D1405225

Microbiological Applications

Laboratory Manual in General Microbiology

Seventh Edition

HAROLD J. BENSON

Professor Emeritus - Pasadena City College

WCB McGraw-Hill

Boston, Massachusetts   Burr Ridge, Illinois   Dubuque, Iowa
Madison, Wisconsin   New York, New York   San Francisco, California   St. Louis, Missouri

McGraw-Hill

A Division of The McGraw·Hill Companies

Microbiological Applications: Complete Version Lab Manual

 This book is printed on recycled, acid-free paper containing 10% postconsumer waste.

1 2 3 4 5 6 7 8 9 0 QPD/QPD 9 0 9 8 7

ISBN 0–697–34139–9

Cover image: ©Dennis Kunkel/Microvision

Publisher: Kevin T. Kane
Sponsoring Editor: Ronald E. Worthington, Ph.D.
Developmental Editor: Terrance Stanton
Marketing Manager: Thomas C. Lyon
Project Manager: Kay J. Brimeyer
Production Supervisor: Laura Fuller
Designer: Katherine Farmer
Cover and Interior Designer: Kathy Theis
Photo Research Coordinator: Lori Hancock
Art Editor: Joyce Watters
Compositor: Carlisle Communications, Inc.
Typeface: 11/12 Times Roman
Printer: Quebecor Printing Book Group/Dubuque, IA

www: mhhe.com

Contents

Preface

This seventh edition of *Microbiological Applications* is an attempt to upgrade concepts where needed, eliminate experiments or portions of experiments that are no longer appropriate, and to clarify text that seems confusing, To help us decide where changes were needed, we were able to enlist the help of twenty-five individuals from 22 colleges and universities in different regions of the country.

Although no new exercises have been added to this edition, 26 exercises in the manual have been altered in one way or another. Many of the changes are quite minor. In some cases only media changes have been made. In others a change may be as insignificant as providing substitute organisms for an experiment. In many instances the changes simply enable instructors to choose alternate methods of performing experiments.

Our usage of Whittaker's five-kingdom system of classification in Part 2 is justifiably criticized as being outdated. This is acknowledged on the title page of Part 2 that pertains to an overview study of microorganisms. However, instead of abandoning the five-kingdom system for the three-domain system at this time we have elected to alert students to the fact that a new classification system is emerging that will eventually displace Whittaker's system. Doing a complete change over to the three-domain system at this time would have drastically altered the structure of all the exercises in Part 2. Hopefully, the next edition will incorporate the desired information.

The following exercises have been expanded to incorporate new alternate procedures: Ex. 19 (Motility Determination), Ex. 71 (Microbiology of Fermented Milk Products) and Ex. 80 (Urinary Tract Pathogens.).

Exercise 23 (Bacterial Population Counts) has been considerably revised. The procedures involved in pipette handling and spectrophotometry have been improved. Changes have also been made in the manner of recording data on the Laboratory Report. Two other exercises that have been considerably revised are Exercises 53 and 54 that pertain to the Enterotube II and OxiFerm Tube II miniturized multitest systems. The revisions in these two exercises were made specifically to upgrade the techniques to current methods.

A great improvement in Part 8 (Identification of Unknown Bacteria) is the incorporation of *Identibacter interactus,* a computer program developed by Allan Konopka, Paul Furbacher, and Clark Gedney at Purdue University. This program is on a CD-ROM disc and can be purchased from WCB/McGraw-Hill Co. in Dubuque. Appendix F provides an explanation of how this program works.

I am greatly indebted to my editor, Elizabeth Sievers, who conducted a survey of individuals who use this manual in search of suggestions for improvement. In addition, I am deeply indebted to the following respondents who willingly yielded to Liz's requests: Barbara B. Hemmingsen at San Diego State University, Robert W. Phelps at San Diego Mesa College, Donald P. Stahly and Richard Roller at the University of Iowa, Thieron P. Harrison at the University of Central Oklahoma, David W. Essar and Emmanuel E. Brako at Winona State University, Joanne M. Kilpatrick and Angela C. Morrow at Auburn University, Dana Kelly Heiny at the University of Arkansas, Elizabeth Godrick at Boston University, Robert W. Bernlohr at Penn State University, Robyn Haskin at SUNY Delhi, Allen W. Nicholson at Wayne State University, Barbara M. Poole at Bossier Parish Community College, Teresa G. Fischer and Judy Gersony at Indian River Community College (Florida), Kathleen Bobbit at Wagner College, Staten Island. Fredrick M. Krick at Des Moines Area Community College, Jeffrey N. Lee at Essex County College (N.J.), Joseph Michalewicz at Holy Family College, Philadelphia, Fred H. Schindler at Indian Hills Community College, Iowa, Alejandro Viera at Passaic County Community College, Elzbet F. Diaz de Leon at Ventura College and Jeffrey C. Burne at Macon College.

Some of the laboratory experiments included in this test may be hazardous if materials are handled improperly or if procedures are conducted incorrectly. Safety precautions are necessary when you are working with chemicals, glass test tubes, hot water baths, sharp instruments, and the like, or for any procedures that generally require caution. Your school may have set regulations regarding safety procedures that your instructor will explain to you. Should you have any problems with materials or procedures, please ask your instructor for help.

These laboratory exercises have been developed to guide you in your daily experiences in microbiology so that you will understand fully the principles involved. Since you will be working with unseen living forms, it will be necessary that you develop a set of techniques that are new to you. Not only will these techniques determine the success or failure of your scientific probing, but they will also be essential to protect you and others around you against potentially harmful forms.

Scheduling During the first week of this course your instructor will provide you with a schedule of laboratory exercises arranged in the order of their performance. Before attending laboratory each day, check the schedule to see what experiment or experiments will be performed and prepare yourself so that you understand what will be done.

Each laboratory session will begin with a short discussion to brief you on the availability of materials and procedures. *Since the preliminary instructions start promptly at the beginning of the period, it is extremely important that you are not late to class.*

Protection of Self and Clothing A lab coat or apron must be worn at all times in the laboratory. Not only will it protect your clothing from accidental contamination, but it will also shield expensive blouses, sweaters, etc., against stains used daily in the lab. When leaving the laboratory, remove coat or apron.

Long hair must be secured in a ponytail to prevent injury from Bunsen burners and contamination of culture material.

To avoid burns, beware of hot Bunsen burners and tripods. Report all accidents and injuries as soon as possible.

Storage of Items Lunches, coats, and books that are not required for this lab, should be stored in some out of the way place (locker, drawer, perimeter counter, or cupboard). Desk space is minimal and must be reserved for essential equipment and your laboratory manual.

At the end of the period all equipment, such as beakers, Bunsen burners, graduates, etc., must be returned to places of origin.

Sanitary Precautions Since we are often working with potentially pathogenic microorganisms, the following protective measures must be adhered to:

1. Before you start your work at the beginning of the period, scrub down the top of your laboratory table with a disinfectant. This will reduce the danger of contaminating your bacterial cultures with dustborn microorganisms and remove any

organisms that may have been left by an inconsiderate student in a previous period.

2. Repeat this scrub-down procedure at the end of the period to remove any microorganisms that might have been unknowingly spilled on the tabletop.

3. Don't smoke or eat food in the laboratory. Make it a habit to keep your hands away from your mouth. Gummed labels should never be moistened with your tongue; use tap water instead.

4. Always use a mechanical device for pipetting fluids or bacterial cultures. *Pipetting by mouth is prohibited in this laboratory.*

5. Observe strict sanitary procedures with respect to the handling of pipettes. In most laboratories pipette suction is provided by mechanical devices, rather than by mouth. Follow the procedures recommended by your instructor.

6. Place old cultures in receptacles that are to be autoclaved. Do not allow your desk drawer or locker to become filled with cultures that have ceased to be of value in your work.

7. Whenever bacterial cultures are accidentally spilled on the floor, notify the instructor so that proper disinfection procedures can be assured.

   In many college laboratories, failure to report a spill may be sufficient reason to prevent you from returning to the laboratory.

8. Do not remove cultures, reagents, or other materials from the laboratory at any time unless specific permission has been granted.

9. Before leaving the laboratory at the end of the period, wash your hands with soap and water.

Biologicals and Body Fluids If the occasion arises where you are handling plasma, test sera, or blood samples in the laboratory, *avoid skin contact with them.* Although suppliers of these substances screen them for the presence of the AIDS virus and other infectious agents, don't allow them to contact your skin. If any blood testing is done in this laboratory, observe strict avoidance of skin contact to blood and test sera.

Planning and Records Always plan your work to avoid serious time-consuming mistakes. Keep an accurate record of what you do. An orderly notebook will pay dividends in the long run.

Before you start to record data on the Laboratory Report sheets, which are located at the back of the manual, remove them from the binding. Trying to shift from the front of the book to the back is inconvenient and time-consuming. Before handing in the Laboratory Reports, trim the perforations from the binding edge with a pair of scissors. These ragged edges make handling of the sheets very difficult.

PART 1 Microscopy

Although there are many kinds of microscopes available to the microbiologist today, only four types will be described here for our use: the brightfield, darkfield, phase-contrast, and fluorescence microscopes. If you have had extensive exposure to microscopy in previous courses, this unit may not be of great value to you; however, if the study of microorganisms is a new field of study for you, there is a great deal of information that you need to acquire about the proper use of these instruments.

Microscopes in a college laboratory represent a considerable investment and require special care to prevent damage to the lenses and mechanicals. The fact that a laboratory microscope may be used by several different individuals during the day and moved around from one place to another results in a much greater chance for damage and wear to occur than if the instrument were used by only one individual.

The complexity of some of the more expensive microscopes also requires that certain adjustments be made periodically. Knowing how to make these adjustments to get the equipment to perform properly is very important. An attempt is made in the five exercises of this unit to provide the necessary assistance in getting the most out of the equipment.

Microscopy should be as fascinating to the beginner as it is to the professional of long standing; however, only with intelligent understanding can the beginner approach the achievement that occurs with years of experience.

Brightfield Microscopy

A microscope that allows light rays to pass directly through to the eye without being deflected by an intervening opaque plate in the condenser is called a *brightfield microscope.* This is the conventional type of instrument encountered by students in beginning courses in biology; it is also the first type to be used in this laboratory.

All brightfield microscopes have certain things in common, yet they differ somewhat in mechanical operation. An attempt will be made in this exercise to point out the similarities and differences of various makes so that you will know how to use the instrument that is available to you. Before attending the first laboratory session in which the microscope will be used, read over this exercise and answer all the questions on the Laboratory Report. Your instructor may require that the Laboratory Report be handed in prior to doing any laboratory work.

Care of the Instrument

Microscopes represent considerable investment and can be damaged rather easily if certain precautions are not observed. The following suggestions cover most hazards.

Transport When carrying your microscope from one part of the room to another, use both hands when holding the instrument, as illustrated in figure 1.1. If it is carried with only one hand and allowed to dangle at your side, there is always the danger of collision with furniture or some other object. And, incidentally, *under no circumstances should one attempt to carry two microscopes at one time.*

Clutter Keep your workstation uncluttered while doing microscopy. Keep unnecessary books, lunches, and other unneeded objects away from your work area. A clear work area promotes efficiency and results in fewer accidents.

Electric Cord Microscopes have been known to tumble off of tabletops when students have entangled a foot in a dangling electric cord. Don't let the light cord on your microscope dangle in such a way as to hazard foot entanglement.

Lens Care At the beginning of each laboratory period check the lenses to make sure they are clean. At the end of each lab session be sure to wipe any immersion oil off the immersion lens if it has been used. More specifics about lens care are provided on page 6.

Dust Protection In most laboratories dustcovers are used to protect the instruments during storage. If one is available, place it over the microscope at the end of the period.

Components

Before we discuss the procedures for using a microscope, let's identify the principal parts of the instrument as illustrated in figure 1.2.

Framework All microscopes have a basic frame structure, which includes the **arm** and **base.** To this framework all other parts are attached. On many of the older microscopes the base is not rigidly attached to the arm as is the case in figure 1.2; instead, a pivot point is present that enables one to tilt the arm backward to adjust the eyepoint height.

Stage The horizontal platform that supports the microscope slide is called the *stage.* Note that it has a clamping device, the **mechanical stage,** which is used for holding and moving the slide

Figure 1.1 The microscope should be held firmly with both hands while carrying it.

around on the stage. Note, also, the location of the **mechanical stage control** in figure 1.2.

Light Source In the base of most microscopes is positioned some kind of light source. Ideally, the lamp should have a **voltage control** to vary the intensity of light. The microscope in figure 1.2 has a knurled wheel on the right side of its base to regulate the voltage supplied to the light bulb. The microscope base in figure 1.4 has a knob (the left one) that controls voltage.

Most microscopes have some provision for reducing light intensity with a **neutral density filter.** Such a filter is often needed to reduce the intensity of light below the lower limit allowed by the voltage control. On microscopes such as the Olympus CH-2, one can simply place a neutral density filter over the light source in the base. On some microscopes a filter is built into the base.

Lens Systems All microscopes have three lens systems: the oculars, the objectives, and the con-

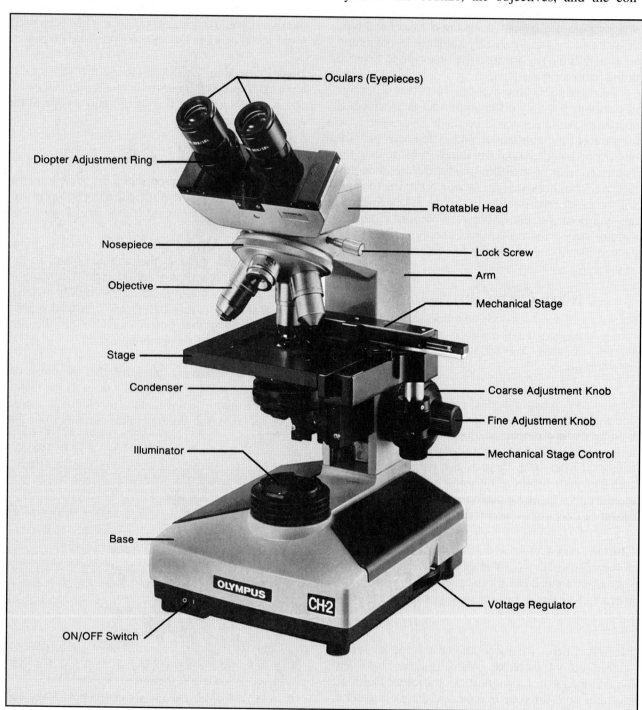

Figure 1.2   **The compound microscope**

Courtesy of the Olympus Corporation, Lake Success, N.Y.

denser. Figure 1.3 illustrates the light path through these three systems.

The **ocular,** or eyepiece, is a complex piece, located at the top of the instrument, that consists of two or more internal lenses and usually has a magnification of 10×. Although the microscope in figure 1.2 has two oculars (binocular), a microscope often has only one.

Three or more **objectives** are usually present. Note that they are attached to a rotatable **nosepiece,** which makes it possible to move them into position over a slide. Objectives on most laboratory microscopes have magnifications of 10×, 45×, and 100×, designated as **low power, high-dry,** and **oil immersion,** respectively. Some microscopes will have a fourth objective for rapid scanning of microscopic fields that is only 4×.

The third lens system is the **condenser,** which is located under the stage. It collects and directs the light from the lamp to the slide being studied. The condenser can be moved up and down by a knob under the stage. A **diaphragm** within the condenser regulates the amount of light that reaches the slide. Microscopes that lack a voltage control on the light source rely entirely on the diaphragm for controlling light intensity. On the Olympus microscope in figure 1.2 the diaphragm is controlled by turning a knurled ring. On some microscopes a diaphragm lever is present. Figure 1.3 illustrates the location of the condenser and diaphragm.

Focusing Knobs The concentrically arranged **coarse adjustment** and **fine adjustment knobs** on the side of the microscope are used for bringing objects into focus when studying an object on a slide. On some microscopes these knobs are not positioned concentrically as shown here.

Ocular Adjustments On binocular microscopes one must be able to change the distance between the oculars and to make diopter changes for eye differences. On most microscopes the interocular distance is changed by simply pulling apart or pushing together the oculars.

To make diopter adjustments, one focuses first with the right eye only. Without touching the focusing knobs, diopter adjustments are then made on the left eye by turning the knurled **diopter adjustment ring** (figure 1.2) on the left ocular until a sharp image is seen. One should now be able to see sharp images with both eyes.

Resolution

The resolution limit, or **resolving power,** of a microscope lens system is a function of its numerical aperture, the wavelength of light, and the design of

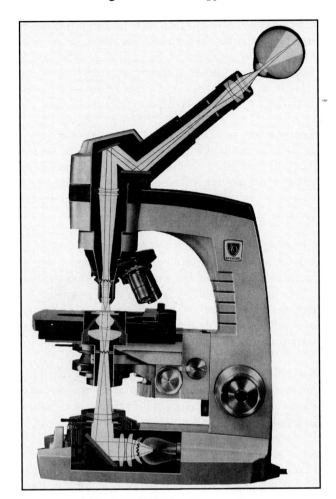

Figure 1.3 The light pathway of a microscope.

the condenser. The optimum resolution of the best microscopes with oil immersion lenses is around 0.2 µm. This means that two small objects that are 0.2 µm apart will be seen as separate entities; objects closer than that will be seen as a single object.

To get the maximum amount of resolution from a lens system, the following factors must be taken into consideration:

- A **blue filter** should be in place over the light source because the short wavelength of blue light provides maximum resolution.
- The **condenser** should be kept at its highest position where it allows a maximum amount of light to enter the objective.
- The **diaphragm** should not be stopped down too much. Although stopping down improves contrast, it reduces the numerical aperture.
- **Immersion oil** should be used between the slide and the 100× objective.

Of significance is the fact that, as magnification is increased, the resolution must also increase. Simply increasing magnification by using a 20× ocular won't increase the resolution.

Lens Care

Keeping the lenses of your microscope clean is a constant concern. Unless all lenses are kept free of dust, oil, and other contaminants, they are unable to achieve the degree of resolution that is intended. Consider the following suggestions for cleaning the various lens components:

Cleaning Tissues   Only lint-free, optically safe tissues should be used to clean lenses. Tissues free of abrasive grit fall in this category. Booklets of lens tissue are most widely used for this purpose. Although several types of boxed tissues are also safe, *use only the type of tissue that is recommended by your instructor.*

Solvents   Various liquids can be used for cleaning microscope lenses. Green soap with warm water works very well. Xylene is universally acceptable. Alcohol and acetone are also recommended, but often with some reservations. Acetone is a powerful solvent that could possibly dissolve the lens mounting cement in some objective lenses if it were used too liberally. When it is used it should be used sparingly. Your instructor will inform you as to what solvents can be used on the lenses of your microscope.

Oculars   The best way to determine if your eyepiece is clean is to rotate it between the thumb and forefinger as you look through the microscope. A rotating pattern will be evidence of dirt.

If cleaning the top lens of the ocular with lens tissue fails to remove the debris, one should try cleaning the lower lens with lens tissue and blowing off any excess lint with an air syringe or gas cannister. *Whenever the ocular is removed from the microscope, it is imperative that a piece of lens tissue be placed over the open end of the microscope as illustrated in figure 1.5.*

Objectives   Objective lenses often become soiled by materials from slides or fingers. A piece of lens tissue moistened with green soap and water, or one of the acceptable solvents mentioned above, will usually remove whatever is on the lens. Sometimes a cotton swab with a solvent will work better than lens tissue. At any time that the image on the slide is unclear or cloudy, assume at once that the objective you are using is soiled.

Condenser   Dust often accumulates on the top surface of the condenser; thus, wiping it off occasionally with lens tissue is desirable.

Procedures

If your microscope has three objectives you have three magnification options: (1) low-power, or 100× total magnification, (2) high-dry magnification, which is 450× total with a 45× objective, and (3) 1000× total magnification with a 100× oil immersion objective. Note that the total magnification seen through an objective is calculated by simply multiplying the power of the ocular by the power of the objective.

Whether you use the low-power objective or the oil immersion objective will depend on how much magnification is necessary. Generally speaking, however, it is best to start with the low-power objective

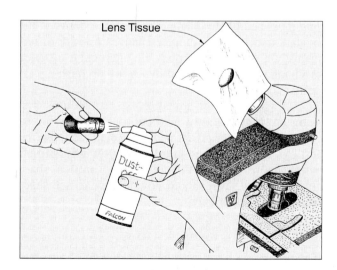

Figure 1.4   On this microscope, the left knob controls voltage. The other knob is used for moving a neutral density filter into position.

Figure 1.5   When oculars are removed for cleaning, cover the ocular opening with lens tissue. A blast from an air syringe or gas cannister removes dust and lint.

and progress to the higher magnifications as your study progresses. Consider the following suggestions for setting up your microscope and making microscopic observations.

Viewing Setup If your microscope has a rotatable head, such as the ones being used by the two students in figure 1.6, there are two ways that you can use the instrument. Note that the student on the left has the arm of the microscope *near* him, and the other student has the arm *away from* her. With this type of microscope, the student on the right has the advantage in that the stage is easier to observe. Note, also that when focusing the instrument she is able to rest her arm on the table. The manufacturer of this type of microscope intended that the instrument be used in the way demonstrated by the young lady. If the microscope head is not rotatable, it will be necessary to use the other position.

Low-Power Examination The main reason for starting with the low-power objective is to enable you to explore the slide to look for the object you are planning to study. Once you have found what you are looking for, you can proceed to higher magnifications. Use the following steps when exploring a slide with the low-power objective:

1. Position the slide on the stage with the material to be studied on the *upper* surface of the slide. Figure 1.7 illustrates how the slide must be held in place by the mechanical stage retainer lever.
2. Turn on the light source, using a *minimum* amount of voltage. If necessary, reposition the slide so that the stained material on the slide is in the *exact center* of the light source.
3. Check the condenser to see that it has been raised to its highest point.
4. If the low-power objective is not directly over the center of the stage, rotate it into position. Be

sure that as you rotate the objective into position it clicks into its locked position.
5. Turn the coarse adjustment knob to lower the objective until it stops. A built-in stop will prevent the objective from touching the slide.
6. While looking down through the ocular (or oculars), bring the object into focus by turning the fine adjustment focusing knob. Don't readjust the coarse adjustment knob. If you are using a binocular microscope it will also be necessary to adjust the interocular distance and diopter adjustment to match your eyes.
7. Manipulate the diaphragm lever to reduce or increase the light intensity to produce the clearest, sharpest image. Note that as you close down the diaphragm to reduce the light intensity, the contrast improves and the depth of field increases. Stopping down the diaphragm when using the low-power objective does not decrease resolution.
8. Once an image is visible, move the slide about to search out what you are looking for. The slide is moved by turning the knobs that move the mechanical stage.
9. Check the cleanliness of the ocular, using the procedure outlined earlier.
10. Once you have identified the structures to be studied and wish to increase the magnification, you may proceed to either high-dry or oil immersion magnification. However, before changing objectives, *be sure to center the object you wish to observe.*

High-Dry Examination To proceed from low-power to high-dry magnification, all that is necessary is to rotate the high-dry objective into position and open up the diaphragm somewhat. It may be necessary to make a minor adjustment with the fine adjustment knob to sharpen up the image, but *the coarse adjustment knob should not be touched.*

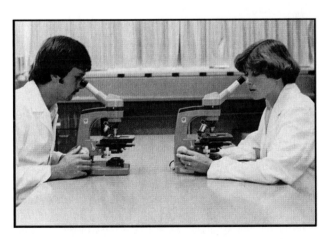

Figure 1.6 The microscope position on the right has the advantage of stage accessibility.

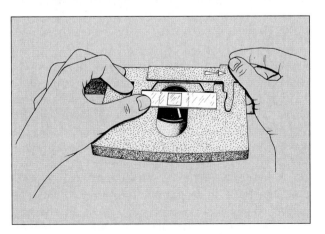

Figure 1.7 The slide must be properly positioned as the retainer lever is moved to the right.

If a microscope is of good quality, only minor focusing adjustments are needed when changing from low power to high-dry because all the objectives will be **parfocalized.** Nonparfocalized microscopes do require considerable refocusing when changing objectives.

High-dry objectives should be used only on slides that have cover glasses; without them, images are usually unclear. When increasing the lighting, be sure to open up the diaphragm first instead of increasing the voltage on your lamp; reason: *lamp life is greatly extended when used at low voltage.* If the field is not bright enough after opening the diaphragm, feel free to increase the voltage. A final point: Keep the condenser at its highest point.

Oil Immersion Techniques The oil immersion lens derives its name from the fact that a special mineral oil is interposed between the lens and the microscope slide. The oil is used because it has the same refractive index as glass, which prevents the loss of light due to the bending of light rays as they pass through air. The use of oil in this way enhances the resolving power of the microscope. Figure 1.8 reveals this phenomenon.

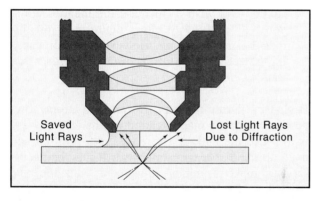

Figure 1.8 Immersion oil, having the same refractive index as glass, prevents light loss due to diffraction.

With parfocalized objectives one can go to oil immersion from either low power or high-dry. On some microscopes, however, going from low power to high power and then to oil immersion is better. Once the microscope has been brought into focus at one magnification, the oil immersion lens can be rotated into position without fear of striking the slide.

Before rotating the oil immersion lens into position, however, a drop of immersion oil must be placed on the slide. An oil immersion lens should never be used without oil. Incidentally, if the oil appears cloudy it should be discarded.

When using the oil immersion lens it is best to open the diaphragm as much as possible . Stopping down the diaphragm tends to limit the resolving power of the optics. In addition, the condenser must be kept at its highest point. If different colored filters are available for the lamp housing, it is best to use blue or greenish filters to enhance the resolving power.

Since the oil immersion lens will be used extensively in all bacteriological studies, it is of paramount importance that you learn how to use this lens properly. Using this lens takes a little practice due to the difficulties usually encountered in manipulating the lighting. A final comment of importance: At the end of the laboratory period remove all immersion oil from the lens tip with lens tissue.

Putting It Away

When you take a microscope from the cabinet at the beginning of the period, you expect it to be clean and in proper working condition. The next person to use the instrument after you have used it will expect the same consideration. A few moments of care at the end of the period will ensure these conditions. Check over this list of items at the end of each period before you return the microscope to the cabinet.

1. Remove the slide from the stage.
2. If immersion oil has been used, wipe it off the lens and stage with lens tissue. (Do not wipe oil off slides you wish to keep. Simply put them into a slide box and let the oil drain off.)
3. Rotate the low-power objective into position.
4. If the microscope has been inclined, return it to an erect position.
5. If the microscope has a built-in movable lamp, raise the lamp to its highest position.
6. If the microscope has a long attached electric cord, wrap it around the base.
7. Adjust the mechanical stage so that it does not project too far on either side.
8. Replace the dustcover.
9. If the microscope has a separate transformer, return it to its designated place.
10. Return the microscope to its correct place in the cabinet.

Laboratory Report

Before the microscope is to be used in the laboratory, answer all the questions on Laboratory Report 1,2 that pertain to brightfield microscopy. Preparation on your part prior to going to the laboratory will greatly facilitate your understanding. Your instructor may wish to collect this report at the *beginning of the period* on the first day that the microscope is to be used in class.

Darkfield Microscopy

Delicate transparent living organisms can be more easily observed with darkfield microscopy than with conventional brightfield microscopy. This method is particularly useful when one is attempting to identify spirochaetes in the exudate from a syphilitic lesion. Figure 2.1 illustrates the appearance of these organisms under such illumination. This effect may be produced by placing a darkfield stop below the regular condenser or by replacing the condenser with a specially constructed one.

Another application of darkfield microscopy is in the fluorescence microscope (Exercise 4). Although fluorescence may be seen without a dark field, it is greatly enhanced with this application.

To achieve the darkfield effect it is necessary to alter the light rays that approach the objective in such a way that only oblique rays strike the objects being viewed. The obliquity of the rays must be so extreme that if no objects are in the field, the background is completely light-free. Objects in the field become brightly illuminated, however, by the rays that are reflected up through the lens system of the microscope.

Although there are several different methods for producing a dark field, only two devices will be described here: the star diaphragm and the cardioid condenser. The availability of equipment will determine the method to be used in this laboratory.

The Star Diaphragm

One of the simplest ways to produce the darkfield effect is to insert a star diaphragm into the filter slot of the condenser housing as shown in figure 2.2. This device has an opaque disk in the center that blocks the central rays of light. Figure 2.3 reveals the effect of this stop on the light rays passing through the condenser. If such a device is not available, one can be made by cutting round disks of opaque paper of different sizes that are cemented to transparent celluloid disks that will fit into the slot. If the microscope normally has a diffusion disk in this slot, it is best to replace it with rigid clear celluloid or glass.

An interesting modification of this technique is to use colored celluloid stops instead of opaque paper. Backgrounds of blue, red, or any color can be produced in this way.

In setting up this type of darkfield illumination it is necessary to keep these points in mind:

1. Limit this technique to the study of large organisms that can be seen easily with low-power magnification. *Good resolution with higher powered objectives is difficult with this method.*
2. Keep the diaphragm wide open and use as much light as possible. If the microscope has a voltage

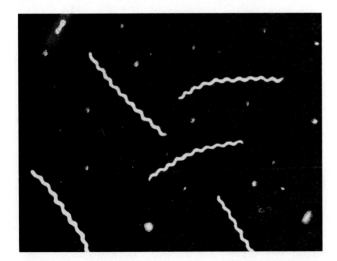

Figure 2.1 Transparent living microorganisms, such as the syphilis spirochaete, can be seen much more easily when observed in a dark field.

Figure 2.2 The insertion of a star diaphragm into the filter slot of the condenser will produce a dark field suitable for low magnifications.

regulator, you will find that the higher voltages will produce better results.

3. Be sure to center the stop as precisely as possible.

4. Move the condenser up and down to produce the best effects.

The Cardioid Condenser

The difficulty that results from using the star diaphragm or opaque paper disks with high-dry and oil immersion objectives is that the oblique rays are not as carefully metered as is necessary for the higher magnifications. Special condensers such as the cardioid or paraboloid types must be used. Since the cardioid type is the most frequently used type, its use will be described here.

Figure 2.4 illustrates the light path through such a condenser. Note that the light rays entering the lower element of the condenser are reflected first off a convex mirrored surface and then off a second concave surface to produce the desired oblique rays of light. Once the condenser has been installed in the microscope, the following steps should be followed to produce ideal illumination.

Materials:

slides and cover glasses of excellent quality (slides of 1.15–1.25 mm thickness and No. 1 cover glasses)

1. Adjust the upper surface of the condenser to a height just below stage level.

2. Place a clear glass slide in position over the condenser.

3. Focus the 10× objective on the top of the condenser until a bright ring comes into focus.

4. Center the bright ring so that it is concentric with the field edge by adjusting the centering screws on the darkfield condenser. If the condenser has a light source built into it, it will also be necessary to center it as well to achieve even illumination.

5. Remove the clear glass slide.

6. If a funnel stop is available for the oil immersion objective, remove this object and insert this unit. (This stop serves to reduce the numerical aperture of the oil immersion objective to a value that is less than the condenser.)

7. Place a drop of immersion oil on the upper surface of the condenser and place the slide on top of the oil. The following preconditions in slide usage must be adhered to:

   • Slides and cover glasses should be optically perfect. Scratches and imperfections will cause annoying diffractions of light rays.

   • Slides and cover glasses must be free of dirt or grease of any kind.

   • A cover glass should always be used.

8. If the oil immersion lens is to be used, place a drop of oil on the cover glass.

9. If the field does not appear dark and lacks contrast, return to the 10× objective and check the ring concentricity and light source centration. If contrast is still lacking after these adjustments, the specimen is probably too thick.

10. If sharp focus is difficult to achieve under oil immersion, try using a thinner cover glass and adding more oil to the top of the cover glass and bottom of the slide.

Laboratory Report

This exercise may be used in conjunction with Part 2 when studying the various types of organisms. After reading over this exercise and doing any special assignments made by your instructor, answer the questions on the last portion of Laboratory Report 1,2 that pertain to darkfield microscopy.

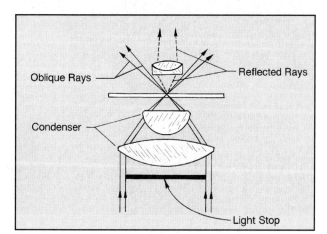

Figure 2.3 The star diaphragm allows only peripheral light rays to pass up through the condenser. This method requires maximum illumination.

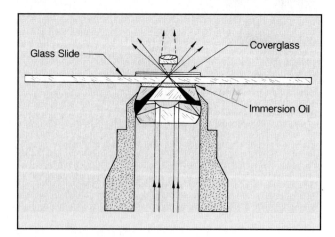

Figure 2.4 A cardioid condenser provides greater light concentration for oblique illumination than the star diaphragm.

Phase-Contrast Microscopy

The difficulty that one encounters in trying to examine cellular organelles is that most protoplasmic material is completely transparent and defies differentiation. It is for this reason that stained slides are usually used in brightfield cytological studies. Since the staining of slides results in cellular death, it is obvious that when we study stained microorganisms on a slide, we are observing artifacts rather than living cells.

A microscope that is able to differentiate transparent protoplasmic structures without staining and killing them is the *phase-contrast microscope.* The first phase-contrast microscope was developed in 1933 by Frederick Zernike and was originally referred to as the *Zernike microscope*. It is the instrument of choice for studying living protozoans and other types of transparent cells. Figure 3.1 illustrates the differences between brightfield and phase-contrast images. Note the greater degree of differentiation that can be seen inside cells when they are observed with phase-contrast optics. In this exercise we will study the principles that govern this type of microscope; we will also see how different manufacturers have met the design challenges of these principles.

Image Contrast

Objects in a microscopic field may be categorized as being either amplitude or phase objects. **Amplitude objects** (illustration 1, figure 3.2) show up as dark objects under the microscope because the amplitude (intensity) of light rays is reduced as the rays pass through the objects. **Phase objects** (illustration 2, figure 3.2), on the other hand, are completely transparent since light rays pass through them unchanged with respect to amplitude. As some of the light rays pass through phase objects, however, they are retarded by ¼ wavelength.

This retardation, known as *phase shift*, occurs with no amplitude diminution; thus, the objects appear transparent rather than opaque. Since most

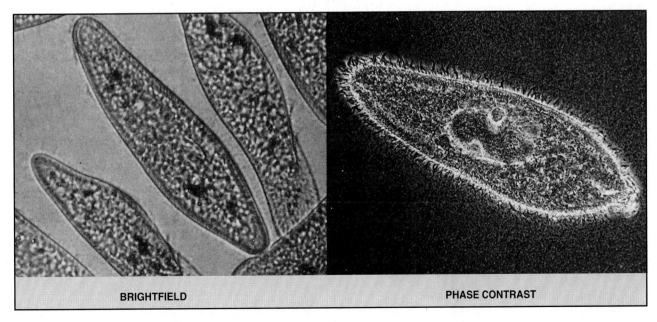

BRIGHTFIELD　　　　　**PHASE CONTRAST**

Figure 3.1　Comparison of brightfield and phase-contrast images

biological specimens are phase objects, lacking in contrast, it becomes necessary to apply dyes of various kinds to cells that are to be studied with a brightfield microscope. To understand how Zernike took advantage of the ¼ wavelength phase shift in developing his microscope we must understand the difference between direct and diffracted light rays.

Two Types of Light Rays

Light rays passing through a transparent object emerge as either direct or diffracted rays. Those rays that pass straight through unaffected by the medium are called **direct rays.** They are unaltered in amplitude and phase. The balance of the rays that are bent by their slowing through the medium (due to density differences) emerge from the object as **diffracted rays.** It is these rays that are retarded ¼ wavelength. Illustration 3, figure 3.2, illustrates these two types of light rays.

An important characteristic of these light rays is that if the direct and diffracted rays of an object can be brought into exact phase, or *coincidence,* with each other, the resultant amplitude of the converged rays is the sum of the two waves. This increase in

amplitude will produce increased brightness of the object in the field. On the other hand, if two rays of equal amplitude are in reverse phase (½ wavelength off), their amplitudes cancel each other to produce a dark object. This phenomenon is called *interference.* Illustration 4, figure 3.2, shows these two conditions.

The Zernike Microscope

In constructing his first phase-contrast microscope, Zernike experimented with various configurations of diaphragms and various materials that could be used to retard or advance the direct light rays. Figure 3.3 illustrates the optical system of a typical modern phase-contrast microscope. It differs from a conventional brightfield microscope by having (1) a different type of diaphragm and (2) a phase plate.

The diaphragm consists of an **annular stop** that allows only a hollow cone of light rays to pass up through the condenser to the object on the slide. The **phase plate** is a special optical disk located at the rear focal plane of the objective. It has a **phase ring** on it that advances or retards the direct light rays ¼ wavelength.

Note in figure 3.3 that the direct rays converge on the phase ring to be advanced or retarded ¼

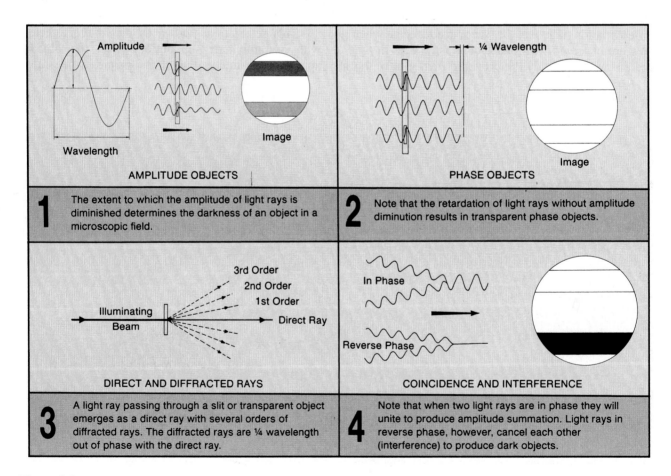

Figure 3.2 The utilization of light rays in phase-contrast microscopy

wavelength. These rays emerge as solid lines from the object on the slide. This ring on the phase plate is coated with a material that will produce the desired phase shift. The diffracted rays, on the other hand, which have already been retarded ¼ wavelength by the phase object on the slide, completely miss the phase ring and are not affected by the phase plate. It should be clear, then, that depending on the type of phase-contrast microscope, the convergence of diffracted and direct rays on the image plane will result in either a brighter image (*amplitude summation*) or a darker image (*amplitude interference* or *reverse*

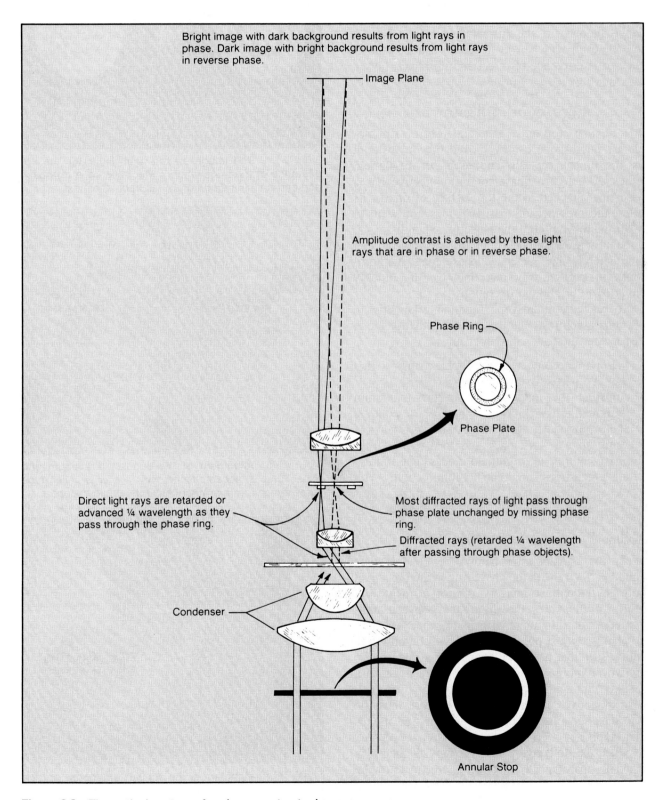

Bright image with dark background results from light rays in phase. Dark image with bright background results from light rays in reverse phase.

Image Plane

Amplitude contrast is achieved by these light rays that are in phase or in reverse phase.

Phase Ring

Phase Plate

Direct light rays are retarded or advanced ¼ wavelength as they pass through the phase ring.

Most diffracted rays of light pass through phase plate unchanged by missing phase ring.

Diffracted rays (retarded ¼ wavelength after passing through phase objects).

Condenser

Annular Stop

Figure 3.3 The optical system of a phase-contrast microscope

13

phase). The former is referred to as **bright phase** microscopy; the latter as **dark phase** microscopy. The apparent brightness or darkness, incidentally, is proportional to the square of the amplitude; thus, the image will be four times as bright or dark as seen through a brightfield microscope.

It should be added here, parenthetically, that the phase plates of some microscopes have coatings to change the phase of the diffracted rays. In any event the end result will be the same: to achieve coincidence or interference of direct and diffracted rays.

Microscope Adjustments

If the annular stop under the condenser of a phase-contrast microscope can be moved out of position, this instrument can also be used for brightfield studies. Although a phase-contrast objective has a phase ring attached to the top surface of one of its lenses, the presence of that ring does not seem to impair the resolution of the objective when it is used in the brightfield mode. It is for this reason that manufacturers have designed phase-contrast microscopes in such a way that they can be quickly converted to brightfield operation.

To make a microscope function efficiently in both phase-contrast and brightfield situations one must master the following procedures:

- lining up the annular ring and phase rings so that they are perfectly concentric,
- adjusting the light source so that maximum illumination is achieved for both phase-contrast and brightfield usage, and
- being able to shift back and forth easily from phase-contrast to brightfield modes. The following suggestions should be helpful in coping with these problems.

Alignment of Annulus and Phase Ring

Unless the annular ring below the condenser is aligned perfectly with the phase ring in the objective, good phase-contrast imagery cannot be achieved. Figure 3.4 illustrates the difference between non-alignment and alignment. If a microscope has only one phase-contrast objective, there will be only one annular stop that has to be aligned. If a microscope has two or more phase objectives, there must be a substage unit with separate annular stops for each phase objective, and alignment procedure must be performed separately for each objective and its annular stop.

Since the objective cannot be moved once it is locked in position, all adjustments are made to the annular stop. On some microscopes the adjustment may be made with tools, as illustrated in figure 3.5.

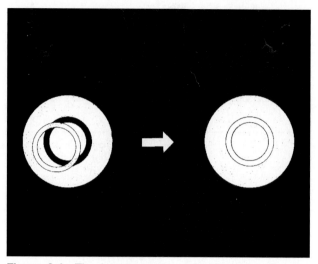

Figure 3.4 The image on the right illustrates the appearance of the rings when perfect alignment of phase ring and annulus diaphragm has been achieved.

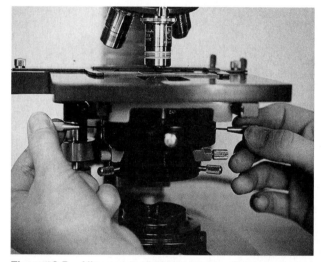

Figure 3.5 Alignment of the annulus diaphragm and phase ring is accomplished with a pair of Allen-type screwdrivers on this American Optical microscope.

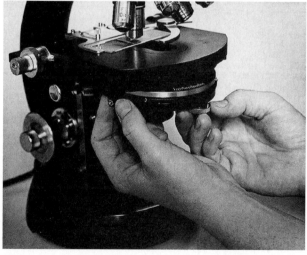

Figure 3.6 Alignment of the annulus and phase ring on this Zeiss microscope is achieved by adjusting the two knobs as shown.

Figure 3.7 If the ocular of a phase-contrast microscope is replaced with a centering telescope, the orientation of the phase ring and annular ring can be viewed.

Figure 3.8 Some microscopes have an aperture viewing unit that can be used instead of a centering telescope for observing the orientation of the phase ring and annular ring.

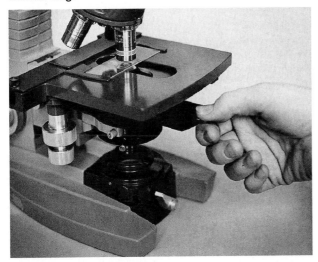

Figure 3.9 The annular stop on this American Optical microscope has the annular stop located on a slideway. When pushed in, the annular stop is in position.

On other microscopes, such as the Zeiss in figure 3.6 which has five phase-contrast objectives, the annular rings are moved into position with special knobs on the substage unit. Since the method of adjustment varies from one brand of microscope to another, one has to follow the instructions provided by the manufacturer. Once the adjustments have been made, they are rigidly set and needn't be changed unless someone inadvertently disturbs them.

To observe ring alignment, one can replace the eyepiece with a **centering telescope** as shown in figure 3.7. With this unit in place, the two rings can be brought into sharp focus by rotating the focusing ring on the telescope. Refocusing is necessary for each objective and its matching annular stop. Some manufacturers, such as American Optical, provide an aperture viewing unit (figure 3.8), which enables one to observe the rings without using a centering telescope. Zeiss microscopes have a unit called the ***Optovar,*** which is located in a position similar to the American Optical unit that serves the same purpose.

Light Source Adjustment

For both brightfield and phase-contrast modes it is essential that optimum lighting be achieved. This is no great problem for a simple setup such as the American Optical instrument shown in figure 3.9. For multiple phase objective microscopes, however, (such as the Zeiss in figure 3.6) there are many more adjustments that need to be made. A few suggestions that highlight some of the problems and solutions follow:

1. Since blue light provides better images for both phase-contrast and brightfield modes, make certain that a blue filter is placed in the filter holder that is positioned in the light path. If the microscope has no filter holder, placing the filter over the light source on the base will help.
2. Brightness of field under phase-contrast is controlled by adjusting the voltage or the iris diaphragm on the base. Considerably more light is required for phase-contrast than for brightfield since so much light is blocked out by the annular stop.
3. The evenness of illumination on some microscopes, such as the Zeiss seen on these pages, can be adjusted by removing the lamp housing from the microscope and focusing the light spot on a piece of translucent white paper. For the detailed steps in this procedure, one should consult the instruction manual that comes with the microscope. Light source adjustments of this nature are not necessary for the simpler types of microscopes.

4. Since each phase-contrast objective must be used with a matching annular stop, make certain that the proper annular stop is being used with the objective that is over the microscope slide. If image quality is lacking, check first to see if the matching annular stop is in position.

Working Procedures

Once the light source is correct and the phase elements are centered you are finally ready to examine slide preparations. Keep in mind that from now on most of the adjustments described earlier should not be altered; however, if misalignment has occurred due to mishandling, it will be necessary to refer back to alignment procedures. The following guidelines should be adhered to in all phase-contrast studies:

- Use only optically perfect slides and cover glasses (no bubbles or striae in the glass).
- Be sure that slides and cover glasses are completely free of grease or chemicals.
- Use wet mount slides instead of hanging drop preparations. The latter leave much to be desired. Culture broths containing bacteria or protozoan suspensions are ideal for wet mounts.
- In general, limit observations to living cells. In most instances stained slides are not satisfactory.

The first time you use phase-contrast optics to examine a wet mount, follow these suggestions:

1. Place the wet mount slide on the stage and bring the material into focus, *using brightfield optics* at low-power magnification.
2. Once the image is in focus, switch to phase optics at the same magnification. Remember, it is necessary to place in position the matching annular stop.
3. Adjust the light intensity, first with the base diaphragm and then with the voltage regulator. In most instances you will need to increase the amount of light for phase-contrast.
4. Switch to higher magnifications, much in the same way you do for brightfield optics, except that you have to rotate a matching annular stop into position.
5. If an oil immersion phase objective is used, add immersion oil to the top of the condenser as well as to the top of the cover glass.
6. Don't be disturbed by the "halo effect" that you observe with phase optics. Halos are normal.

Laboratory Report

This exercise may be used in conjunction with Part 2 in studying various types of organisms. Organelles in protozoans and algae will show up more distinctly than with brightfield optics. After reading this exercise and doing any special assignments made by your instructor, answer the questions on combined Laboratory Report 3–5 that pertain to this exercise.

Fluorescence Microscopy

4

The fluorescence microscope is a unique instrument that is indispensible in certain diagnostic and research endeavors. Differential dyes and immunofluorescence techniques have made laboratory diagnosis of many diseases much simpler with this type of microscope than with the other types described in Exercises 1, 2, and 3. If you are going to prepare and study any differential fluorescence slides that are described in certain exercises in this manual, you should have a basic understanding of the microscope's structure, its capabilities, and its limitations. In addition, it is important that one be aware of the potential of experiencing eye injury if one of these instruments is not used in a safe manner.

A fluorescence microscope differs from an ordinary brightfield microscope in several respects. First of all, it utilizes a powerful mercury vapor arc lamp for its light source. Secondly, a darkfield condenser is usually used in place of the conventional Abbé brightfield condenser. The third difference is that it employs three sets of filters to alter the light that passes up through the instrument to the eye. Some general principles related to its operation will follow an explanation of the principle of fluorescence.

The Principle of Fluorescence

It was pointed out in the last exercise that light exists as a form of energy propagated in wave form. An interesting characteristic of such an electromagnetic wave is that it can influence the electrons of molecules that it encounters, causing significant interaction. Those electrons within a molecule that are not held too securely may be set in motion by the oscillations of the light beam. Not only are these electrons interrupted from their normal pathways, but they are also forced to oscillate in resonance with the passing light wave. This excitation, caused by such oscillation, requires energy that is supplied by the light beam. When we say that a molecule absorbs light, this is essentially what is taking place.

Whenever a physical body absorbs energy, as in the case of the activated molecule, the energy doesn't just disappear; it must reappear again in some other form. This new manifestation of the energy may be in the form of a chemical reaction, heat, or light. If light is emitted by the energized molecules, the phenomenon is referred to as **photoluminescence.** In photoluminescence there is always a certain time lapse between the absorption and emission of light. If the time lag is greater than 1/10,000 of a second it is generally called **phosphorescence.** On the other hand, if the time lapse is less than 1/10,000 of a second, it is known as **fluorescence.**

Thus, we see that fluorescence is initiated when a molecule absorbs energy from a passing wave of light. The excited molecule, after a brief period of time, will return to its fundamental energy state after emitting fluorescent light. It is significant that *the wavelength of fluorescence is always longer than the exciting light.* This follows Stokes' law, which applies to liquids but not to gases. This phenomenon is due to the fact that energy loss occurs in the process so that the emitting light has to be of a longer wavelength. This energy loss, incidentally, occurs as a result of the mobilization of the comparatively heavy atomic nuclei of the molecules rather than the displacement of the lighter electrons.

Microbiological material that is to be studied with a fluorescence microscope must be coated with special compounds that possess this quality of fluorescence. Such compounds are called **fluorochromes.** Auramine O, acridine orange, and fluorescein are well-known fluorochromes. Whether a compound will fluoresce will depend on its molecular structure, the temperature, and the pH of the medium. The proper preparation and use of fluorescent materials for microbiological work must take all these factors into consideration.

Microscope Components

Figure 4.2 illustrates, diagrammatically, the light pathway of a fluorescence microscope. The essential components are the light source, heat filter, exciter filter, condenser, and barrier filter. The characteristics and functions of each item follow.

Light Source The first essential component of a fluorescence microscope is its bright mercury vapor arc lamp. Such a bulb is preferred over an incandescent one because it produces an ample supply of shorter wavelengths of light (ultraviolet, violet, and blue) that are needed for good fluorescence. To produce the arc in one of these lamps, voltages as high as 18,000 volts are required; thus, a power supply transformer is always used.

The wavelengths produced by these lamps include the ultraviolet range of 200–400 nm, the visible range of 400–780 nm, and the long infrared rays that are above 780 nm.

Mercury vapor arc lamps are expensive and potentially dangerous. Certain precautions must be taken, not only to promote long bulb life, but to protect the user as well. One of the hazards of these bulbs is that they are pressurized and can explode. Another hazard exists in direct exposure of the eyes to harmful rays. Knowledge of these hazards is essential to safe operation. If one follows certain precautionary measures, there is little need for anxiety. However, one should not attempt to use one of these instruments without a complete understanding of its operation.

Heat Filter The infrared rays generated by the mercury vapor arc lamp produce a considerable amount of heat. These rays serve no useful purpose in fluorescence and place considerable stress on the filters within the system. To remove these rays, a heat-absorbing filter is the first element in front of the condensers. Ultraviolet rays, as well as most of the visible spectrum, pass through this filter unimpeded.

Exciter Filter After the light has been cooled down by the heat filter it passes through the exciter filter, which absorbs all the wavelengths except the short ones needed to excite the fluorochrome on the slide. These filters are very dark and are designed to let through only the green, blue, violet, or ultraviolet rays. If the exciter filter is intended for visible light (blue, green, or violet) transmission, it will also allow ultraviolet transmittance.

Condenser To achieve the best contrast of a fluorescent object in the microscopic field, a darkfield condenser is used. It must be kept in mind that weak fluorescence of an object in a brightfield would be difficult to see. The dark background produced by the darkfield condenser, thus, provides the desired contrast. Another bonus of this type of condenser is that the majority of the ultraviolet light rays are

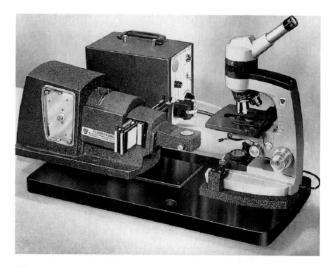

Figure 4.1 An early model American Optical fluorescence illuminator (*Fluorolume*) that could be adapted to an ordinary darkfield microscope.

deflected by the condenser, protecting the observer's eyes. To achieve this, the numerical aperture of the objective is always 0.05 less than that of the condenser.

Barrier Filter This filter is situated between the objective and the eyepiece to remove all remnants of the exciting light so that only the fluorescence is seen. When ultraviolet excitation is employed with its very dark, almost black-appearing exciter filters, the corresponding barrier filters appear almost colorless. On the other hand, when blue exciter filters are used, the matching barrier filters have a yellow to deep orange color. In both instances, the significant fact is that the barrier filter should cut off precisely the shorter exciter wavelengths without affecting the longer fluorescence wavelengths.

Use of the Microscope

As in the case of most sophisticated equipment of this type, it is best to consult the manufacturer's instruction manual before using it. Although different makes of fluorescence microscopes are essentially alike in principle, they may differ considerably in the fine points of operation. Since it is not possible to be explicit about the operation of all makes, all that will be attempted here is to generalize.

Some Precautions To protect yourself and others it is well to outline the hazards first. Keep the following points in mind:

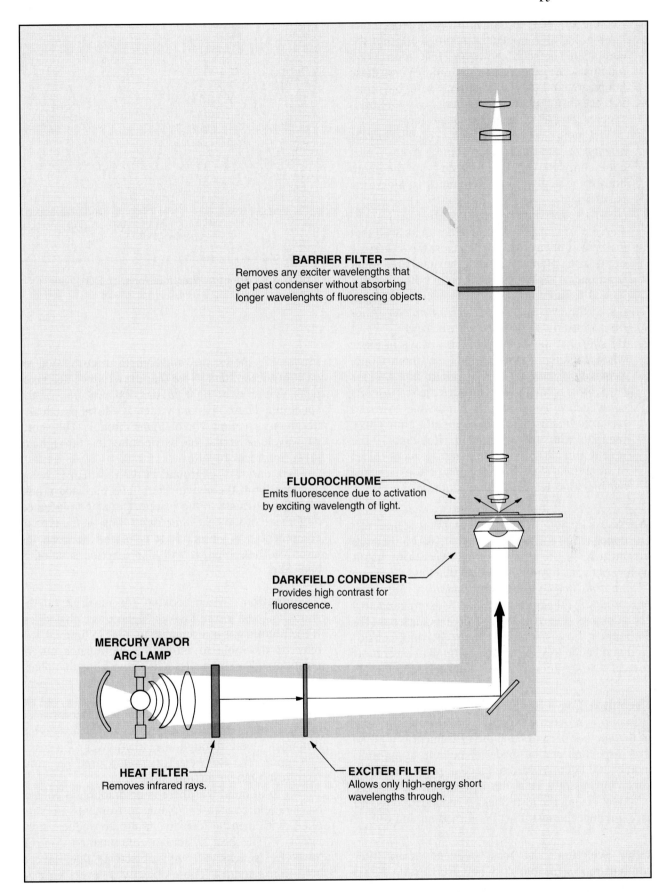

BARRIER FILTER
Removes any exciter wavelengths that get past condenser without absorbing longer wavelenghts of fluorescing objects.

FLUOROCHROME
Emits fluorescence due to activation by exciting wavelength of light.

DARKFIELD CONDENSER
Provides high contrast for fluorescence.

MERCURY VAPOR ARC LAMP

HEAT FILTER
Removes infrared rays.

EXCITER FILTER
Allows only high-energy short wavelengths through.

Figure 4.2 The light pathway of a fluorescence microscope

1. Remember that the pressurized mercury arc lamp is literally a potential bomb. Design of the equipment is such, however, that with good judgment, no injury should result. When these lamps are cold they are relatively safe, but when hot, the inside pressure increases to eight atmospheres, or 112 pounds per square inch.

   The point to keep in mind is this—*never attempt to inspect the lamp while it is hot.* Let it cool completely before opening up the lamp housing. Usually, 15 to 20 minutes cooling time is sufficient.

2. *Never expose your eyes to the direct rays of the mercury arc lamp.* Equipment design is such that the bulb is always shielded against the scattering of its rays. Remember that the unfiltered light from one of these lamps is rich in both ultraviolet and infrared rays—both of which are damaging to the eyes. *Severe retinal burns can result from exposure to the mercury arc rays.*

3. Be sure that the barrier filter is always in place when looking down through the microscope. *Removal of the barrier filter or exciter filter or both filters while looking through the microscope could cause eye injury.* It is possible to make mistakes of this nature if one is not completely familiar with the instrument. Remember, the function of the barrier filter is to prevent traces of ultraviolet light from reaching the eyes without blocking wavelengths of fluorescence.

Warm-up Period The lamps in fluorescence microscopes require a warm-up period. When they are first turned on the illumination is very low, but it increases to maximum in about 2 minutes. *Optimum illumination occurs when the equipment has been operating for 30 minutes or more.* Most manufacturers recommend leaving the instruments turned on for an hour or more when using them. It is not considered good economy to turn the instrument on and off several times within a 2- or 3-hour period.

Keeping a Log The life expectancy of a mercury arc lamp is around 400 hours. A log should be kept of the number of hours that the instrument is used so that inspection can be made of the bulb at approximately 200 hours. A card or piece of paper should be kept conveniently near the instrument so that the individual using the instrument is reminded to record the time that the instrument is turned on and off.

Filter Selection The most frequently used filter combination is the bluish Schott BG12 (AO #702) exciter and the yellowish Schott OG1 barrier filters.

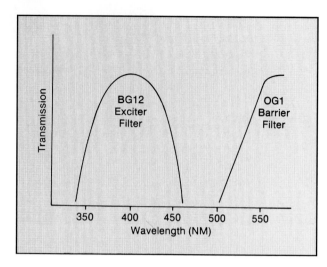

Figure 4.3 Spectral transmissions of BG12 and OG1 filters

Figure 4.3 shows the wavelength transmission of each of these filters. Note that the exciter filter gives peak emission of light in the 400 nm area of the spectrum. These rays are violet. It allows practically no green or yellow wavelengths through. The shortest wavelengths that this barrier filter lets through are green to greenish-yellow.

If a darker background is desired than is being achieved with the above filters, one may add a pale blue Schott BG38 to the system. It may be placed on either side of the heat filter, depending on the type of equipment being used. If it is placed between the lamp and heat filter, it will also function as another heat filter.

Examination When looking for material on the slide, it is best to use low- or high-power objectives. If the illuminator is a separate unit, as in figure 4.1, it may be desirable to move the illuminator out of position and use incandescent lighting for this phase of the work. Once the desirable field has been located, the mercury vapor arc illuminator can be moved into position. One problem with fluorescence microscopes is that most darkfield condensers do not illuminate well through the low-power objectives (exception: the Reichert-Toric setup used on some American Optical instruments).

Keep in mind that there is no diaphragm control on darkfield condensers. Some instruments are supplied with neutral density filters to reduce light intensity. The best system of illumination control, however, is achieved with objectives that have a built-in iris control. These objectives have a knurled ring that can be rotated to control the contrast.

For optimum results it is essential that oil be used between the condenser and the slide. And, of course, if the oil immersion lens is used, the oil must also be interposed between the slide and the objective. It is also important that special low-fluorescing immersion oil be used. *Ordinary immersion oil should be avoided.*

Although the ocular of a fluorescence microscope is usually 10×, one should not hesitate to try other size oculars if they are available. With bright-field microscopes it is generally accepted that noth-ing is gained by going beyond 1000× magnification. In a fluorescence microscope, however, the image is formed in a manner quite different from its bright-field counterpart, obviating the need for following the 1000× rule. The only loss by using the higher magnification is some brightness.

Laboratory Report

Complete all the answers to the questions on Laboratory Report 3–5 that pertain to this exercise.

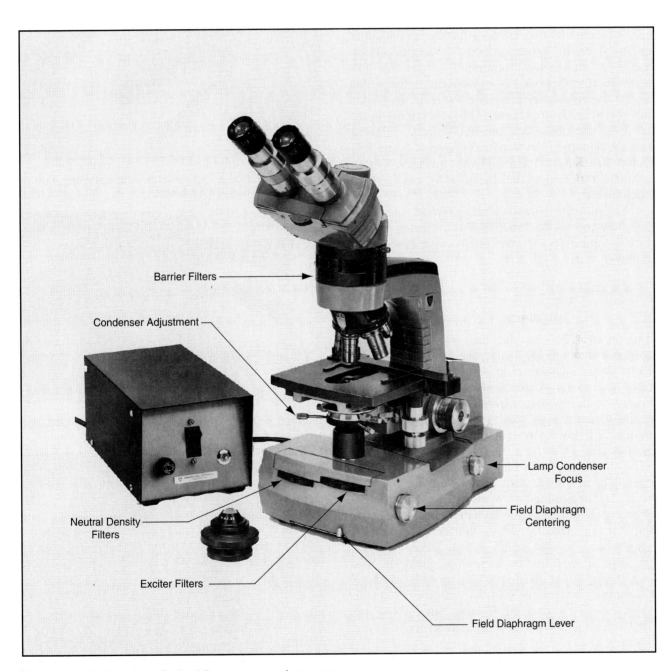

Figure 4.4 An American Optical fluorescence microscope

With an ocular micrometer properly installed in the eyepiece of your microscope, it is a simple matter to measure the size of microorganisms that are seen in the microscopic field. An **ocular micrometer** consists of a circular disk of glass that has graduations engraved on its upper surface. These graduations appear as shown in illustration B, figure 5.4. On some microscopes one has to disassemble the ocular so that the disk can be placed on a shelf in the ocular tube between the two lenses. On most microscopes, however, the ocular micrometer is simply inserted into the bottom of the ocular, as shown in figure 5.1. Before one can use the micrometer it is necessary to calibrate it for each of the objectives by using a stage micrometer.

The principal purpose of this exercise is to show you how to calibrate an ocular micrometer for the various objectives on your microscope. Proceed as follows:

Calibration Procedure

The distance between the lines of an ocular micrometer is an arbitrary value that has meaning only if the ocular micrometer is calibrated for the objective that is being used. A **stage micrometer,** also known as an *objective micrometer,* has lines scribed on it that are exactly 0.01 mm (10 μm) apart. Illustration C, figure 5.4 reveals the appearance of these graduations.

To calibrate the ocular micrometer for a given objective, it is necessary to superimpose the two scales and determine how many of the ocular graduations coincide with one graduation on the scale of the stage micrometer. Illustration A in figure 5.4 shows how the two scales appear when they are properly aligned in the microscopic field. In this case, seven ocular divisions match up with one stage micrometer division of 0.01 mm to give an ocular value of 0.01/7, or 0.00143 mm. Since there are 1000 micrometers in 1 millimeter, these divisions are 1.43 μm apart.

With this information known, the stage micrometer is replaced with a slide of organisms to be measured. Illustration D, figure 5.4, shows how a field of microorganisms might appear with the ocular

Figure 5.1 Ocular micrometer with retaining ring is inserted into base of eyepiece.

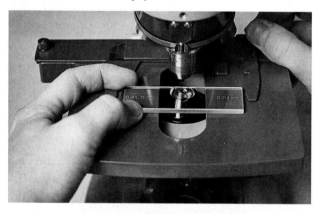

Figure 5.2 Stage micrometer is positioned by centering small glass disk over the light source.

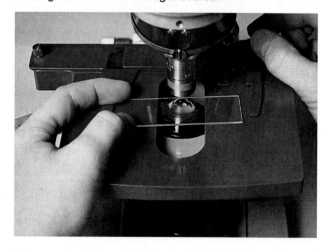

Figure 5.3 After calibration is completed, stage micrometer is replaced with slide for measurements.

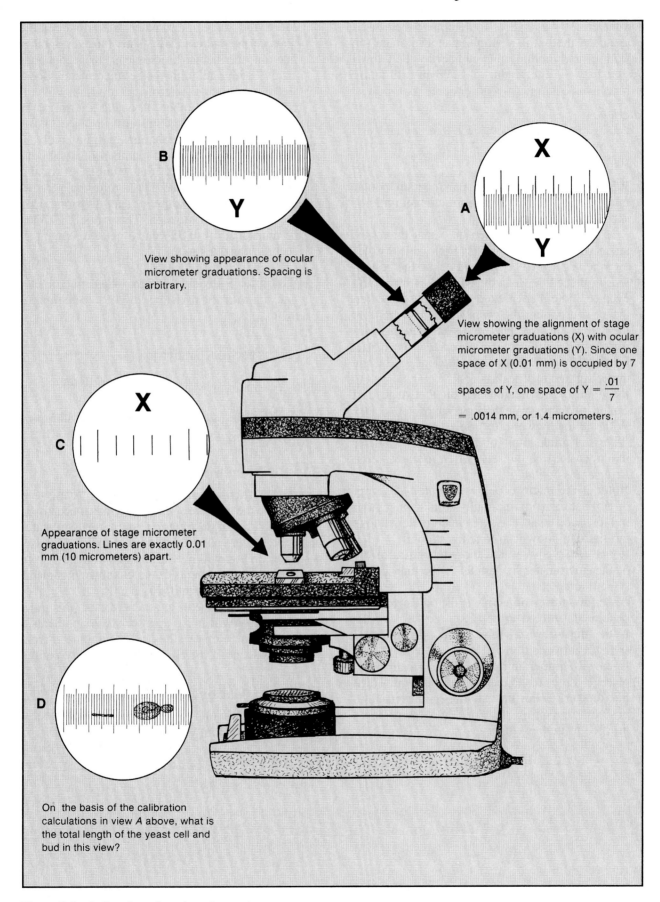

B

Y

View showing appearance of ocular
micrometer graduations. Spacing is
arbitrary.

X

A

Y

View showing the alignment of stage
micrometer graduations (X) with ocular
micrometer graduations (Y). Since one
space of X (0.01 mm) is occupied by 7
spaces of Y, one space of $Y = \dfrac{.01}{7}$

$= .0014$ mm, or 1.4 micrometers.

X

C

Appearance of stage micrometer
graduations. Lines are exactly 0.01
mm (10 micrometers) apart.

D

On the basis of the calibration
calculations in view *A* above, what is
the total length of the yeast cell and
bud in this view?

Figure 5.4   Calibration of ocular micrometer

micrometer in the eyepiece. To determine the size of an organism, then, it is a simple matter to count the graduations and multiply this number by the known distance between the graduations. When calibrating the objectives of a microscope, proceed as follows.

Materials:

ocular micrometer or eyepiece that contains a
   micrometer disk
stage micrometer

1. If eyepieces are available that contain ocular micrometers, replace the eyepiece in your microscope with one of them.

   If it is necessary to insert an ocular micrometer in your eyepiece, find out from your instructor whether it is to be inserted below the bottom lens or placed between the two lenses within the eyepiece. In either case, great care must be taken to avoid dropping the eyepiece or reassembling the lenses incorrectly. *Only with your instructor's prior approval shall eyepieces be disassembled.* Be sure that the graduations are on the upper surface of the glass disk.

2. Place the stage micrometer on the stage and center it exactly over the light source.

3. With the low-power (10×) objective in position, bring the graduations of the stage micrometer into focus, *using the coarse adjustment knob. Reduce the lighting.*

   Note: If the microscope has an automatic stop, do not use it as you normally would for regular microscope slides. The stage micrometer slide is too thick to allow it to function properly.

4. Rotate the eyepiece until the graduations of the ocular micrometer lie parallel to the lines of the stage micrometer.

5. If the **low-power objective** is the objective to be calibrated, proceed to step 8.

6. If the **high-dry objective** is to be calibrated, swing it into position and proceed to step 8.

7. If the **oil immersion lens** is to be calibrated, place a drop of immersion oil on the stage micrometer, swing the oil immersion lens into position, and bring the lines into focus; then, proceed to the next step.

8. Move the stage micrometer laterally until the lines at one end coincide. Then look for another line on the ocular micrometer that coincides *exactly* with one on the stage micrometer. Occasionally one stage micrometer division will include an even number of ocular divisions, as shown in illustration A. In most instances, however, several stage graduations will be involved. In this case, divide the number of stage micrometer divisions by the number of ocular divisions that coincide. The figure you get will be that part of a stage micrometer division that is seen in an ocular division. This value must then be multiplied by 0.01 mm to get the amount of each ocular division.

Example: 3 divisions of the stage micrometer line up with 20 divisions of the ocular micrometer.

$$\text{Each ocular division} = \frac{3}{20} \times 0.01$$
$$= 0.0015 \text{ mm}$$
$$= 1.5 \ \mu\text{m}$$

9. Replace the stage micrometer with slides of organisms to be measured.

Measuring Assignments

Organisms such as protozoans, algae, fungi, and bacteria in the next few exercises may need to be measured. If your instructor requires that measurements be made, you will be referred to this exercise.

Later on you will be working with unknowns. In some cases measurements of the unknown organisms will be pertinent to identification.

If trial measurements are to be made at this time, your instructor will make appropriate assignments.

Important: Remove the ocular micrometer from your microscope at the end of the laboratory period.

Laboratory Report

Answer the questions on combined Laboratory Report 3–5 that pertain to this exercise.

PART 2 Survey of Microorganisms

The four exercises of this unit pertain to a study of the overall population of the microbial world. Too often, in our serious concern with the direct applications of microbiology to human welfare, we neglect the large number of interesting free-living microorganisms that abound in the water, soil, and air. In this unit we will study representatives of three kingdoms (Monera, Protista, and Myceteae) that include the majority of microorganisms in fresh water and air. Although there are no phyla listed below under Kingdom Animalia, certain species of invertebrates are usually encountered when studying pond-water populations. It is for this reason that Exercise 7 (Microscopic Invertebrates) has been included to satisfy student curiosity pertaining to the identity of some of these behemoths of the microbial world. The organisms to be studied at this time are shown in **bold face type** in the following outline:

Kingdom MONERA
    Subkingdom I: **Cyanobacteria**
    Subkingdom II: **Bacteria**

Kingdom PROTISTA
    Subkingdom I: **Protozoa**
    Subkingdom II: **Algae**

Kingdom MYCETEAE
    Division 1: Gymnomycota
    Division 2: Mastigomycota
    Division 3: **Amastigomycota**

Kingdom PLANTAE
Kingdom ANIMALIA

The principal source of protozoa, algae, and cyanobacteria in Exercise 6 will be pond waters. Water and bottom debris from various ponds will be available in specimen bottles on the demonstration table. Wet mount slides from these samples will be studied with brightfield or phase-contrast microscopes to identify the various genera. It is necessary to study these microorganisms simultaneously because the organisms will be identified as they are encountered, not in any particular sequence.

The fungi (Amastigomycota) for Exercise 8 will be collected from the air. Plates of media that have been exposed to the air will be the principal source of these organisms. Microscope slides will be made from plate cultures so that detailed structure of molds and yeasts can be observed. For our study of bacteria in Exercise

9 we will expose nutrient agar plates and tubes of nutrient broth to the air, the hands, and objects in the environment. After the plates and broth tubes have been incubated, microscope slides will be made from them for microscopic examination.

Classification Systems

Although Whittaker's five-kingdom system of classification as presented on the previous page, has been used for some time, the "three-domain" system discussed on the next page presents a truer picture of the relationships between microorganisms. Since 1977 when Carl Woese and Raplp S. Wolfe at the University of Illinois reported the discovery of domain *Archaea,* many subsequent studies of the nucleotide sequence have reinforced the concept that there are three major groups of life on this planet. It has also been established that the archaea are more like the eucaryotes than the bacteria. The similarity to eucaryotes relates to the type of proteins coded by the genes in *Methanoccoccus jannaschii.* The transcription apparatus, which synthesizes RNA in this archaeal organism, is completely different form that seen in the bacteria, and seems to be a simpler version of that found in eucaryotes. The evolutionary tree that is illustrated below visualizes the present concept of the three-domain system.

The student should be aware that although it is convenient for us to use the five-kingdom system at the present time, this new system of three domains is really more relevent. However, since there are still some problems to be solved in developing and using the new arrangement, we feel more comfortable at this time utilizing the five-kingdom system.

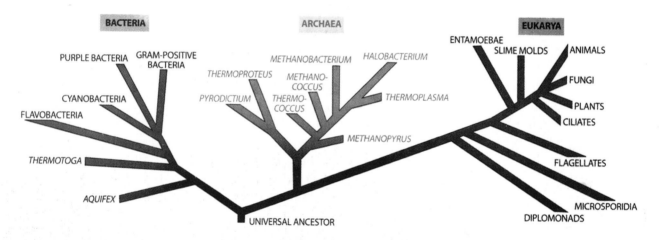

From: Extremophiles. Michael T. Madigan and Barry L. Marrs in *Scientific American* Vol. 276, Number 4, pages 82–87, April 1997.

The Three-Domain System

Instead of five kingdoms, this system separates all living forms into three DOMAINS. Note that Kingdom Monera (procaryotes) of the five-kingdom system is split into two domains: Archaea and Bacteria. All remaining forms of life (nine kingdoms) are grouped under Domain Eucarya. Note, also, that **bold face type** signifies those groups that contain only microorganisms. Groups that are underlined are the ones that we will study in this part.

Domain Archaea (all are procaryotes)

Division The **methanogens** (obligate anaerobes)
Division The **extreme halophiles** (require at least 1.5M NaCl)
Division The **hyperthermophiles** (require temperatures of >70° C)

Domain Bacteria (all are procaryotes)

Division **Firmicutes** (gram-positive bacteria)
Division **Cyanobacteria**
Division **Proteobacteria**
Division **Spirochaetes**
Division *Flavobacterium, Bacteriodes,* and relatives
Division **Chlamydias**
Division *Deinococcus* and relatives
Division **Chloroflexaceae** (green non-sulfur bacteria)
Division **Chlorobiaceae** (green sulfur bacteria)
Division *Plantomyces* and relatives
Division **Thermotogales**
Division *Verrucomicrobium* and *Prosthecobacter*
Division **Aquificales**
Division *Fibrobacter*

Domain Eucarya (all have nuclei)

Group 1. Anaerobes, 70S ribosomes, no Golgi, no mitochondria: *Giardia, Microsporidia*
Group 2. Aerobes or facultative, 80S ribosomes, Golgi, mitochondria, chloroplasts if photosynthetic
    Kingdom **Euglenids**
    Kingdom **Entamebas**
    Kingdom Cellular Slime Molds
    Kingdom **Dinoflagellates** and **Ciliates**
    Kingdom Stramenopiles **Diatoms,** Brown Seaweeds, Xanthophytes, Silicoflagellate Bicosoecid
        Flagellates, Oomycetes)
    Kingdom Red Seaweeds
    Kingdom **Myceteae** (fungi)
    Kingdom Plants and **Green Algae**
    Kingdom Animalia (includes microscopic invertebrates and acellular slime molds)

References:

Madigan, Michael T. et al. 1997 *Brock Biology of Microorganisms,* 8th ed. Englewood Cliffs, NJ: Prentice-Hall
Madigan, Michael T. et al, 1997 *Extremophiles.* Scientific American Vol 276, Number 4: pp 82–87. New York.
Knoll, A. H. 1992. *The early evolution of eukaryotes: A geological perspective.* Science 256: 622–627.
Woese, C. R. et al. 1994 *Towards a natural system of organisms: Proposal for the domains Archae, Bacteria, and Eucarya.* Proc. Natl. Acad Sci. 87: pp 4576–4579.
Internet: http://phylogeny. arizona.edu/tree/phylogeny.html

Contributor:

Barbara B. Hemmingsen, Ph.D.
San Diego State University.

6

Protozoa, Algae, and Cyanobacteria

In this exercise a study will be made of protozoans, algae, and cyanobacteria that are found in pond water. Bottles that contain water and bottom debris from various ponds will be available for study. Illustrations and text provided in this exercise will be used to assist you in an attempt to identify the various types that are encountered. Unpigmented, moving microorganisms will probably be protozoans. Greenish or golden-brown organisms are usually algae. Organisms that appear blue-green will be cyanobacteria. Supplementary books on the laboratory bookshelf will also be available for assistance in identifying organisms that are not described in the short text of this exercise. If you encounter invertebrates and are curious as to their identification, you may refer to Exercise 7; however, keep in mind that our prime concern here is only with protozoans, algae, and cyanobacteria.

The purpose of this exercise is, simply, to provide you with an opportunity to become familiar with the differences between the three groups by comparing their characteristics. The extent to which you will be held accountable for the names of various organisms will be determined by your instructor. The amount of time available for this laboratory exercise will determine the depth of scope to be pursued.

To study the microorganisms of pond water, it will be necessary to make wet mount slides. The procedure for making such slides is relatively simple. All that is necessary is to place a drop of suspended organisms on a microscope slide and cover it with a cover glass. If several different cultures are available, the number of the bottle should be recorded on the slide with a china marking pencil. As you prepare and study your slides, observe the following guidelines:

Materials:
bottles of pond-water samples
microscope slides and cover glasses
rubber-bulbed pipettes and forceps
china marking pencil
reference books

1. Clean the slide and cover glass with soap and water, rinse thoroughly, and dry. Do not attempt to study a slide that lacks a cover glass.
2. When using a pipette, insert it into the bottom of the bottle to get a maximum number of organisms. Very few organisms will be found swimming around in mid-depth of the bottle.
3. To remove filamentous algae from a specimen bottle use forceps. Avoid putting too much material on the slides.
4. Explore the slide first with the low-power objective. Reduce the lighting with the iris diaphragm. Keep the condenser at its highest point.
5. When you find an organism of interest, swing the high-dry objective into position and adjust the lighting to get optimum contrast. If your microscope has phase-contrast elements, use them.
6. Refer to Figures 6.1 through 6.6 and the text on these pages to identify the various organisms that you encounter.
7. Record your observations on the Laboratory Report.

Protozoa

The Subkingdom **Protozoa** includes all the animal-like microorganisms of the Kingdom Protista. All of the representatives in this subkingdom are single-celled organisms; however, some of them do form colonial aggregates.

Externally, the cells are covered with a cell membrane, or pellicle; cell walls are absent, and distinct nuclei with nuclear membranes are present. Specialized organelles, such as contractile vacuoles, cytostomes, mitochondria, ribosomes, flagella, and cilia may also be present.

All protozoa produce *cysts,* which are resistant dormant stages that enable them to survive drought, heat, and freezing. They reproduce asexually by cell division and exhibit various degrees of sexual reproduction.

The Subkingdom Protozoa is divided into three phyla: Sarcomastigophora, Ciliophora, and Apicomplexa. Type of locomotion plays an important role in classification here. A brief description of each phylum follows:

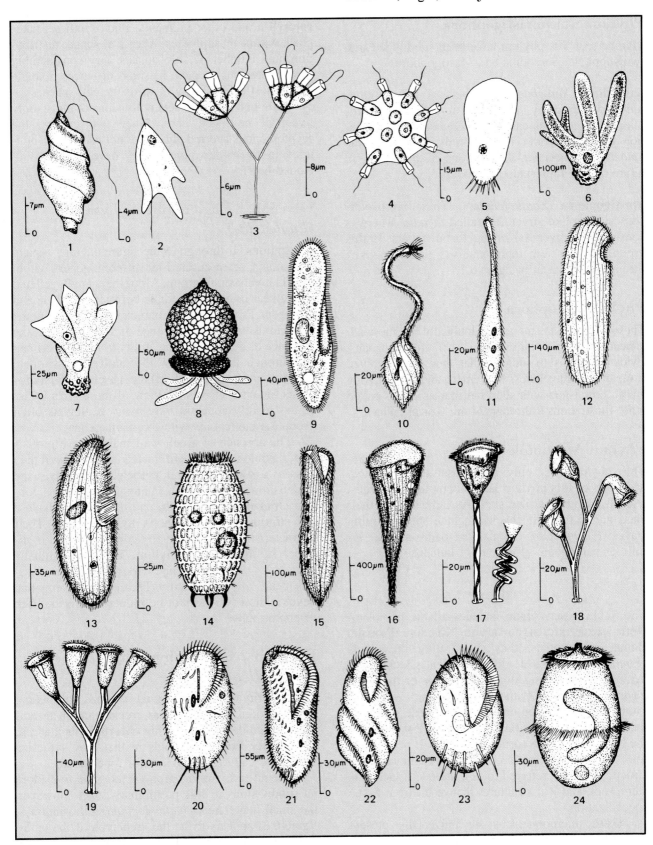

Figure 6.1 Protozoans

1. *Heteronema*
2. *Cercomonas*
3. *Codosiga*
4. *Protospongia*
5. *Trichamoeba*
6. *Amoeba*
7. *Mayorella*
8. *Diffugia*
9. *Paramecium*
10. *Lacrymaria*
11. *Lionotus*
12. *Loxodes*
13. *Blepharisma*
14. *Coleps*
15. *Condylostoma*
16. *Stentor*
17. *Vorticella*
18. *Carchesium*
19. *Zoothamnium*
20. *Stylonychia*
21. *Onychodromos*
22. *Hypotrichidium*
23. *Euplotes*
24. *Didinium*

Phylum Sarcomastigophora

Members of this phylum have been subdivided into two subphyla: Sarcodina and Mastigophora.

Sarcodina (*Amoebae*) Members of this subphylum move about by the formation of flowing protoplasmic projections called *pseudopodia.* The formation of pseudopodia is commonly referred to as *amoeboid movement.* Illustrations 5 through 8 in figure 6.1 are representative amoebae.

Mastigophora (*Zooflagellates*) These protozoans possess whiplike structures called *flagella.* There is considerable diversity among the members of this group. Only a few representatives (illustrations 1 through 4) are seen in figure 6.1.

Phylum Ciliophora

These microorganisms are undoubtedly the most advanced and structurally complex of all protozoans. Evidence seems to indicate that they have evolved from the zooflagellates. Movement and food-getting are accomplished with short hairlike structures called *cilia.* Illustrations 9 through 24 are typical ciliates.

Phylum Apicomplexa

This phylum has only one class, the *Sporozoa.* Members of this phylum lack locomotor organelles, and all are internal parasites. As indicated by their class name, their life cycles include spore-forming stages. *Plasmodium,* the malarial parasite, is a significant pathogenic sporozoan of humans.

Algae

The Subkingdom **Algae** includes all the photosynthetic eukaryotic organisms in Kingdom Protista. Being true protists, they differ from the plants (*Plantae*) in that tissue differentiation is lacking. In Whittaker's five-kingdom system some of the algae have been included with protozoans.

The algae may be unicellular, as those shown in the top row of figure 6.2; colonial, like the four in the lower right-hand corner of figure 6.2; or filamentous, as those in figure 6.3. The undifferentiated algal structure is often referred to as a *thallus.* It lacks the stem, root, and leaf structures that result from tissue specialization.

These microorganisms are universally present where ample moisture, favorable temperature, and sufficient sunlight exist. Although a great majority of them live submerged in water, some grow on soil; others grow on the bark of trees or on the surfaces of rocks.

Algae have distinct, visible nuclei and chloroplasts. **Chloroplasts** are organelles that contain chlorophyll **a** and other pigments. Photosynthesis takes place within these bodies. The size, shape, distribution, and number of chloroplasts vary considerably from species to species. In some instances a single chloroplast may occupy most of the cell space.

Although there are seven divisions of algae, only five will be listed here. Since two groups, the cryptomonads and red algae, are not usually encountered in freshwater ponds, they have not been included here.

Division 1 Euglenophycophyta (*Euglenoids*)

Illustrations 1 through 6 in figure 6.2 are typical euglenoids, representing four different genera within this relatively small group. All of them are flagellated and appear to be intermediate between the algae and protozoa. Protozoanlike characteristics seen in the euglenoids are: (1) the absence of a cell wall, (2) the presence of a gullet, (3) the ability to ingest food, but not through the gullet, (4) the ability to assimilate organic substances, and (5) the absence of chloroplasts in some species. In view of these facts, it becomes readily apparent why many zoologists often group the euglenoids with the zooflagellates.

The absence of a cell wall makes these protists very flexible in movement. Instead of a cell wall they possess a semirigid outer **pellicle,** which gives the organism a definite form. Photosynthetic types contain **chlorophylls a** and **b,** and they always have a red **stigma** (eyespot) that is light-sensitive. Their characteristic food storage compound is a lipopolysaccharide, **paramylum.** The photosynthetic euglenoids can be bleached experimentally by various means in the laboratory. The colorless forms that develop, however, cannot be induced to revert back to phototrophy.

Division 2 Chlorophycophyta (*Green Algae*)

The majority of algae observed in ponds will belong to this group. They are grass-green in color, resembling the euglenoids in having **chlorophylls a** and **b.** They differ from euglenoids in that they sythesize **starch** instead of paramylum for food storage.

The diversity of this group is too great to explore its subdivisions in this preliminary study; however, the small flagellated *Chlamydomonas* (illustration 8, figure 6.2) appears to be the archetype of the entire group and has been extensively studied. Many colonial forms, such as *Pandorina, Eudorina, Gonium,* and *Volvox* (illustrations 14, 15, 19, and 20, figure 6.2) consist of organisms similar to *Chlamydomonas.* It is the consensus that all the filamentous algae have evolved from this flagellated form.

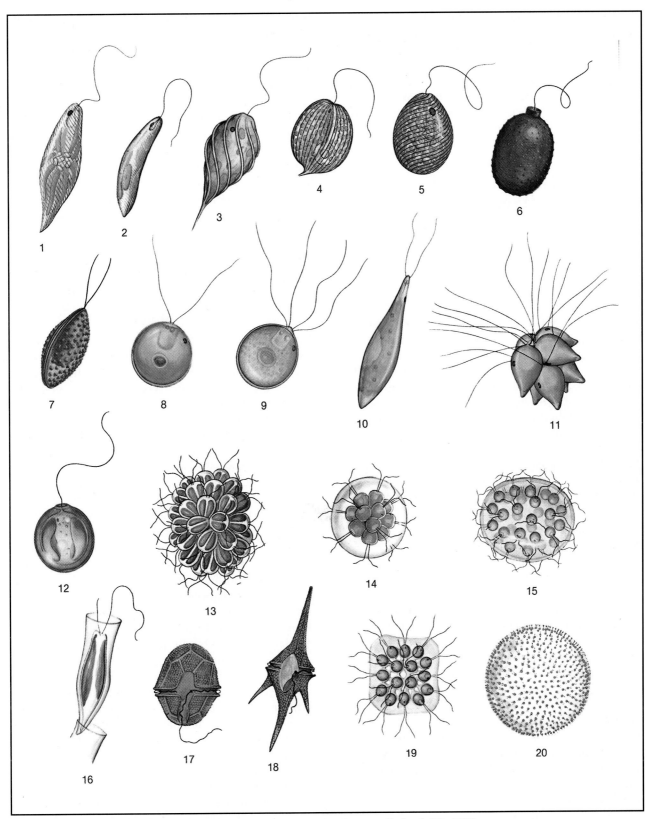

Courtesy of the U.S. Environmental Protection Agency, Office of Research & Development, Cincinnati, Ohio 45268.

1. *Euglena* (700X)
2. *Euglena* (700X)
3. *Phacus* (1000X)
4. *Phacus* (350X)
5. *Lepocinclis* (350X)

6. *Trachelomonas* (1000X)
7. *Phacotus* (1500X)
8. *Chlamydomonas* (1000X)
9. *Carteria* (1500X)
10. *Chlorogonium* (1000X)

11. *Pyrobotrys* (1000X)
12. *Chrysococcus* (3000X)
13. *Synura* (350X)
14. *Pandorina* (350X)
15. *Eudorina* (175X)

16. *Dinobyron* (1000X)
17. *Peridinium* (350X)
18. *Ceratium* (175X)
19. *Gonium* (350X)
20. *Volvox* (100X)

Figure 6.2 Flagellated algae

Except for *Vaucheria* and *Tribonema,* all of the filamentous forms in figure 6.3 are Chlorophycophyta. All of the nonfilamentous, nonflagellated algae in figure 6.4 also are green algae.

A unique group of green algae are the **desmids** (illustrations 16 through 20, figure 6.4). With the exceptions of a few species, the cells of desmids consist of two similar halves, or semicells. The two halves usually are separated by a constriction, the *isthmus.*

Division 3 Chrysophycophyta (*Golden Brown Algae*)

This large diversified division consists of over 6000 species. They differ from the euglenoids and green algae in that: (1) food storage is in the form of **oils** and **leucosin,** a polysaccharide, (2) **chlorophylls a** and **c** are present, and (3) **fucoxanthin,** a brownish pigment, is present. It is the combination of fucoxanthin, other yellow pigments, and the chlorophylls that causes most of these algae to appear golden brown.

Representatives of this division are seen in figures 6.2, 6.3, and 6.5. In figure 6.2, *Chrysococcus, Synura,* and *Dinobyron* are typical flagellated chrysophycophytes. *Vaucheria* and *Tribonema* are the only filamentous chrysophycophytes shown in figure 6.3.

All of the organisms in figure 6.5 are chrysophycophytes and fall into a special category of algae called the **diatoms.** The diatoms are unique in that they have hard cell walls of pectin, cellulose, or silicon oxide that are constructed in two halves. The two halves fit together like lid and box.

Skeletons of dead diatoms accumulate on the ocean bottom to form *diatomite,* or "diatomaceous earth," that is commercially available as an excellent polishing compound. It is postulated by some that much of our petroleum reserves may have been formulated by the accumulation of oil from dead diatoms over millions of years.

Division 4 Phaeophycophyta (*Brown Algae*)

With the exception of three freshwater species, all algal protists of this division exist in salt water (marine); thus, it is unlikely that you will encounter any phaeophycophytes in this laboratory experience. These algae have essentially the same pigments seen in the chrysophycophytes, but they appear brown because of the masking effect of the greater amount of fucoxanthin. Food storage in the brown algae is in the form of **laminarin,** a polysaccharide, and **mannitol,** a sugar alcohol. All species of brown algae are multicellular and sessile. Most seaweeds are brown algae.

Division 5 Pyrrophycophyta (*Fire Algae*)

The principal members of this division are the **dinoflagellates.** Since the majority of these protists are marine, only two freshwater forms are shown in figure 6.2: *Peridinium* and *Ceratium* (illustrations 17 and 18). Most of these protists possess celulose walls of interlocking armor plates, as in *Ceratium.* Two flagella are present: one is directed backward when swimming, and the other moves within a transverse groove. Many marine dinoflagellates are bioluminescent. Some species of marine *Gymnodinium,* when present in large numbers, produce the **red tides** that cause water discoloration and unpleasant odors along our coastal shores.

These algae have **chlorophylls a** and **c** and several xanthophylls. Foods are variously stored in the form of **starch, fats,** and **oils.**

Cyanobacteria

The *Cyanobacteria* (blue-green bacteria) constitute Subkingdom I of the Kingdom Monera. Although these microorganisms were formerly referred to as algae, their prokaryotic type of nucleus definitely sets them apart from the eukaryotic algae.

Although some bacteria are phototrophic, the difference between phototrophic bacteria and cyanobacteria is that the cyanobacteria have **chlorophyll a,** and the phototrophic bacteria do not. Bacteriochlorophyll is the photosynthetic pigment in the phototrophic bacteria.

Over 1000 species of cyanobacteria have been reported. They are present in almost all moist environments from the tropics to the poles, including both fresh water and marine. Figure 6.6 illustrates only a random few that are frequently seen.

The designation of these bacteria as "blue-green" is somewhat misleading in that many cyanobacteria are actually black, purple, red, and various shades of green instead of blue-green. These different colors are produced by the varying proportions of the numerous pigments present. These pigments are **chlorophyll a, carotene, xanthophylls,** blue **c-phycocyanin,** and red **c-phycoerythrin.** The last two pigments are unique to the cyanobacteria and red algae.

Cellular structure is considerably different from the eukaryotic algae. As stated earlier, nuclear membranes in cyanobacteria are absent. The nuclear material consists of DNA granules in a more or less colorless area in the center of the cell.

Unlike the algae, the pigments of the cyanobacteria are not contained in chloroplasts; instead, they are located in granules (**phycobilisomes**) that are attached to membranes (**thylakoids**) that permeate the cytoplasm.

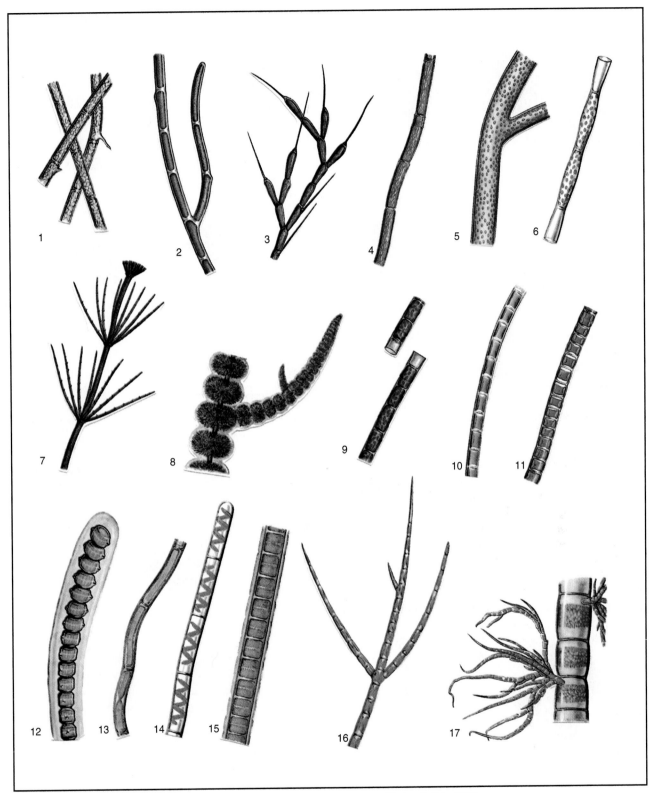

1. *Rhizoclonium* (175X)
2. *Cladophora* (100X)
3. *Bulbochaete* (100X)
4. *Oedogonium* (350X)

5. *Vaucheria* (100X)
6. *Tribonema* (300X)
7. *Chara* (3 X)
8. *Batrachospermum* (2 X)

9. *Microspora* (175X)
10. *Ulothrix* (175X)
11. *Ulothrix* (175X)
12. *Desmidium* (175X)

13. *Mougeotia* (175X)
14. *Spirogyra* (175X)
15. *Zygnema* (175X)
16. *Stigeoclonium* (300X)
17. *Draparnaldia* (100X)

Figure 6.3 Filamentous algae

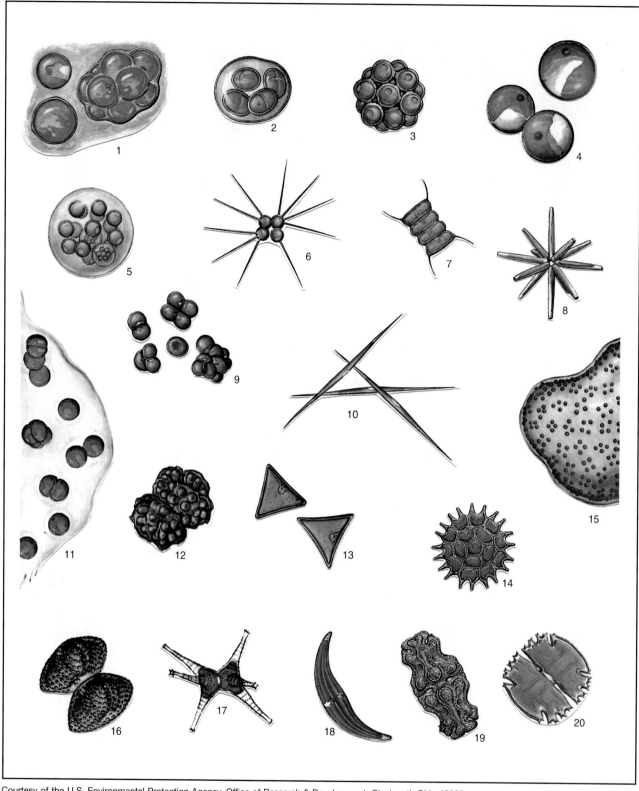

Courtesy of the U.S. Environmental Protection Agency, Office of Research & Development, Cincinnati, Ohio 45268.

1. *Chlorococcum* (700X)
2. *Oocystis* (700X)
3. *Coelastrum* (350X)
4. *Chlorella* (350X)
5. *Sphaerocystis* (350X)

6. *Micractinium* (700X)
7. *Scendesmus* (700X)
8. *Actinastrum* (700X)
9. *Phytoconis* (700X)
10. *Ankistrodesmus* (700X)

11. *Pamella* (700X)
12. *Botryococcus* (700X)
13. *Tetraedron* (1000X)
14. *Pediastrum* (100X)
15. *Tetraspora* (100X)

16. *Staurastrum* (700X)
17. *Staurastrum* (350X)
18. *Closterium* (175X)
19. *Euastrum* (350X)
20. *Micrasterias* (175X)

Figure 6.4 **Nonfilamentous and nonflagellated algae**

Courtesy of the U.S. Environmental Protection Agency, Office of Research & Development, Cincinnati, Ohio 45268.

1. *Diatoma* (1000X)
2. *Gomphonema* (175X)
3. *Cymbella* (175X)
4. *Cymbella* (1000X)
5. *Gomphonema* (2000X)
6. *Cocconeis* (750X)

7. *Nitschia* (1500X)
8. *Pinnularia* (175X)
9. *Cyclotella* (1000X)
10. *Tabellaria* (175X)
11. *Tabellaria* (1000X)
12. *Synedra* (350X)

13. *Synedra* (175X)
14. *Melosira* (750X)
15. *Surirella* (350X)
16. *Stauroneis* (350X)
17. *Fragillaria* (750X)
18. *Fragillaria* (750X)

19. *Asterionella* (175X)
20. *Asterionella* (750X)
21. *Navicula* (750X)
22. *Stephanodiscus* (750X)
23. *Meridion* (750X)

Figure 6.5 Diatoms

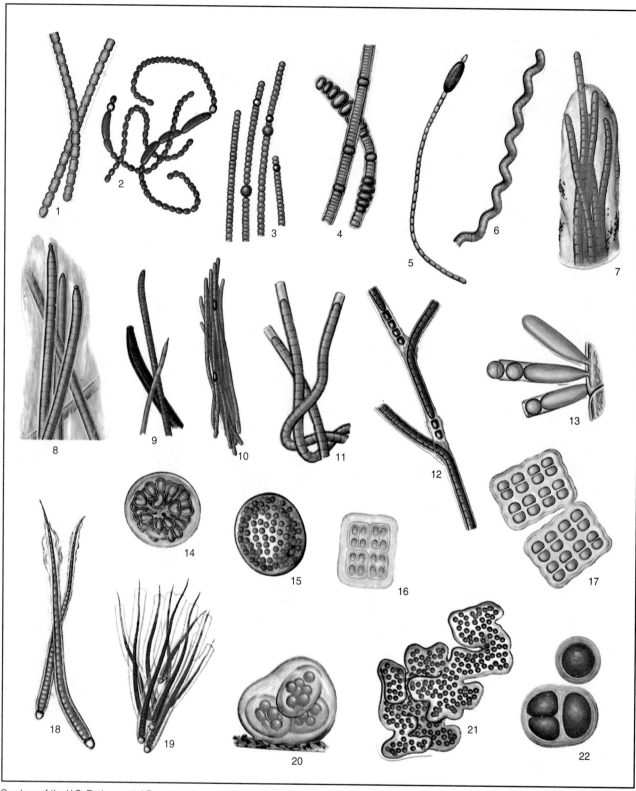

Courtesy of the U.S. Environmental Protection Agency, Office of Research & Development, Cincinnati, Ohio 45268.

1. *Anabaena* (350X)
2. *Anabaena* (350X)
3. *Anabaena* (175X)
4. *Nodularia* (350X)
5. *Cylindrospermum* (175X)
6. *Arthrospira* (700X)

7. *Microcoleus* (350X)
8. *Phormidium* (350X)
9. *Oscillatoria* (175X)
10. *Aphanizomenon* (175X)
11. *Lyngbya* (700X)
12. *Tolypothrix* (350X)

13. *Entophysalis* (1000X)
14. *Gomphosphaeria* (1000X)
15. *Gomphosphaeria* (350X)
16. *Agmenellum* (700X)
17. *Agmenellum* (175X)
18. *Calothrix* (350X)

19. *Rivularia* (175X)
20. *Anacystis* (700X)
21. *Anacystis* (175X)
22. *Anacystis* (700X)

Figure 6.6 Cyanobacteria

While looking for protozoa, algae, and cyanobacteria in pond water, one invariably encounters large, transparent, complex microorganisms that, to the inexperienced, appear to be protozoans. In most instances these moving "monsters" are rotifers (illustrations 13 through 17, figure 7.1); in some cases they are copepods, daphnia, or any one of the other forms illustrated in figure 7.1.

All of the organisms illustrated in figure 7.1 are multicellular with organ systems. If organ systems are present, then the organisms cannot be protists, because organs indicate the presence of tissue differentiation. Collectively, these microscopic forms are designated as "invertebrates." It is to prevent you from misinterpreting some of these invertebrates as protozoans that they are described here.

In using figure 7.1 to identify what you consider might be an invertebrate, keep in mind that there are considerable size differences. A few invertebrates, such as *Dugesia* and *Hydra,* are macroscopic in adult form but microscopic when immature. Be sure to judge size differences by reading the scale beside each organism. The following phyla are listed according to the degree of complexity, the simplest first.

Phylum Coelenterata
(Illustration 1)

Members of this phylum are almost exclusively marine. The only common freshwater form shown in figure 7.1 is *Hydra.* In addition, there are a few less-common freshwater genera similar to the marine hydroids.

The hydras are quite common in ponds and lakes. They are usually attached to rocks, twigs, or other substrata. Around the mouth at the free end are five tentacles of variable length, depending on the species. Smaller organisms, such as *Daphnia,* are grasped by the tentacles and conveyed to the mouth. These animals have a digestive cavity that makes up the bulk of the interior. Since no anus is present, undigested remains of food are expelled through the mouth.

Phylum Platyhelminthes
(Illustrations 2, 3, 4, 5)

The invertebrates of this phylum are commonly referred to as **flatworms.** The phylum contains two parasitic classes and one class of free-living organisms, the *Turbellaria.* It is the organisms in this class that are encountered in fresh water. The four genera of this class shown in figure 7.1 are *Dugesia, Planaria, Macrostomum,* and *Provortex.* The characteristics common to all these organisms are dorsoventral flatness, a ciliated epidermis, a ventral mouth, and eyespots on the dorsal surface near the anterior end. As in the coelenterates, undigested food must be ejected through the mouth since no anus is present. Reproduction may be asexual by fission or fragmentation; generally, however, reproduction is sexual, each organism having both male and female reproductive organs. Species identification of the turbellarians is exceedingly difficult and is based to a great extent on the details of the reproductive system.

Phylum Nematoda
(Illustration 6)

The members of this phylum are the **roundworms.** They are commonly referred to as *nemas* or *nematodes.* They are characteristically round in cross section, have an external cuticle without cilia, lack eyes, and have a tubular digestive system complete with mouth, intestine, and anus. The males are generally much smaller than the females and have a hooked posterior end. The number of named species is only a fraction of the total nematodes in existence. Species identification of these invertebrates requires very detailed study of many minute anatomical features, which requires complete knowledge of anatomy.

Phylum Aschelminthes

This phylum includes classes *Gastrotricha* and *Rotifera.* Most of the members of this phylum are microscopic. Their proximity to the nematodes in classification is due to the type of body cavity (*pseudocoel*) that is present in both phyla.

The **gastrotrichs** (illustrations 7, 8, 9, 10) range from 10 to 540 μm in size. They are very similar to the ciliated protozoans in size and habits. The typical gastrotrich is elongate, flexible, forked at the posterior end, and covered with bristles. The digestive system consists of an anterior mouth surrounded by bristles, a pharynx, intestine, and posterior anus. Species identification is based partially on the shape of the head, tail structure and size, and distribution of spines. Overall length is also an important identification characteristic. They feed primarily on unicellular algae.

The **rotifers** (illustrations 13, 14, 15, 16, and 17) are most easily differentiated by the wheellike arrangement of cilia at the anterior end and the presence of a chewing pharynx within the body. They are considerably diversified in food habits: some feed on algae and protozoa, others on juices of plant cells, and some are parasitic. They play an important role in keeping waters clean. They also serve as food for small worms and crustaceans, being an important link in the food chain of fresh waters.

Phylum Annelida
(Illustration 18)

This phylum includes three classes: *Oligochaeta, Polychaeta,* and *Hirudinea*. Since polychaetes are primarily marine and the leeches (Hirudinea) are mostly macroscopic and parasitic, only the oligochaete is represented in figure 7.1. Some oligochaetes are marine, but the majority are found in fresh water and soil. These worms are characterized by body segmentation, bristles (*setae*) on each segment, an anterior mouth, and a roundish protrusion—the *prostomium*—anterior to the mouth. Although most oligochaetes breathe through the skin, some aquatic forms possess gills at the posterior end or along the sides of the segments. Most oligochaetes feed on vegetation; some feed on the muck of the bottoms of polluted waters, aiding in purifying such places.

Phylum Tardigrada
(Illustrations 11 and 12)

These invertebrates are of uncertain taxonomic position. They appear to be closely related to both the Annelida and Arthropoda. They are commonly referred to as the **water bears.** They are generally no more than 1 mm long, with a head, four trunk segments, and four pairs of legs. The ends of the legs may have claws, fingers, or disklike structures. The anterior end has a retractable snout with teeth. Eyes are often present. Sexes are separate, and females are oviparous. They are primarily herbivorous. Locomotion is by crawling, not swimming. During desiccation of their habitat they contract to form barrel-shaped *tuns* and are able to survive years of dryness, even in extremes of heat and cold. Widespread distribution is due to dispersal of the tuns by the wind.

Phylum Arthropoda
(Illustrations 19, 20, 21, 22, 23)

This phylum contains most of the known Animalia, almost a million species. Representatives of three groups of the Class *Crustacea* are shown in figure 7.1: *Cladocera, Ostracoda,* and *Copepoda*. The characteristics these three have in common are jointed appendages, an exoskeleton, and gills.

The **cladocera** are represented by *Daphnia* and *Latonopsis* in figure 7.1. They are commonly known as **water fleas.** All cladocera have a distinct head. The body is covered by a bivalvelike carapace. There is often a distinct cervical notch between the head and body. A compound eye may be present; when present, it is movable. They have many appendages: antennules, antennae, mouth parts, and four to six pairs of legs.

The **ostracods** are bivalved crustaceans that are distinguished from minute clams by the absence of lines of growth on the shell. Their bodies are not distinctly segmented. They have seven pairs of appendages. The end of the body terminates with a pair of *caudal furca*.

The **copepods** represented here are *Cyclops* and *Canthocamptus*. They lack the shell-like covering of the ostracods and cladocera; instead, they exhibit distinct body segmentation. They may have three simple eyes or a single median eye. Eggs are often seen attached to the abdomen on females.

Laboratory Report

There is no Laboratory Report for this exercise.

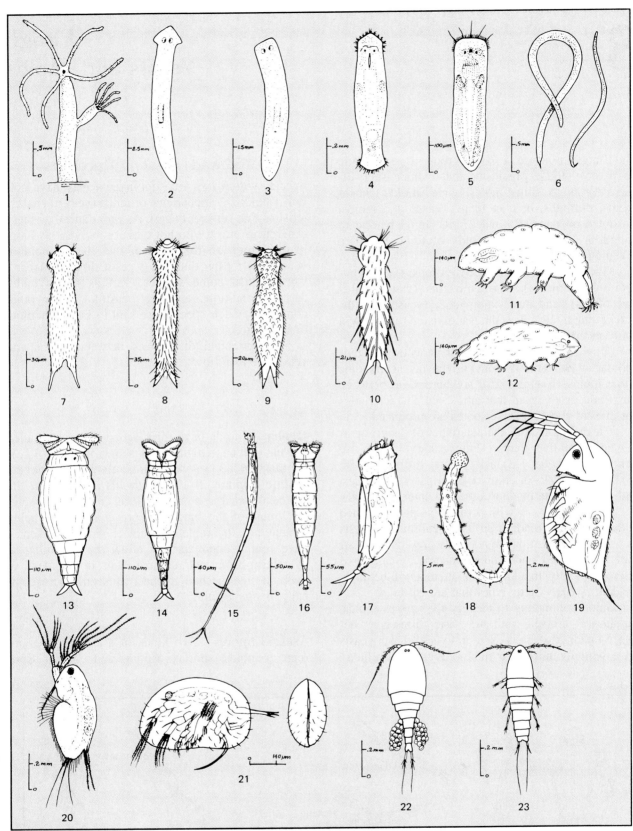

1. *Hydra*
2. *Dugesia*
3. *Planaria*
4. *Macrostomum*
5. *Provortex*
6. *Nematodes*
7. *Lepidermella*
8, 9, 10. *Chaetonotus*
11, 12. *Hypsibius*
13, 14. *Philodina*
15, 16. *Rotaria*
17. *Euchlanis*
18. *Oligochaete*
19. *Daphnia*
20. *Latonopis*
21. *Ostracod*
22. *Cyclops*
23. *Canthocamptus*

Figure 7.1 Microscopic invertebrates

Fungi: Yeasts and Molds

The fungi comprise a large group of eukaryotic nonphotosynthetic organisms that include such diverse forms as slime molds, water molds, mushrooms, puffballs, bracket fungi, yeasts, and molds. As noted on pages 25 and 27, the fungi belong to Kingdom **Myceteae.** The study of fungi is called **mycology.**

Myceteae consist of three divisions: Gymnomycota (slime molds), Mastigomycota (water molds and others), and Amastigomycota (yeasts, molds, bracket fungi, etc.). It is the last division that we will study in this exercise.

Fungi may be saprophytic or parasitic and unicellular or filamentous. Some organisms, such as the slime molds (Exercise 26), are borderline between fungi and protozoa in that amoeboid characteristics are present and fungi-like spores are produced.

The distinguishing characteristics of the group as a whole are that they: (1) are eukaryotic, (2) are nonphotosynthetic, (3) lack tissue differentiation, (4) have cell walls of chitin or other polysaccharides, and (5) propagate by spores (sexual and/or asexual).

In this study we will examine prepared stained slides and slides made from living cultures of yeasts and molds. Molds that are normally present in the air will be cultured and studied macroscopically and microscopically. In addition, an attempt will be made to identify the various types that are cultured.

Before attempting to identify the various molds, familiarize yourself with the basic differences between molds and yeasts. Note in figure 8.1 that yeasts are essentially unicellular and molds are multicellular.

Mold and Yeast Differences

Species within the Amastigomycota may have cottony (moldlike) appearance or moist (yeasty) characteristics that set them apart. As pronounced as these differences are, we do not classify the various fungi in this group on the basis of their being mold or yeast. The reason that this type of division doesn't work is that some species exist as molds under certain conditions and are yeastlike under other conditions. Such species are said to be **dimorphic,** or **biphasic.**

The principal differences between molds and yeasts are as follows:

Molds

Hyphae Molds have microscopic filaments called *hyphae* (hypha, singular). As shown in figure 8.1, if the filament has crosswalls, it is referred to as having **septate hyphae.** If no crosswalls are present, the filament is said to be **nonseptate,** or **aseptate.** Actually, most of the fungi that are classified as being septate are incompletely septate since the septae have central openings that allow the streaming of cytoplasm from one compartment to the next. A mass of intermeshed hyphae, as seen macroscopically, is a *mycelium.*

Asexual Spores Two kinds of asexual spores are seen in molds: sporangiospores and conidia. **Spo-**

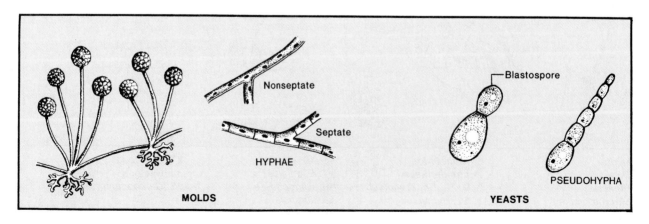

Figure 8.1 Structural differences between molds and yeasts

rangiospores are spores that form within a sac called a *sporangium.* The sporangia are attached to stalks called *sporangiophores.* See illustration 1, figure 8.2.

Conidia are asexual spores that form on specialized hyphae called *conidiophores.* If the conidia are small they are called *microconidia;* large multicellular conidia are known as *macroconidia.* The following four types of conidia are shown in figure 8.2:

- **Phialospores:** Conidia of this type are produced by vase-shaped cells called *phialides.* Note in figure 8.2 that *Penicillium* and *Gliocadium* produce this type.
- **Blastoconidia:** Conidia of this type are produced by budding from cells of preexisting conidia, as in *Cladosporium,* which typically has lemon-shaped spores.
- **Arthrospores:** This type of conidia forms by separation from preexisting hyphal cells. Example: *Oospora.*
- **Chlamydospores:** These spores are large, thick-walled, round or irregular structures formed within or on the ends of a hypha. Common to most fungi, they generally form on old cultures. Example: *Candida albicans.*

Sexual Spores Three kinds of sexual spores are seen in molds: zygospores, ascospores, and basidiospores. Figure 8.3 illustrates the three types.

Zygospores are formed by the union of nuclear material from the hyphae of two different strains. **Ascospores,** on the other hand, are sexual spores produced in enclosures, which may be oval sacs or elongated tubes. **Basidiospores** are sexually produced on club-shaped bodies called *basidia.* A basidium is considered by some to be a modified type of ascus.

Yeasts

Hyphae Unlike molds, yeasts do not have true hyphae. Instead they form multicellular structures called **pseudohyphae.** See figure 8.1.

Asexual Spores The only asexual spore produced by yeasts is called a **blastospore,** or **bud.** These spores form as an outpouching of a cell by a budding process. It is easily differentiated from the parent cell by its small size. It may separate from the original cell or remain attached. If successive buds remain attached in the budding process, the result is the formation of a pseudohypha.

Subdivisions of the Amastigomycota

Division Amastigomycota consists of four subdivisions: Zygomycotina, Ascomycotina, Basidiomycotina, and Deuteromycotina. They are separated on the basis of the type of sexual reproductive spores as follows:

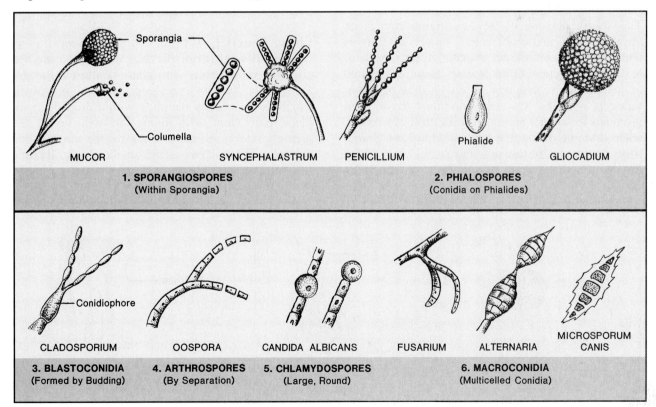

Figure 8.2 Types of asexual spores seen in fungi

Zygomycotina

These fungi have nonseptate hyphae and produce zygospores. They also produce sporangiospores. *Rhizopus, Mucor,* and *Syncephalastrum* are representative genera of this subdivision.

Ascomycotina

Since all the fungi in this subdivision produce ascospores, they are grouped into one class, the *Ascomycetes.* They are commonly referred to as the "ascomycetes" and are also called "sac fungi." All of them have septate hyphae and most of them have chitinous walls.

Fungi in this group that produce a single ascus are called *ascomycetes yeasts.* Other ascomycetes produce numerous asci in complex flask-shaped fruiting bodies called **perithecia** or **pseudothecia,** in cup-shaped structures, or in hollow spherical bodies, as in powdery mildews, *Eupenicillium* or *Talaromyces* (the sexual stages for *Penicillium*).

Basidiomycotina

All fungi in this subdivision belong to one class, the Basidiomycetes. Puffballs, mushrooms, smuts, rust, and shelf fungi on tree branches are also basidiomycetes. The sexual spores of this class are **basidiospores.**

Deuteromycotina

This fourth division of the Amastigomycota is an artificial group that was created to place any fungus that has not been shown to have some means of sexual reproduction. Often, species that are relegated to this division remain here for only a short period of time: as soon as the right conditions have been provided for sexual spores to form, they are reclassified into one of the first three subdivisions. Sometimes, however, the asexual and sexual stages of a fungus are discovered and named separately by different mycologists, with the result that a single species acquires two different names. Although, generally there is a switch over to the sexual stage name, not all mycologists conform to this practice.

Members of this group are commonly referred to as the *fungi imperfecti* or *deuteromycetes.* It is a large group, containing over 15,000 species.

Laboratory Procedures

Several options are provided here for the study of molds and yeasts. The procedures to be followed will be outlined by your instructor.

Yeast Study

The organism *Saccharomyces cerevisiae,* which is used in bread making and alcohol fermentation, will be used for this study. Either prepared slides or living organisms may be used.

Materials:
　　prepared slides of *Saccharomyces cerevisiae*
　　broth cultures of *Saccharomyces cerevisiae*
　　methylene blue stain
　　microscope slides and cover glasses

Prepared Slides　If prepared slides are used, they may be examined under high-dry or oil immersion. One should look for typical **blastospores** and **ascospores.** Space is provided on the Laboratory Report for drawing the organisms.

Living Material　If broth cultures of *Saccharomyces cerevisiae* are available they should be examined on a wet mount slide with phase-contrast or bright-field optics. Two or three loopfuls of the organisms should be placed on the slide with a drop of methylene blue stain. Oil immersion will reveal the greatest amount of detail. Look for the **nucleus** and **vacuole.** The nucleus is the smaller body. Draw a few cells on the Laboratory Report.

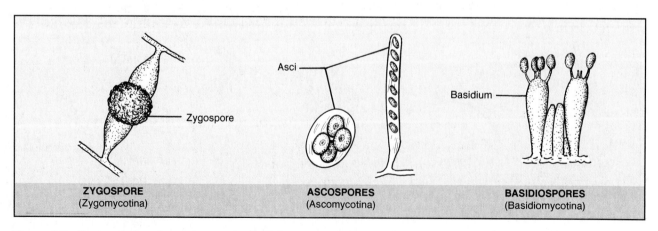

ZYGOSPORE	**ASCOSPORES**	**BASIDIOSPORES**
(Zygomycotina)	(Ascomycotina)	(Basidiomycotina)

Figure 8.3　Types of sexual spores seen in the Amastigomycota

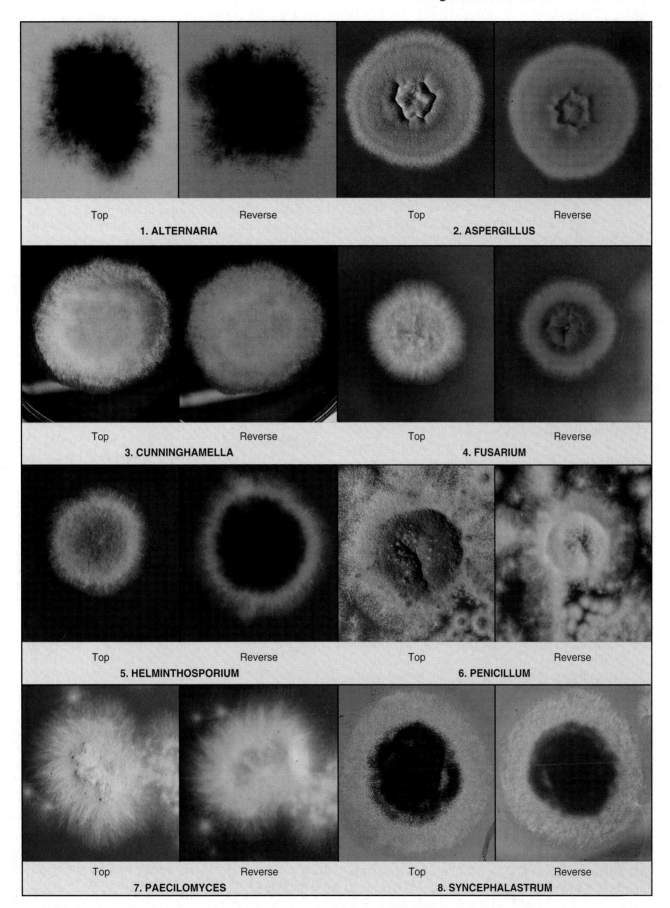

Figure 8.4 Colony characteristics of some of the more common molds

Mold Study

Examine a Petri plate of Sabouraud's agar that has been exposed to the air for about an hour and incubated at room temperature for 3–5 days. This medium has a low pH, which makes it selective for molds. A good plate will have many different-colored colonies. Note the characteristic "cottony" nature of the colonies. Also, look at the bottom of the plate and observe how the colonies differ in color here. The identification of molds is based on surface color, backside color, hyphal structure, and types of spores.

Figure 8.4 reveals how some of the more common molds appear when grown on Sabouraud's agar. Keep in mind when using figure 8.4 that the appearance of a mold colony can change appreciably as it gets older. The photographs in figure 8.4 are of colonies that are 10 to 21 days old.

Conclusive identification cannot be made unless a microscope slide is made to determine the type of hyphae and spores that are present. Figure 8.5 reveals, diagrammatically, the microscopic differences that one looks for when identifying mold genera.

Two Options   In making slides from mold colonies one can make either wet mounts directly from the colonies by the procedure outlined here or cultured slides as outlined in Exercise 24. The following steps should be used for making stained slides directly from the colonies. Your instructor will indicate the number of identifications that are to be made.

Materials:
> mold cultures on Sabouraud's agar
> microscope slides and cover glasses
> lactophenol cotton blue stain
> sharp-pointed scalpels or dissecting needles

1. Place the uncovered plate on the stage of your microscope and examine the edge of a colored colony with the low-power objective. Look for hyphal structure and spore arrangement. Ignore the white colonies since they generally lack spores and are difficult to identify.
2. Consult figures 8.4 and 8.5 to make a preliminary identification based on colony characteristics and low-power magnification of hyphae and spores.
3. Make a wet mount slide by transferring a small amount of the culture with a sharp scalpel or dissecting needle to a drop of lactophenol cotton blue stain on a slide. Cover with a cover glass and examine under low-power and high-dry objectives. Refer again to figure 8.5 to confirm any conclusions drawn from your previous examination of the edge of the colony.
4. Repeat the above procedure for each different colony.

Laboratory Report

After recording your results on the Laboratory Report, answer all the questions.

Figure 8.5   Legend

1. *Penicillium*—bluish-green; brush arrangement of phialospores.
2. *Aspergillus*—bluish-green with sulfur-yellow areas on the surface. *Aspergillus niger* is black.
3. *Verticillium*—pinkish-brown, elliptical microconidia.
4. *Trichoderma*—green, resemble *Penicillium* macroscopically.
5. *Gliocadium*—dark green; conidia (phialospores) borne on phialides, similar to *Penicillium*; grows faster than *Penicillium.*
6. *Cladosporium (Hormodendrum)*—light green to grayish surface; gray to black back surface; blastoconidia.
7. *Pleospora*—tan to green surface with brown to black back; ascospores shown are produced in sacs borne within brown, flask-shaped fruiting bodies called pseudothecia.
8. *Scopulariopsis*—light brown; rough-walled microconidia.
9. *Paecilomyces*—yellowish-brown; elliptical microconidia.
10. *Alternaria*—dark greenish-black surface with gray periphery; black on reverse side; chains of macroconidia.
11. *Bipolaris*—black surface with grayish periphery; macroconidia shown.
12. *Pullularia*—black, shiny, leathery surface; thick-walled; budding spores.
13. *Diplosporium*—buff-colored wooly surface; reverse side has red center surrounded by brown.
14. *Oospora (Geotrichum)*—buff-colored surface; hyphae break up into thin-walled rectangular arthrospores.
15. *Fusarium*—variants of yellow, orange, red, and purple colonies; sickle-shaped macroconidia.
16. *Trichothecium*—white to pink surface; two-celled conidia.
17. *Mucor*—a zygomycete; sporangia with a slimy texture; spores with dark pigment.
18. *Rhizopus*—a zygomycete; spores with dark pigment.
19. *Syncephalastrum*—a zygomycete; sporangiophores bear rod-shaped sporangioles, each containing a row of spherical spores.
20. *Nigrospora*—conidia black, globose, one-celled, borne on a flattened, colorless vesicle at the end of a conidiophore.
21. *Montospora*—dark gray center with light gray periphery; yellow-brown conidia.

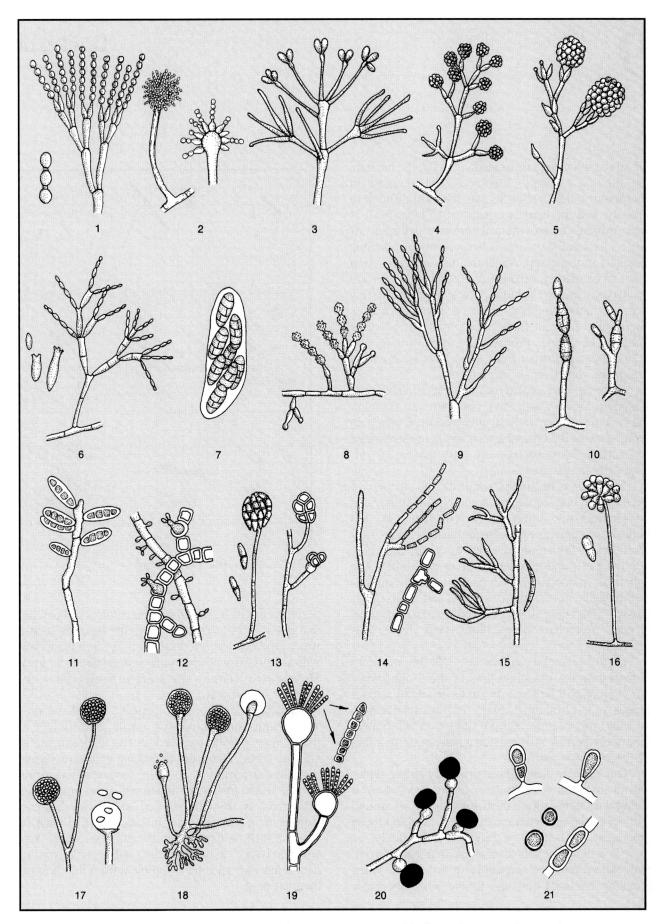

Figure 8.5 Microscopic appearance of some of the more common molds (legend on opposite page)

9

Bacteria

Of all the microorganisms studied so far, the bacteria are the most widely distributed, the simplest in morphology, the smallest in size, the most difficult to classify, and the hardest organisms to identify. It is even difficult to provide a descriptive definition of what a bacterial organism is because of considerable diversity in the group. About the only generalization that can be made for the entire group is that they are prokaryotic and are seldom photosynthetic. The few that are photosynthetic utilize a pigment that is chemically different from chlorophyll a. It is called *bacteriochlorophyll*. Probably the simplest definition that one can construct from these facts is: *Bacteria are prokaryons without chlorophyll a.*

Since they are prokaryons, the bacteria share the Kingdom Monera with the Cyanobacteria. Although the bacteria are generally smaller than the Cyanobacteria, some of the Cyanobacteria are in the size range of bacteria. Most bacteria are only 0.5 to 2.0 micrometers in diameter.

Figure 9.1 illustrates most of the shapes of bacteria that one would encounter. Note that they can be grouped into three types: rod, spherical, and helical or curved. Rod-shaped bacteria may vary considerably in length; may have square, round, or pointed ends; and may be motile or nonmotile. The spherical, or coccus-shaped, bacteria may occur singly, in pairs, in tetrads, in chains, and in irregular masses. The helical and curved bacteria exist as slender spirochaetes, spirillum, and bent rods (vibrios).

In this exercise an attempt will be made to demonstrate the ubiquitousness of these organisms. No attempt will be made to study detailed bacterial anatomy or physiology. Many exercises related to staining, microscopy, and physiology in subsequent laboratory periods will provide a clear understanding of these microorganisms.

Our concern here relates primarily to the widespread distribution of bacteria in our environment. Being thoroughly aware of their existence all around us is of prime importance if we are to develop those laboratory skills that we refer to, collectively, as *aseptic technique.* The awareness that bacteria are everywhere must be constantly in our minds when handling bacterial cultures. In the next laboratory

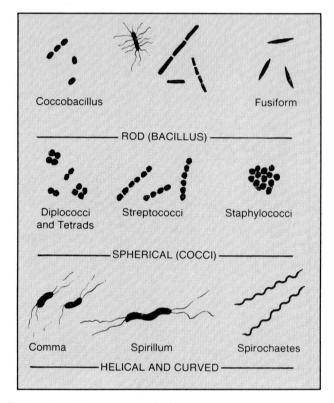

Figure 9.1 Bacterial morphology

period you will be handling tube cultures of bacteria, and unless you learn how to handle them in such a way as to keep foreign bacteria away from them, you will always be working with contaminated cultures. *Without pure cultures the study of bacteriology becomes a hopeless endeavor.*

During this laboratory period you will be provided with three kinds of sterile bacteriological media that will be exposed to the environment in various ways. To ensure that these exposures cover as wide a spectrum as possible, specific assignments will be made for each student. In some instances a moistened swab will be used to remove bacteria from some object; in other instances a Petri plate of medium will be exposed to the air or a cough. You will be issued a number that will enable you to determine your specific assignment from the chart on the next page.

Materials:

per student:
1 tube of nutrient broth
1 Petri plate of trypticase soy agar (TSA)
1 sterile cotton swab
china marking pencil

per 2 or more students:
1 Petri plate of blood agar

1. Expose your TSA plate according to your assignment in the table below. *Label the bottom* of your plate with your initials, your assignment number, and the date.
2. Moisten a sterile swab by immersing it into a tube of nutrient broth and expressing most of the broth out of it by pressing the swab against the inside wall of the tube.
3. Rub the moistened swab over a part of your body such as a finger, ear, etc., or some object such as a doorknob, telephone mouthpiece, etc., and return the swab to the tube of broth. It may be necessary to break off the stick end of the swab so that you can replace the cap on the tube.
4. Label the tube with your initials and the source of the bacteria.
5. Expose the blood agar plate by coughing onto it. Label the bottom of the plate with the initials of the individuals that cough onto it. Be sure to date the plate also.
6. Incubate the plates and tube at 37° C for 48 hours.

Evaluation

After 48 hours incubation, examine the tube of nutrient broth and two plates. Shake the tube vigorously without wetting the cap. Is it cloudy or clear? Compare it with an uncontaminated tube of broth. What is the significance of cloudiness? Do you see any colonies growing on the blood agar plate? Are the colonies all the same size and color? If not, what does this indicate? Group together a set of TSA plates representing all nine types of exposures. Record your results on the Laboratory Report.

Your instructor will indicate whether these tubes and plates are to be used for making slides in Exercise 12 (Simple Staining). If the plates and tubes are to be saved, containers will be provided for their storage in the refrigerator. Place the plates and tubes in the designated containers.

Laboratory Report

Complete the Laboratory Report for this exercise.

Exposure Method for TSA Plate	Student Number
1. To the air in laboratory for 30 minutes	1, 10, 19, 28
2. To the air in room other than laboratory for 30 minutes	2, 11, 20, 29
3. To the air outside of building for 30 minutes	3, 12, 21, 30
4. Blow dust onto exposed medium	4, 13, 22, 31
5. Moist lips pressed against medium	5, 14, 23, 32
6. Fingertips pressed lightly on medium	6, 15, 24, 33
7. Several coins pressed temporarily on medium	7, 16, 25, 34
8. Hair is combed over exposed medium (10 strokes)	8, 17, 26, 35
9. Optional: Any method not listed above	9, 18, 27, 36

PART 3

Microscope Slide Techniques (Bacterial Morphology)

The ten exercises in this unit include the procedures for eleven slide techniques that one might employ in morphological studies of bacteria. A culture method in Exercise 19 also is included as a substitute for slide techniques when pathogens are encountered.

These exercises are intended to serve two equally important functions: (1) to help you to develop the necessary skills in making slides and (2) to introduce you to the morphology of bacteria. Although the title of each exercise pertains to a specific technique, the organisms chosen for each method have been carefully selected so that you can learn to recognize certain morphological features. For example, in the exercise on simple staining (Exercise 12) the organisms selected exhibit metachromatic granules, pleomorphism, and palisade arrangement of cells. In Exercise 14 (Gram Staining) you will observe the differences between cocci and bacilli, as well as learn how to execute the staining routine.

The importance of the mastery of these techniques cannot be overemphasized. Although one is seldom able to make species identification on the basis of morphological characteristics alone, it is a very significant starting point. This fact will become increasingly clear with subsequent experiments.

Although the steps in the various staining procedures may seem relatively simple, student success is often quite unpredictable. Unless your instructor suggests a variation in the procedure, try to follow the procedures exactly as stated, without improvisation. Photomicrographs in color have been provided for many of the techniques; use them as a guide to evaluate the slides you have prepared. Once you have mastered a specific technique feel free to experiment.

*no a bacterial stain
the background is stained*

The simplest way to make a slide of bacteria is to prepare a wet mount, much in the same manner that was used for studying protozoa and algae. Although this method will quickly produce a slide, finding the bacteria on the slide may be difficult, especially for a beginner. The problem one encounters is that bacteria are quite colorless and transparent, and unless the diaphragm is carefully adjusted, the beginner usually has considerable difficulty bringing the organisms into focus.

A better way to observe bacteria for the first time is to prepare a slide by a process called *negative,* or *background staining.* This method consists of mixing the microorganisms in a small amount of nigrosine or india ink and spreading the mixture over the surface of a slide. (Incidentally, nigrosine is far superior to india ink.)

Since these two pigments are not really bacterial stains, they do not penetrate the microorganisms; instead they obliterate the background, leaving the organisms transparent and visible in a darkened field.

Although this technique has limitations, it can be useful for determining cell morphology and size. Since no heat is applied to the slide, there is no shrinkage of the cells, and, consequently, more accurate cell size determinations result than with some other methods. This method is also useful for studying spirochaetes that don't stain readily with ordinary dyes.

Three Methods

Negative staining can be done by one of three different methods. Figure 10.1 illustrates the more commonly used method in which the organisms are mixed in a drop of nigrosine and spread over the slide with another slide. The goal is to produce a smear that is thick at one end and feather-thin at the other end.

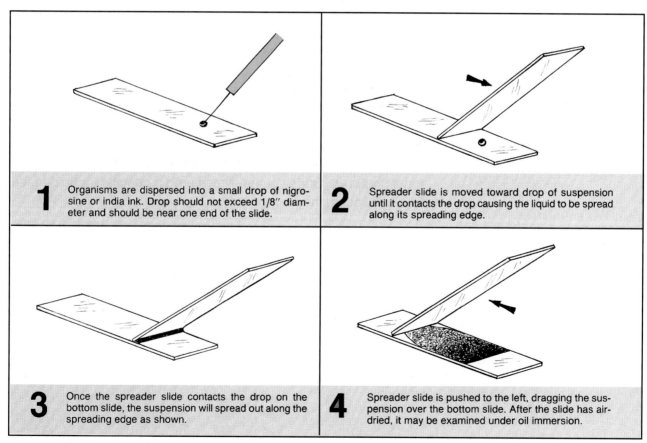

1 Organisms are dispersed into a small drop of nigrosine or india ink. Drop should not exceed 1/8″ diameter and should be near one end of the slide.

2 Spreader slide is moved toward drop of suspension until it contacts the drop causing the liquid to be spread along its spreading edge.

3 Once the spreader slide contacts the drop on the bottom slide, the suspension will spread out along the spreading edge as shown.

4 Spreader slide is pushed to the left, dragging the suspension over the bottom slide. After the slide has air-dried, it may be examined under oil immersion.

Figure 10.1 Negative staining technique, using a spreader slide

Somewhere between the too thick and too thin areas will be an ideal spot to study the organisms.

Figure 10.2 illustrates a second method, in which organisms are mixed in only a loopful of nigrosine instead of a full drop. In this method the organisms are spread over a smaller area in the center of the slide with an inoculating needle. No spreader slide is used in this method.

The third procedure (Woeste-Demchick's method), which is not illustrated here, involves applying ink to a conventional smear with a black felt marking pen. If this method is used, it should be done on a smear prepared in the manner described in the next exercise. Simply put, the technique involves applying a *single coat* of felt pen ink over a smear.

Note in the procedure below that slides may be made from organisms between your teeth or from specific bacterial cultures. Your instructor will indicate which method or methods you should use and demonstrate some basic aseptic techniques. Various options are provided here to ensure success.

Materials:

> microscope slides (with polished edges)
> nigrosine solution or india ink
> slant cultures of *S. aureus* and *B. megaterium*
> inoculating straight wire and loop
> Bunsen burner

china marking pencil
felt marking pen (see *Instructor's Handbook*)

1. Clean two or three microscope slides with Bon Ami to rid them of all dirt and grease.
2. By referring to figure 10.1 or 10.2, place the proper amount of stain on the slide.
3. **Oral Organisms:** Remove a small amount of material from between your teeth with a sterile straight wire, and mix it into the stain on the slide. Be sure to break up any clumps of organisms with the wire. When using a wire, *be sure to flame it first to make it sterile.*
4. **From Cultures:** With a *sterile* straight wire transfer a very small amount of bacteria from the slant to the center of the stain on the slide.
5. Spread the mixture over the slide according to the procedure used in figure 10.1 or 10.2.
6. Allow the slide to air-dry and examine with an oil immersion objective.

[handwritten margin note: procedure Pg 50]

Laboratory Report

Draw a few representative types of organisms on Laboratory Report 10–13. If slide is of oral organisms, look for yeasts and hyphae as well as bacteria. Spirochaetes may also be present.

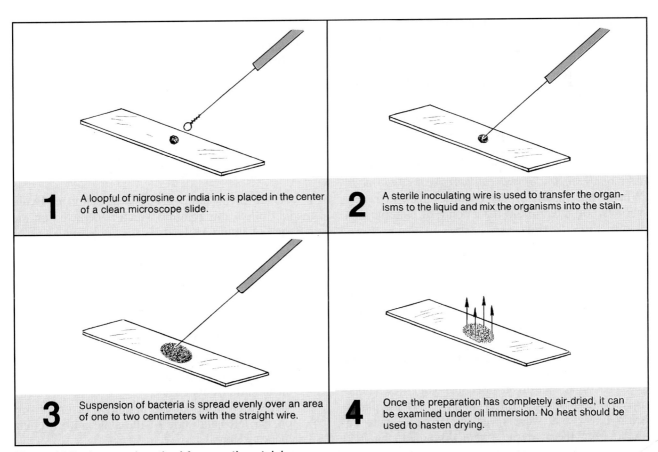

1 A loopful of nigrosine or india ink is placed in the center of a clean microscope slide.

2 A sterile inoculating wire is used to transfer the organisms to the liquid and mix the organisms into the stain.

3 Suspension of bacteria is spread evenly over an area of one to two centimeters with the straight wire.

4 Once the preparation has completely air-dried, it can be examined under oil immersion. No heat should be used to hasten drying.

Figure 10.2   A second method for negative staining

While negative staining is a simple enough process to make bacteria more visible with a brightfield microscope, it is of little help when one attempts to observe anatomical microstructures such as flagella, granules, and endospores. Only by applying specific bacteriological stains to organisms can such organelles be seen. However, success at bacterial staining depends first of all on the preparation of a suitable *smear* of the organisms. A properly prepared bacterial smear is one that withstands one or more washings during staining without loss of organisms; is not too thick; and does not result in excessive distortion due to cell shrinkage. The procedure for making such a smear is illustrated in figure 11.1.

The first step in preparing a bacteriological smear differs according to the source of the organisms. If the bacteria are growing in a liquid medium (broths, milk, saliva, urine, etc.), one starts by placing one or two loopfuls of the liquid medium directly on the slide.

From solid media such as nutrient agar, blood agar, or some part of the body, one starts by placing one or two loopfuls of water on the slide and then uses a straight inoculating wire to disperse the organisms in the water. Bacteria growing on solid media tend to cling to each other and must be dispersed sufficiently by dilution in water; unless this is done, the smear will be too thick. *The most difficult concept for students to understand about making slides from solid media is that it takes only a very small amount of material to make a good smear.* When your instructor demonstrates this step, pay very careful attention to the amount of material that is placed on the slide.

Another hurdle to overcome in making bacterial smears is to learn the proper procedure for handling cultures. Figure 11.2 visualizes the various steps that must become routine whenever you remove organisms from a test tube. This is the beginning of a series of aseptic techniques that must become as automatic as breathing. Tabletop disinfection must also be performed at the beginning of the period and at the close of the laboratory session.

The organisms to be used for your first slides may be from several different sources. If the plates from Exercise 9 were saved, some slides may be made from them. If they were discarded, the first slides may be made for Exercise 12, which pertains to simple staining. Your instructor will indicate which cultures are to be used.

Tabletop Preparation

At the beginning of every laboratory period from now on, it is essential that your laboratory tabletop be cleared of all paraphernalia and wiped clean with a disinfectant. This procedure is necessary to minimize the possibility of contaminating your cultures. Loose dust particles (and bacteria) on your desk are removed to prevent them from getting into your cultures. In addition, sanitizing your work area protects yourself against microorganisms that might have been carelessly left on the tabletop by someone in a previous period.

The procedure is simple: a little disinfectant is poured onto the tabletop and spread over the entire surface with a sponge. The disinfectant may be Roccal, Zephiran, Betadine, or some other acceptable agent. Your instructor will show you where the disinfectant and sponges are kept. As a consideration for those students that use your station in the next class period, it is expected that you will repeat this scrub-down procedure at the end of the period.

From Liquid Media
(From Broths, Saliva, Milk, etc.)

If you are preparing a bacterial smear from liquid media, follow this routine, which is depicted on the left side of figure 11.1.

Materials:
    microscope slides
    Bunsen burner
    wire loop
    china marking pencil
    slide holder (clothespin), optional

(handwritten note: — Not to much Saline — Don't cook the bacteria)

1. Wash a slide with soap or Bon Ami and hot water, removing all dirt and grease. Handle the clean slide by its edges.
2. Write the initials of the organism or organisms on the left-hand side of the slide with a china marking pencil.

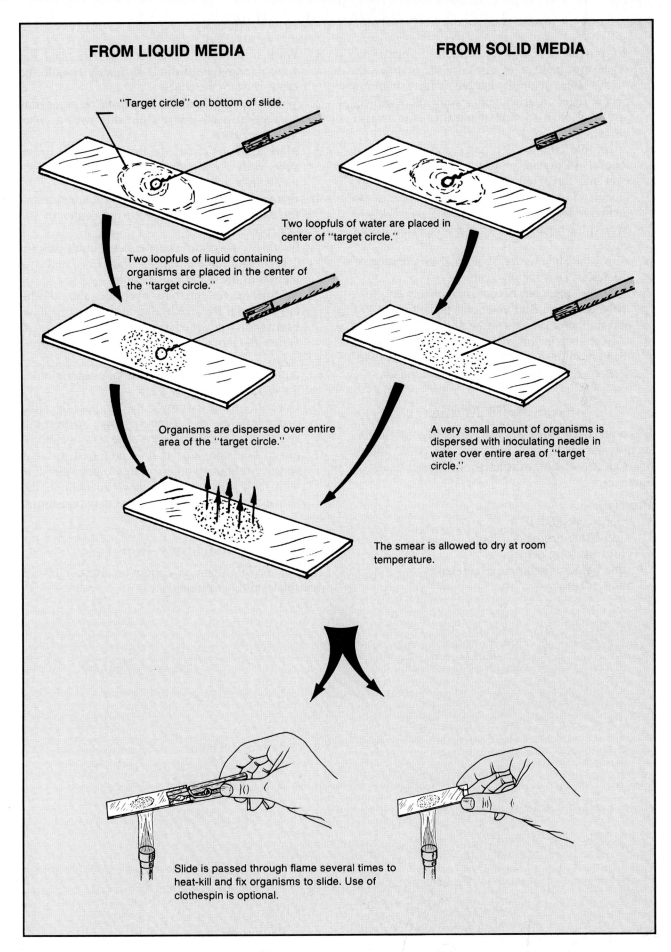

FROM LIQUID MEDIA

FROM SOLID MEDIA

"Target circle" on bottom of slide.

Two loopfuls of water are placed in center of "target circle."

Two loopfuls of liquid containing organisms are placed in the center of the "target circle."

Organisms are dispersed over entire area of the "target circle."

A very small amount of organisms is dispersed with inoculating needle in water over entire area of "target circle."

The smear is allowed to dry at room temperature.

Slide is passed through flame several times to heat-kill and fix organisms to slide. Use of clothespin is optional.

Figure 11.1 Procedure for making a bacterial smear

3. To provide a target on which to place the organisms, make a ½″ circle on the *bottom* side of the slide, centrally located, with a marking pencil. Later on, when you become more skilled, you may wish to omit the use of this "target circle."

4. Shake the culture vigorously and transfer two loopfuls of organisms to the center of the slide over the target circle. Follow the routine for inoculations shown in figure 11.2. *Be sure to flame the loop after it has touched the slide.*

5. Spread the organisms over the area of the target circle.

6. Allow the slide to dry by normal evaporation of the water. Don't apply heat.

7. After the smear has become completely dry, pass the slide over a Bunsen burner flame to heat-kill and fix the organisms to the slide.

    Note that in this step one has the option of using or not using a clothespin to hold the slide. *Use the option preferred by your instructor.*

From Solid Media

use saline

When preparing a bacterial smear from solid media, such as nutrient agar or a part of the body, follow this routine, which is depicted on the right side of figure 11.1.

Materials:

> microscope slides
> inoculating needle and loop
> china marking pencil
> slide holder (clothespin), optional
> Bunsen burner

1. Wash a slide with soap or Bon Ami and hot water, removing all dirt and grease. Handle the clean slide by its edges.

2. Write the initials of the organism or organisms on the left-hand side of the slide with a china marking pencil.

3. Mark a "target circle" on the bottom side of the slide with a china marking pencil. (See comments in step 3 at left.)

4. Flame an inoculating loop, let it cool, and transfer two loopfuls of water to the center of the target circle.

5. Flame an inoculating needle, let it cool, pick up *a very small amount of the organisms,* and mix them into the water on the slide.

    Disperse the mixture over the area of the target circle. Be certain that the organisms have been well emulsified in the liquid. *Be sure to flame the inoculating needle before placing it aside.*

6. Allow the slide to dry by normal evaporation of the water. Don't apply heat.

7. Once the smear is completely dry, pass the slide over the flame of a Bunsen burner to heat-kill and fix the organisms to the slide. Use a clothespin to hold the slide if it is preferred by your instructor. Some workers prefer to hold the slide with their fingers so that they can monitor the temperature of the slide (to prevent overheating).

Laboratory Report

Answer the questions on Laboratory Report 10–13 that relate to this exercise.

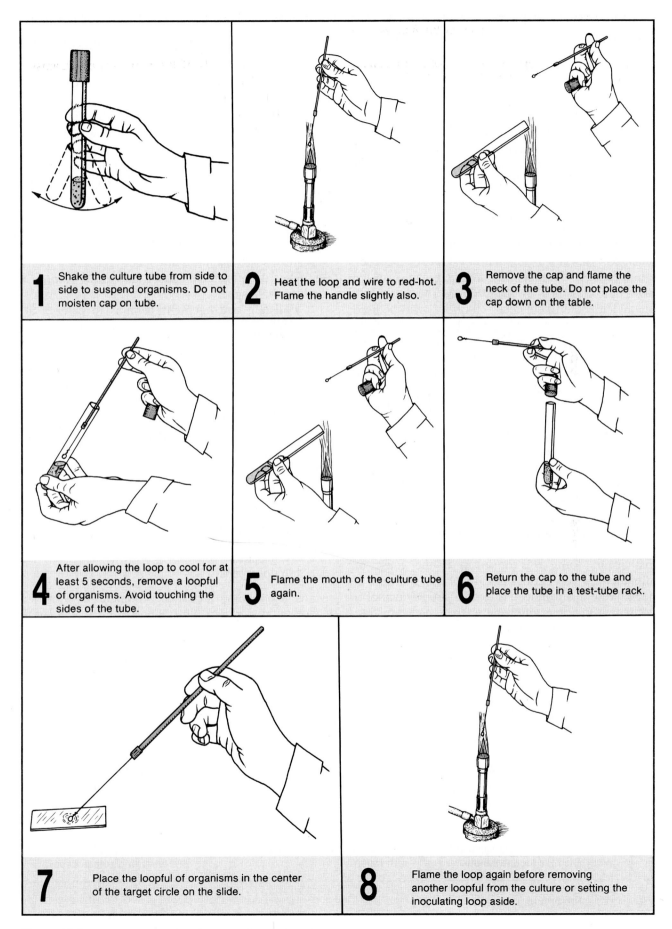

1 Shake the culture tube from side to side to suspend organisms. Do not moisten cap on tube.

2 Heat the loop and wire to red-hot. Flame the handle slightly also.

3 Remove the cap and flame the neck of the tube. Do not place the cap down on the table.

4 After allowing the loop to cool for at least 5 seconds, remove a loopful of organisms. Avoid touching the sides of the tube.

5 Flame the mouth of the culture tube again.

6 Return the cap to the tube and place the tube in a test-tube rack.

7 Place the loopful of organisms in the center of the target circle on the slide.

8 Flame the loop again before removing another loopful from the culture or setting the inoculating loop aside.

Figure 11.2 Aseptic procedure for organism removal

12

Simple Staining

(use of a single stain)

many
shape

The use of a single stain to color a bacterial organism is commonly referred to as *simple staining.* Some of the most commonly used dyes for simple staining are methylene blue, basic fuchsin, and crystal violet. All of these dyes work well on bacteria because they have color-bearing ions (*chromophores*) that are positively charged (cationic).

The fact that bacteria are slightly negatively charged produces a pronounced attraction between these cationic chromophores and the organism. Such dyes are classified as **basic dyes.** The basic dye methylene blue (methylene$^+$ chloride$^-$) will be used in this exercise. Those dyes that have anionic chromophores are called **acidic dyes.** Eosin (sodium$^+$ eosinate$^-$) is such a dye. The anionic chromophore, eosinate$^-$, will not stain bacteria because of the electrostatic repelling forces that are involved.

The staining times for most simple stains are relatively short, usually from 30 seconds to 2 minutes, depending on the affinity of the dye. After a smear has been stained for the required time, it is washed off gently, blotted dry, and examined directly under oil immersion. Such a slide is useful in determining basic morphology and the presence or absence of certain kinds of granules.

An avirulent strain of *Corynebacterium diphtheriae* will be used here for simple staining. In its pathogenic form, this organism is the cause of diphtheria, a very serious disease. One of the steps in identifying this pathogen is to do a simple stain of it to demonstrate the following unique characteristics: pleomorphism, metachromatic granules, and palisade arrangement of cells.

Pleomorphism pertains to irregularity of form: i.e., demonstrating several different shapes. While *C. diphtheriae* is basically rod-shaped, it also appears club-shaped, spermlike, or needle-shaped. *Bergey's Manual* uses the terms "pleomorphic" and "irregular" interchangeably.

Metachromatic granules are distinct reddish-purple granules within cells that show up when the organisms are stained with methylene blue. These granules are considered to be masses of *volutin,* a polymetaphosphate.

Palisade arrangement pertains to parallel arrangement of rod-shaped cells. This characteristic, also called "picket fence" arrangement, is common to many corynebacteria.

Procedure

Prepare a slide of *C. diphtheriae,* using the procedure outlined in figure 12.1. It will be necessary to refer back to Exercise 11 for the smear preparation procedure.

Materials:
    slant culture of avirulent strain of
    ~~Corynebacterium diphtheriae~~ *- has*
        └ Klebsiella
    methylene blue (Loeffler's)
    wash bottle
    bibulous paper

After examining the slide, compare it with the photomicrograph in illustration 1, figure 14.3 (page 60). Record your observations on Laboratory Report 10–13.

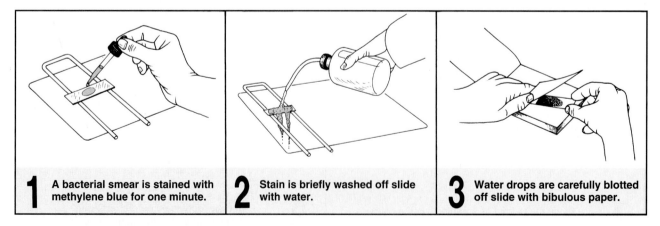

1 A bacterial smear is stained with methylene blue for one minute.

2 Stain is briefly washed off slide with water.

3 Water drops are carefully blotted off slide with bibulous paper.

Figure 12.1   Procedure for simple staining

Capsular Staining

13

Some bacterial cells are surrounded by a pronounced gelatinous or slimy layer called a *capsule.* There is considerable evidence to support the view that all bacteria have some amount of slime material surrounding their cells. In most instances, however, the layer is not of sufficient magnitude to be readily discernible. Although some capsules appear to be made of glycoprotein, others contain polypeptides. All appear to be water-soluble.

Staining the bacterial capsule cannot be accomplished by ordinary simple staining procedures. The problem with trying to stain capsules is that if you prepare a heat-fixed smear of the organism by ordinary methods, you will destroy the capsule; and, if you do not heat-fix the slide, the organism will slide off the slide during washing. In most of our bacteriological studies our principal concern is simply to

demonstrate the presence or absence of a pronounced capsule. This can be easily achieved by combining negative and simple staining techniques, as in figure 13.1. To learn about this technique prepare a capsule "stained" slide of *Klebsiella pneumoniae,* using the procedure outlined in figure 13.1.

Materials:
    36–48 hour milk culture of *Klebsiella*
        pneumoniae
    india ink
    crystal violet

Observation: Examine the slide under oil immersion and compare your slide with illustration 2, figure 14.3 on page 60. Record your results on Laboratory Report 10–13.

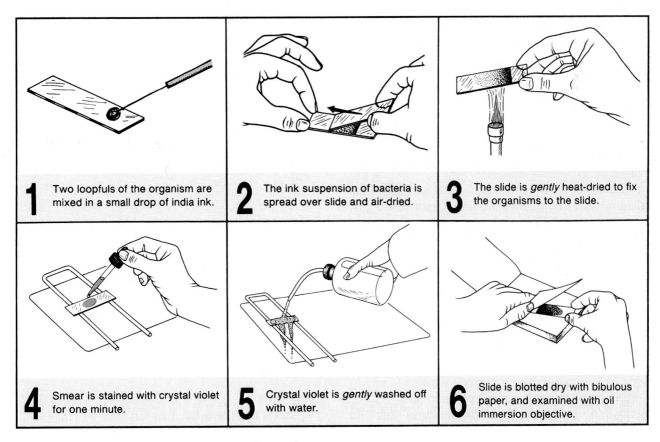

1 Two loopfuls of the organism are mixed in a small drop of india ink.

2 The ink suspension of bacteria is spread over slide and air-dried.

3 The slide is *gently* heat-dried to fix the organisms to the slide.

4 Smear is stained with crystal violet for one minute.

5 Crystal violet is *gently* washed off with water.

6 Slide is blotted dry with bibulous paper, and examined with oil immersion objective.

Figure 13.1   **Procedure for demonstration of capsule presence**

Gram Staining

In 1884 the Danish bacteriologist Christian Gram developed a staining technique that separates bacteria into two groups: those that are gram-positive and those that are gram-negative. The procedure is based on the ability of microorganisms to retain the purple color of crystal violet during decolorization with alcohol. Gram-negative bacteria are decolorized by the alcohol, losing the purple color of crystal violet. Gram-positive bacteria are not decolorized and remain purple. After decolorization, safranin, a red counterstain, is used to impart a pink color to the decolorized gram-negative organisms.

Figure 14.1 illustrates the effects of the various reagents on bacterial cells at each stage in the process. Note that crystal violet, the **primary stain,** causes both gram-positive and gram-negative organisms to become purple after 20 seconds of staining. When Gram's iodine, the **mordant,** is applied to the cells for one minute, the color of gram-positive and gram-negative bacteria remains the same: purple. The function of the mordant here is to combine with crystal violet to form a relatively insoluble compound in the gram-positive bacteria. When the **decolorizing agent,** 95% ethanol, is added to the cells for 10-20 seconds, the gram-negative bacteria are leached colorless, but the gram-positive bacteria remain purple. In the final step a **counterstain,** safranin, adds a pink color to the decolorized gram-negative bacteria without affecting the color of the purple gram-positive bacteria.

Of all the staining techniques you will use in the identification of unknown bacteria, Gram staining is, undoubtedly, the most important tool you will use. Although this technique seems quite simple, performing it with a high degree of reliability is a goal that requires some practice and experience. Here are two suggestions that can be helpful: first, don't make your smears too thick, and second, pay particular attention to the comments in step 4 on the next page that pertain to decolorization.

When working with unknowns keep in mind that old cultures of gram-positive bacteria tend to decolorize more rapidly than young ones, causing them to appear gram-negative instead of gram-positive. For reliable results one should use cultures that are approximately 16 hours old. Another point to remember is that some species of *Bacillus* tend to be

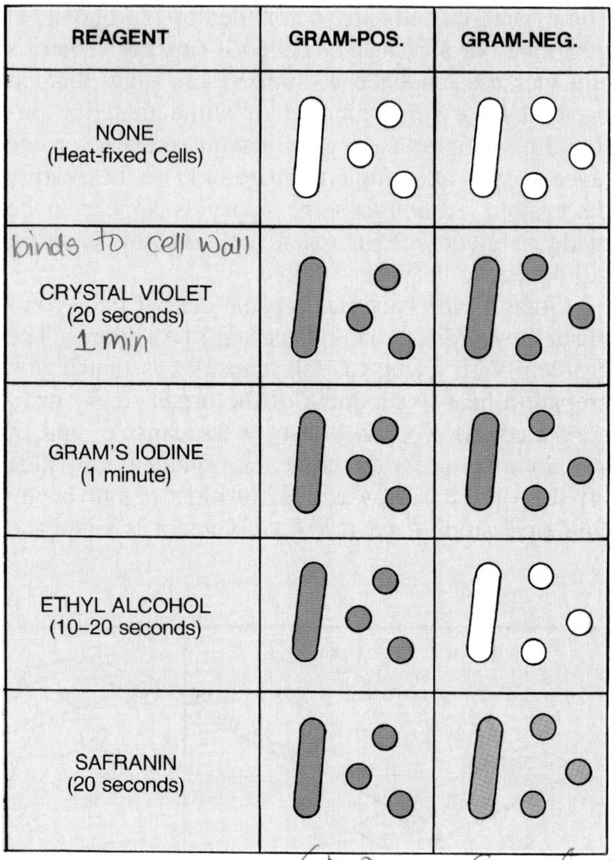

Figure 14.1 Color changes that occur at each step in the gram-staining process

gram-variable; i.e., sometimes positive and sometimes negative.

During this laboratory period you will be provided an opportunity to stain several different kinds of bacteria to see if you can achieve the degree of success that is required. Remember, if you don't master this technique now, you will have difficulty with your unknowns later.

Staining Procedure

Materials:

  slides with heat-fixed smears
  gram-staining kit and wash bottle
  bibulous paper

1. Cover the smear with **crystal violet** and let stand for *20 seconds*.
2. Briefly wash off the stain, using a wash bottle of distilled water. Drain off excess water.
3. Cover the smear with **Gram's iodine** solution and let it stand for *one minute*. (Your instructor may prefer only 30 seconds for this step.)
4. Pour off the Gram's iodine and flood the smear with **95% ethyl alcohol** for *10 to 20 seconds*. This step is critical. Thick smears will require more time than thin ones. *Decolorization has occurred when the solvent flows colorlessly from the slide.*
5. Stop action of the alcohol by rinsing the slide with water from wash bottle for a *few seconds*.
6. Cover the smear with **safranin** for *20 seconds*. (Some technicians prefer more time here.)
7. Wash gently for a few seconds, blot dry with bibulous paper, and air-dry.
8. Examine the slide under oil immersion.

Assignments

The organisms that will be used here for Gram staining represent a diversity of form and staining characteristics. Some of the rods and cocci are

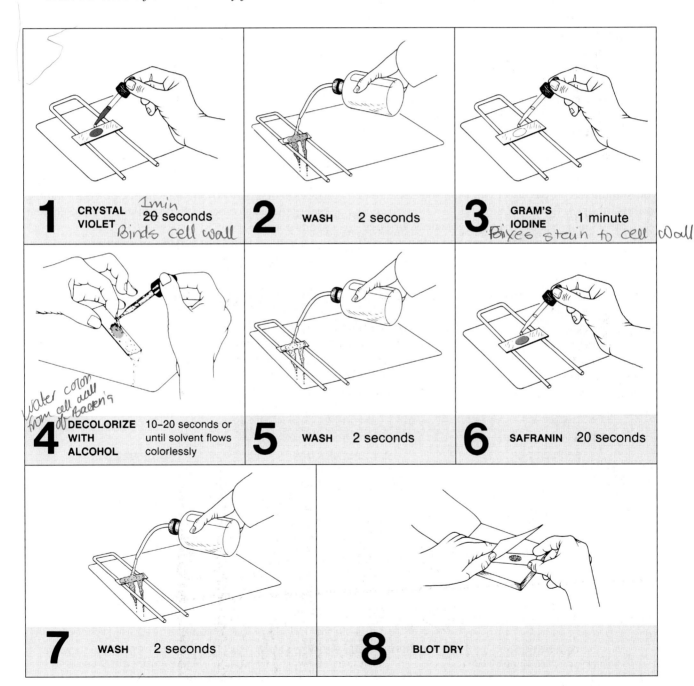

1 CRYSTAL VIOLET 20 seconds   *1 min*   *Binds cell wall*

2 WASH 2 seconds

3 GRAM'S IODINE 1 minute   *Fixes stain to cell wall*

4 DECOLORIZE WITH ALCOHOL 10–20 seconds or until solvent flows colorlessly   *water color from cell wall of bacteria*

5 WASH 2 seconds

6 SAFRANIN 20 seconds

7 WASH 2 seconds

8 BLOT DRY

Figure 14.2 The gram-staining procedure

gram-positive; others are gram-negative. One rod-shaped organism is a spore-former and another is acid-fast. The challenge here is to make gram-stained slides of various combinations that reveal their differences.

Materials:
> broth cultures of *Staphylococcus aureus,
> Pseudomonas aeruginosa,* and *Moraxella
> (Branhamella) catarrhalis*
> nutrient agar slant cultures of *Bacillus
> megaterium* and *Mycobacterium smegmatis*

Mixed Organisms I (Triple Smear Practice Slides) Prepare three slides with three smears on each slide. On the left portion of each slide make a smear of *Staphylococcus aureus.* On the right portion of each slide make a smear of *Pseudomonas aeruginosa.* In the middle of the slide make a smear that is a mixture of both organisms, using two loopfuls of each organism. ***Be sure to flame the loop sufficiently to avoid contaminating cultures.***

Gram stain one slide first, saving the other two for later. Examine the center smear. If done properly, you should see purple cocci and pink rods as shown in illustration 3, figure 14.3.

Call your instructor over to evaluate your slide. If the slide is improperly stained, the instructor will be able to tell what went wrong by examining all three smears. He or she will inform you how to correct your technique when you stain the next triple smear reserve slide.

Record your results on Laboratory Report 14–17 by drawing a few cells in the appropriate circle.

Mixed Organisms II Make a gram-stained slide of a mixture of *Bacillus megaterium* and *Moraxella (Branhamella) catarrhalis.*

This mixture differs from the previous slide in that the rods (*B. megaterium*) will be purple and the cocci (*M.B. catarrhalis*) will be large pink diplococci. See illustration 4, figure 14.3.

As you examine this slide look for clear areas on the rods which represent endospores. Since endospores are refractile and impermeable to crystal violet they will appear as transparent holes in the cells.

Draw a few cells in the appropriate circle on your Laboratory Report sheet.

Acid-Fast Bacteria To see how acid-fast mycobacteria react to Gram's stain, make a gram-stained slide of *Mycobacterium smegmatis.* If your staining technique is correct, the organisms should appear gram-positive.

Draw a few cells in the appropriate circle on your Laboratory Report sheet.

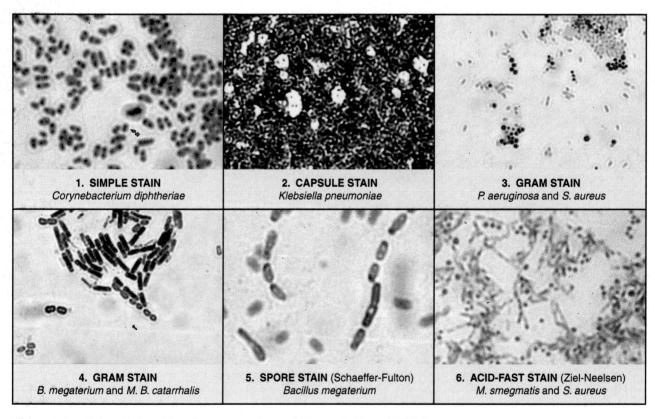

1. SIMPLE STAIN
Corynebacterium diphtheriae

2. CAPSULE STAIN
Klebsiella pneumoniae

3. GRAM STAIN
P. aeruginosa and *S. aureus*

4. GRAM STAIN
B. megaterium and *M. B. catarrhalis*

5. SPORE STAIN (Schaeffer-Fulton)
Bacillus megaterium

6. ACID-FAST STAIN (Ziel-Neelsen)
M. smegmatis and *S. aureus*

Figure 14.3 **Photomicrographs of representative staining techniques (8000×)**

Spore Staining: Two Methods

15

Species of bacteria, belonging principally to the genera *Bacillus* and *Clostridium,* produce extremely heat-resistant structures called **endospores.** In addition to being heat-resistant, they are also very resistant to many chemicals that destroy non-spore-forming bacteria. This resistance to heat and chemicals is due primarily to a thick, tough spore coat.

It was observed in Exercise 14 that Gram staining will not stain endospores. Only if considerable heat is applied to a suitable stain can the stain penetrate the spore coat. Once the stain has entered the spore, however, it is not easily removed with decolorizing agents or water.

Several methods are available that employ heat to provide stain penetration. However, since the Schaeffer-Fulton and Dorner methods are the principal ones used by most bacteriologists, both have been included in this exercise. Your instructor will indicate which procedure is preferred in this laboratory.

Schaeffer-Fulton Method

This method, which is depicted in figure 15.1, utilizes malachite green to stain the endospore and safranin to stain the vegetative portion of the cell. Utilizing this technique, a properly stained spore-former will have a green endospore contained in a pink sporangium. Illustration 5, figure 14.3 on page 60 reveals what such a slide looks like under oil immersion.

After preparing a smear of *Bacillus megaterium,* follow the steps outlined in figure 15.1 to stain the spores.

Materials:

24-36 hour nutrient agar slant culture of
Bacillus megaterium
electric hot plate and small beaker (25 ml size)
spore-staining kit consisting of a bottle each of
5% malachite green and safranin

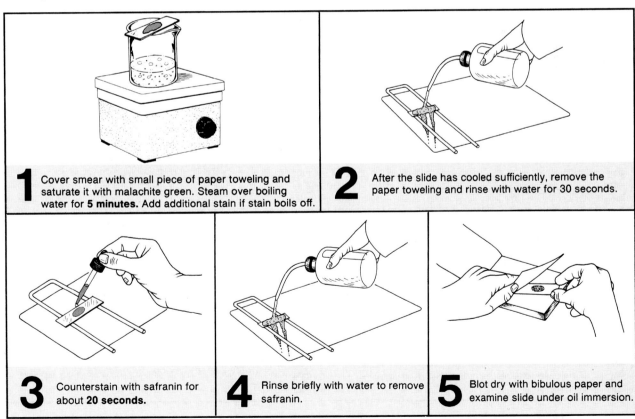

1 Cover smear with small piece of paper toweling and saturate it with malachite green. Steam over boiling water for **5 minutes.** Add additional stain if stain boils off.

2 After the slide has cooled sufficiently, remove the paper toweling and rinse with water for 30 seconds.

3 Counterstain with safranin for about **20 seconds.**

4 Rinse briefly with water to remove safranin.

5 Blot dry with bibulous paper and examine slide under oil immersion.

Figure 15.1   The Schaeffer-Fulton spore stain method

Dorner Method

The Dorner method for staining endospores produces a red spore within a colorless sporangium. Nigrosine is used to provide a dark background for contrast. The six steps involved in this technique are shown in figure 15.2. Although both the sporangium and endospore are stained during boiling in step 3, the sporangium is decolorized by the diffusion of safranin molecules into the nigrosine.

Prepare a slide of *Bacillus megaterium* that utilizes the Dorner method. Follow the steps in figure 15.2.

Materials:
   nigrosine
   electric hot plate and small beaker (25 ml size)
   small test tube (10 × 75 mm size)
   test-tube holder
   24-36 hour nutrient agar slant culture of
      Bacillus megaterium

Laboratory Report

After examining the organisms under oil immersion, draw a few cells in the appropriate circles on Laboratory Report 14–17.

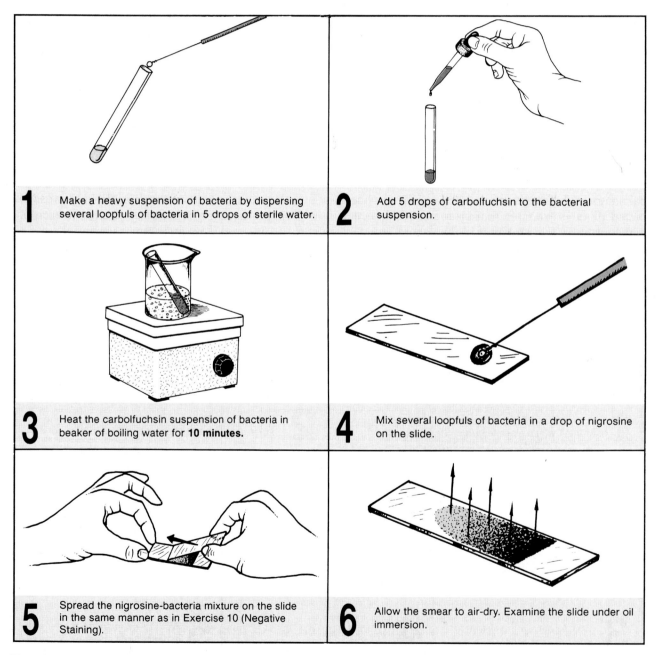

1 Make a heavy suspension of bacteria by dispersing several loopfuls of bacteria in 5 drops of sterile water.

2 Add 5 drops of carbolfuchsin to the bacterial suspension.

3 Heat the carbolfuchsin suspension of bacteria in beaker of boiling water for **10 minutes.**

4 Mix several loopfuls of bacteria in a drop of nigrosine on the slide.

5 Spread the nigrosine-bacteria mixture on the slide in the same manner as in Exercise 10 (Negative Staining).

6 Allow the smear to air-dry. Examine the slide under oil immersion.

Figure 15.2 The Dorner spore stain method

Acid-Fast Staining: Ziehl-Neelsen Method

16

AFB - Acid fast Bacili

Most bacteria in the genus *Mycobacterium* contain considerable amounts of waxlike lipoidal material, which affects their staining properties. Unlike most other bacteria, once they are properly stained with carbolfuchsin, they resist decolorization with acid-alcohol. Since they are not easily decolorized they are said to be **acid-fast.** This property sets them apart from many other bacteria.

This stain is used primarily in the identification of the tuberculosis bacillus, *Mycobacterium tuberculosis,* and the leprosy organism, *Mycobacterium leprae.* After decolorization, methylene blue is added to the organisms to counterstain any material that is not acid-fast; thus, a properly stained slide of a mixture of acid-fast organisms, tissue cells, and non-acid-fast bacteria will reveal red acid-fast rods with bluish tissue cells and bacteria. An example of acid-fast staining is shown in illustration 6 of figure 14.3.

The two organisms used in this staining exercise are *Mycobacterium smegmatis,* a nonpathogenic acid-fast rod found in soil and on external genitalia, and *Staphylococcus aureus,* a non-acid-fast coccus.

Materials:
 nutrient agar slant culture of *Mycobacterium smegmatis* (48-hour culture)
 nutrient broth culture of *S. aureus*
 electric hot plate and small beaker
 acid-fast staining kit (carbolfuchsin, acid alcohol, and methylene blue)

Smear Preparation Prepare a mixed culture smear by placing two loopfuls of *S. aureus* on a slide and transferring a small amount of *M. smegmatis* to the broth on the slide with an inoculating needle. Since the smegma bacilli are waxy and tend to cling to each other in clumps, break up the masses of organisms with the inoculating needle. After air-drying the smear, heat-fix it.

Staining Follow the staining procedure outlined in figure 16.1.

Examination Examine under oil immersion and compare your slide with illustration 6, figure 14.3.

Laboratory Report Record your results on Laboratory Report 14–17.

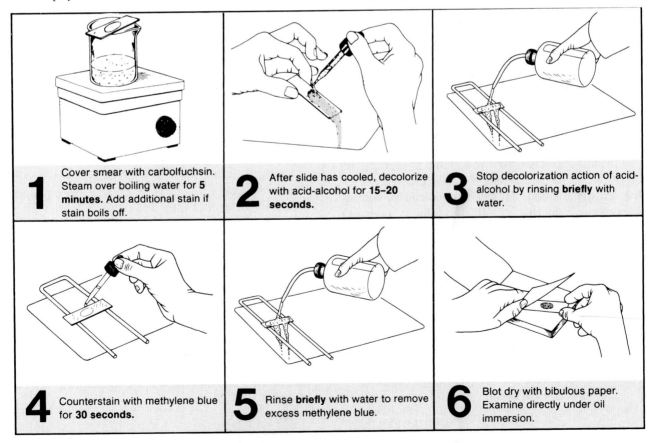

1 Cover smear with carbolfuchsin. Steam over boiling water for **5 minutes.** Add additional stain if stain boils off.

2 After slide has cooled, decolorize with acid-alcohol for **15–20 seconds.**

3 Stop decolorization action of acid-alcohol by rinsing **briefly** with water.

4 Counterstain with methylene blue for **30 seconds.**

5 Rinse **briefly** with water to remove excess methylene blue.

6 Blot dry with bibulous paper. Examine directly under oil immersion.

Figure 16.1 Ziehl-Neelsen acid-fast staining procedure

decolorizer - 95% Eythl & HCl

17 Acid-Fast Staining: Fluorescence Method

In laboratories where large numbers of sputum, gastric washings, urine, and other body fluid samples are tested for pathogenic mycobacteria, **_fluorochrome acid-fast staining_** is used in conjunction with the Ziehl-Neelsen technique. The advantage of using a fluorescence method is that fluorochrome stained slides can be scanned under lower magnification. While a Ziehl-Neelsen prepared slide must be examined under oil immersion (1000× magnification), fluorochrome stained slides can be examined with 60× or 100× magnification. In only a few minutes an entire fluorochrome prepared slide can be scanned. Because of this fact, many laboratories use this faster technique as a screening tool. When they encounter a positive slide with this method, they use a Ziehl-Neelsen prepared slide as a means of confirmation. The fact that dead or noncultivatable mycobacteria may fluoresce makes it necessary to use a confirmatory technique.

The **Truant method** of fluorochrome staining (figure 17.1) consists of staining smears with auramine-rhodamine for 20 minutes, decolorizing with acid-alcohol, and "counterstaining" with potassium permanganate. As soon as the slides are dry, they are examined with a fluorescence microscope. Bacteria that are acid-fast will fluoresce as yellow-orange rods in a dark field. Areas of fluorescence that show up during scanning can be examined more critically under high-dry or oil immersion.

In this exercise you will stain a mixture of *Staphylococcus aureus* and *Mycobacterium phlei* by the Truant method. It will be examined with a fluorescence microscope. If the slide is prepared properly, only the acid-fast rod-shaped mycobacteria will fluoresce.

Materials:
  broth culture of *S. aureus*
  slant culture of *M. phlei* (Lowenstein-Jensen medium)
  auramine-rhodamine stain
  acid-alcohol (for fluorochrome staining)
  postassium permanganate (0.5% solution)
  microscope slides, inoculating loop, Bunsen burner, wash bottle

1. Prepare a mixed smear of *S. aureus* and *M. phlei* by adding a small amount of *M. phlei* to two loopfuls of *S. aureus* on a clean slide. The organisms should be well dispersed on the slide by vigorous manipulation of the inoculating loop on the clumps of organisms.
2. Allow the smear to air-dry completely.
3. Flame-fix the slide over a Bunsen burner. Avoid overheating.

    In diagnostic work where pathogens are being stained, the smear is usually heat-fixed on a slide warmer (65° C) for 2 hours.
4. Cover the smear with **auramine-rhodamine** stain and let stand for **20 minutes** at room temperature.
5. Rinse off the stain with wash bottle.
6. Decolorize with **acid-alcohol** (2.5% HCl in 70% ethanol) for **7–10 seconds.**
7. Rinse thoroughly with wash bottle.
8. Cover the smear with **potassium permanganate** and let stand for **3 minutes.** This solution eliminates background fluorescence ("quenching"). Although this step is often referred to as "counterstaining," in actuality it is not. Excessive counterstaining must be avoided because fluorescence will be completely eliminated.
9. Rinse with water and air-dry.

Observation

The fluorescence microscope should be equipped with a BG12 exciter filter and an OG1 barrier filter. Scan the slide with the lowest magnification that is possible on the microscope. On some instruments this may be high-dry, not low power. If high-dry is used, it is necessary to place a cover glass on the smear with immersion oil between the slide and cover glass. Be sure to use only oil that is specific for fluorescence viewing.

Once you have located areas of fluorescence (yellow-orange spots), add oil and swing the oil immersion lens into position for more critical observation. Record your results on Laboratory Report 14–17.

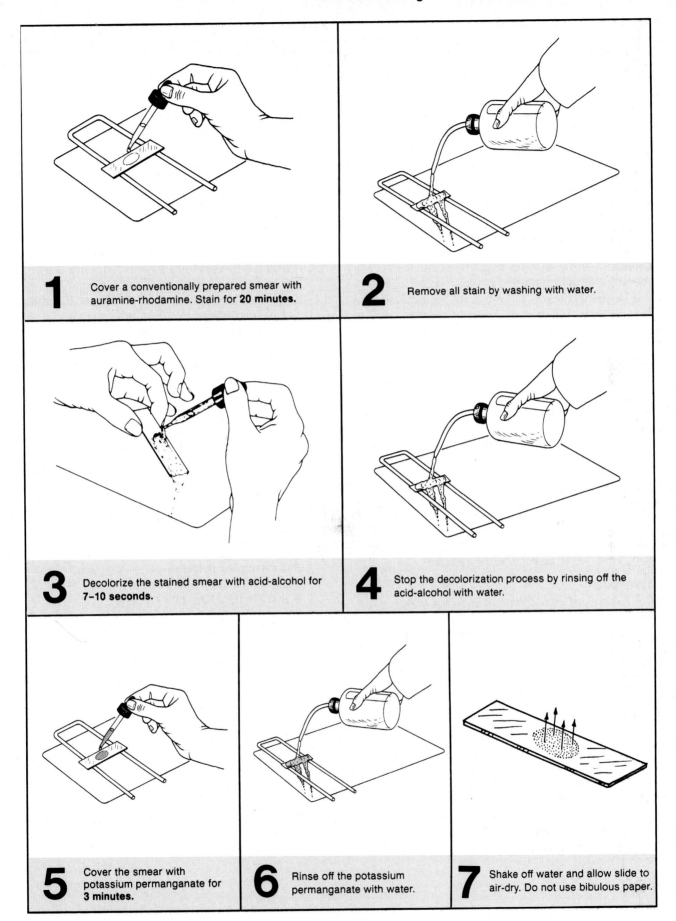

1 Cover a conventionally prepared smear with auramine-rhodamine. Stain for **20 minutes.**

2 Remove all stain by washing with water.

3 Decolorize the stained smear with acid-alcohol for **7–10 seconds.**

4 Stop the decolorization process by rinsing off the acid-alcohol with water.

5 Cover the smear with potassium permanganate for **3 minutes.**

6 Rinse off the potassium permanganate with water.

7 Shake off water and allow slide to air-dry. Do not use bibulous paper.

Figure 17.1   Fluorochrome acid-fast staining routine

Flagellar Staining: Leifson's Method

Of the three types of motility seen in bacteria (gliding, rotary, and flagellar), the latter is most common and best understood. Bacterial flagella may be as long as 70 μm; however, on average, they fall in the range of 10–20 μm long. Although they vary somewhat in thickness, they approximate 20 nm in diameter—a size that places them outside the visibility range of the brightfield microscope.

Visual studies of flagella prior to the advent of the electron microscope necessitated the use of flagellar stains. However, since more reliable information is attainable with electron microscopy, and since flagellar staining is not an easy technique, this procedure is not used a great deal today. Where electron microscopy is not available, however, Leifson's method works quite well.

To make flagella visible, it is necessary to increase flagellar diameters by precipitating a coating of dye over their entire length. Leifson's method accomplishes this by using a single staining reagent that utilizes pararosaniline as a staining agent and tannic acid as a mordant. Staining takes place within 5 to 15 minutes.

Culture Preparation

Materials:
    cultures of *Pseudomonas fluorescens* or
        Morganelli morganii
    tube of brain-heart infusion broth (4 ml)
    1 ml serological pipettes
    formalin
    centrifuge

1. Inoculate a tube of brain-heart infusion broth with the organism and incubate at room temperature for 18 to 20 hours.
2. With a 1 ml pipette add 0.25 ml of formalin to the culture, mix by shaking, and let stand for 15 minutes.
3. Fill the tube to within 1 cm of top with distilled water, mix, and centrifuge for 3 minutes.
4. Pour off supernatant fluid without disturbing organisms in bottom of the tube.
5. Add fresh distilled water, mix, and centrifuge again.
6. Remove supernatant fluid again and resuspend organisms in about 2 ml of distilled water.
7. Dilute the suspension with additional distilled water until suspension is barely turbid.

Staining Procedure

Follow the procedure illustrated in figure 18.1 to produce a bacterial smear and stain it. Once the slide has dried off it can be examined under oil immersion.

Materials:
    suspension of bacteria
    glass slides and china marking pencil
    Leifson's flagellar stain (fresh)

Laboratory Report

There is no Laboratory Report for this exercise.

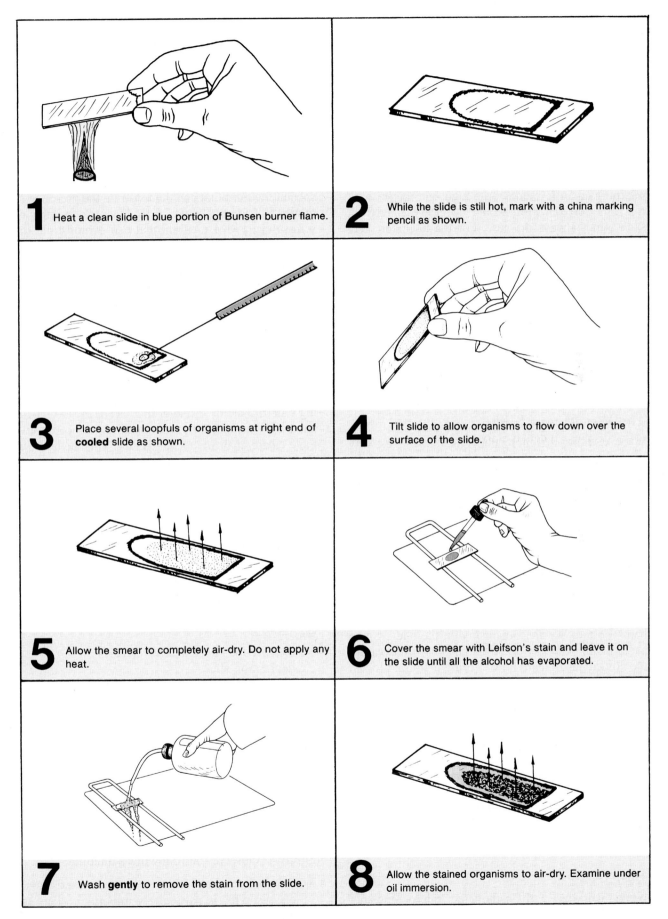

1 Heat a clean slide in blue portion of Bunsen burner flame.

2 While the slide is still hot, mark with a china marking pencil as shown.

3 Place several loopfuls of organisms at right end of **cooled** slide as shown.

4 Tilt slide to allow organisms to flow down over the surface of the slide.

5 Allow the smear to completely air-dry. Do not apply any heat.

6 Cover the smear with Leifson's stain and leave it on the slide until all the alcohol has evaporated.

7 Wash **gently** to remove the stain from the slide.

8 Allow the stained organisms to air-dry. Examine under oil immersion.

Figure 18.1 Leifson's method for staining flagella

When attempting to identify an unknown bacterium it is usually necessary to determine whether the microorganism is motile. Although one might think that this determination would be easily arrived at, such is not always the case. For the beginner there are many opportunities to err.

Four Methods

For nonpathogens there are two slide techniques that one might use. For pathogens one tube and one Petri plate method can be used. Each method has its advantages and limitations. The method you use will depend on which one is most suitable for the situation at hand. A discussion of each procedure follows.

The Wet Mount Slide

When working with nonpathogens, the simplest way to determine motility is to place a few loopfuls of the organism on a clean slide and cover it with a cover glass. In addition to being able to determine the presence or absence of motility, this method is useful in determining cellular shape (rod, coccus, or spiral) and arrangement (irregular clusters, packets, pairs, or long chains). A wet mount is especially useful if phase optics are used. Unlike stained slides that are heat-fixed for staining, there is no distortion of cells on a wet mount.

One problem for beginners is the difficulty of being able to see the organisms on the slide. Since bacteria are generally colorless and very transparent, the novice has to learn how to bring them into focus.

The Hanging Drop Slide

If it is necessary to study viable organisms on a microscope slide for a longer period of time than is possible with a wet mount, one can resort to a hanging drop slide. As shown in illustration 4 of figure 19.1, organisms are observed in a drop that is suspended under a cover glass in a concave depression slide. Since the drop lies within an enclosed glass chamber, drying out occurs very slowly.

Tube Method

When working with pathogenic microorganisms such as the typhoid bacillus, it is too dangerous to attempt to determine motility with slide techniques. A much safer method is to culture the organisms in a special medium that can demonstrate the presence of motility. The procedure is to inoculate a tube of semisolid or SIM medium that can demonstrate the presence of motility. Both media have a very soft consistency that allows motile bacteria to migrate readily through them causing cloudiness. Figure 10.2 illustrates the inoculation procedure.

Soft Agar Plate Method

Although the tube method is the generally accepted procedure for determining motility of pathogens, it is often very difficult for beginners to interpret. Richard Roller at the University of Iowa suggests that incubating a Petri plate of soft agar that has been stab inoculated with a motile organism will show up motility more clearly than an inoculated tube. This method will also be tried here in this laboratory period.

First Period

During the first period you will make wet mount and hanging drop slides of two organisms: *Proteus vulgaris* and *Micrococcus luteus*. Tube media (semisolid medium or SIM medium) and a soft agar plate will also be inoculated. The media inoculations will have to be incubated to be studied in the next period. Proceed as follows:

Materials:
>   microscope slides and cover glasses
>   depression slide
>   2 tubes of semisolid or SIM medium
>   1 Petri plate of soft nutrient agar (20–25 ml of
>   soft agar per plate)
>   nutrient broth cultures of *Micrococcus luteus*
>   and *Proteus vulgaris* (young cultures)
>   inoculating loop and needle
>   Bunsen burner

Wet Mounts   Prepare wet mount slides of each of the organisms, using several loopfuls of the organism on the slides. Examine under an oil immersion objective. Observe the following guidelines:

- Use only scratch-free, clean slides and cover glasses. This is particularly important when using phase-contrast optics.

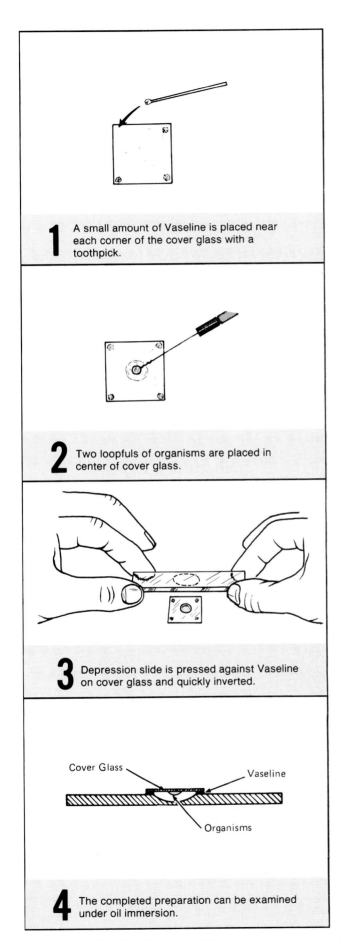

1 A small amount of Vaseline is placed near each corner of the cover glass with a toothpick.

2 Two loopfuls of organisms are placed in center of cover glass.

3 Depression slide is pressed against Vaseline on cover glass and quickly inverted.

Cover Glass

Vaseline

Organisms

4 The completed preparation can be examined under oil immersion.

Figure 19.1 **The hanging drop slide.**

• Label each slide with the name of the organism.
• By manipulating the diaphragm and voltage control, reduce the lighting sufficiently to make the organisms visible. Unstained bacteria are very transparent and difficult to see.
• For proof of true motility, look for directional movement that is several times the long dimension of the bacterium. The movement will also occur in different directions in the same field.
• Ignore Brownian movement. *Brownian movement* is vibrational movement caused by invisible molecules bombarding bacterial cells. If the only movement you see is vibrational and not directional, the organism is nonmotile.
• If you see only a few cells exhibiting motility, consider the organism to be motile. Characteristically, only a few of the cells will be motile at a given moment.
• Don't confuse water current movements with true motility. Water currents are due to capillary action caused by temperature changes and drying out. All objects move in a straight line in one direction.
• And, finally, always *examine a wet mount immediately,* once it has been prepared, because motility decreases with time after preparation.

Hanging Drop Slides By referring to figure 19.1 prepare hanging drop slides of each organism. Be sure to use clean cover glasses and label each slide with a china marking pencil. When placing loopfuls of organisms on the cover glass, be sure to flame the loop between applications. Once the slide is placed on the microscope stage, do as follows:

. Examine the slide first with the low-power objective. If your microscope is equipped with an automatic stop, avoid using the stop; instead, use the coarse adjustment knob for bringing the image into focus. The greater thickness of the depression slide prevents one from being able to focus at the stop point.
. Once the image is visible under low power, swing the high-dry objective into position and readjust the lighting. Since most bacteria are drawn to the edge of the drop by surface tension, **focus near the edge of the drop.**
. If your microscope has phase-contrast optics, switch to high-dry phase. Although a hanging drop does not provide the shallow field desired for phase-contrast, you may find that it works fairly well.
. If you wish to use oil immersion, simply rotate the high-dry objective out of position, add immersion oil to the cover glass, and swing the oil immersion lens into position.

69

5. Avoid delay in using this setup. Water of condensation may develop to decrease clarity and the organisms become less motile with time.

6. Review all the characteristics of bacterial motility that are stated above under wet mounts.

Tube Method   Inoculate tubes of semisolid or SIM media with each organism according to the following instructions:

1. Label the tubes of semisolid (or SIM) media with the names of the organisms. Place your initials on the tubes, also.

2. Flame and cool the inoculating needle, and insert it into the culture after flaming the neck of the tube.

3. Remove the plug from the tube of medium, flame the neck, and stab it ⅔ of the way down to the bottom, as shown in figure 19.2. Flame the neck of the tube again before returning the plug to the tube.

4. Repeat steps 2 and 3 for the other culture.

5. Incubate the tubes at room temperature for 24 to 48 hours.

Plate Method   Mark the bottom of a plate of soft agar with two one-half inch circles about one inch apart. Label one circle ML and the other PV. These circles will be targets for your culture stabs. Put your initials on the plate also.

Using proper aseptic techniques, stab the medium in the center of the ML circle with *M. luteus* and the center of the other circle with *P. vulgaris*. Incubate the plate for 24 to 48 hours at room temperature.

Second Period

Assemble the following materials that were inoculated during the last period and incubated.

Materials:

culture tubes of motility medium that have been incubated
inoculated Petri plate that has been incubated

Compare the two tubes that were inoculated with *M. luteus* and *P. vulgaris*. Look for cloudiness as evidence of motility. *Proteus* should exhibit motility. Does it? Record your results on the Laboratory Report.

Compare the appearance of the two stabs in the soft agar. Describe the differences that exist in the two stabs.

Does the plate method provide any better differentiation of results than the tube method?

Laboratory Report

Complete the Laboratory Report for this exercise.

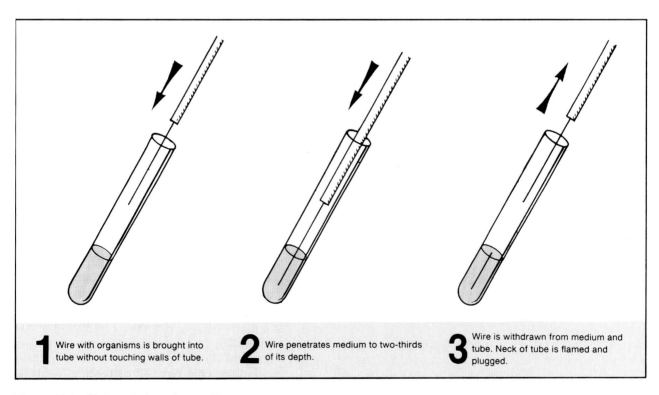

1 Wire with organisms is brought into tube without touching walls of tube.

2 Wire penetrates medium to two-thirds of its depth.

3 Wire is withdrawn from medium and tube. Neck of tube is flamed and plugged.

Figure 19.2   Stab technique for motility test

PART Culture Methods

All nutritional types are represented among the protists. This diversity requires a multiplicity of culture methods. An attempt has been made here to present those techniques that have proven most successful for the culture of autotrophic and heterotrophic bacteria, molds, and slime molds.

The first four exercises (20, 21, 22, and 23) pertain to basic techniques applicable to both autotrophs and heterotrophs. The other four exercises are concerned with the culture of specific types. In performing the last four experiments, you should be just as concerned with understanding the growth conditions as with the successful growth of a particular isolate. For example, the use of an enrichment medium, such as in Exercise 27 (Photosynthetic Bacteria), has direct application in the culture of other autotrophic bacteria as well. The method of employing a synthetic medium to bring a specific type of organism out of mixed populations will be used in several subsequent exercises.

This unit culminates the basic techniques phase of this course. A thorough understanding of microscopy, slide techniques, and culture methods provides a substantial foundation for the remainder of the exercises in this manual. If independent study projects are to be pursued as a part of this course, the completion of this unit will round out the background knowledge and skills for such work.

Culture Media Preparation

From now on, most of the laboratory experiments in this manual will utilize bacteriological media. In most instances it will be provided for you. However, circumstances may arise when you will need some special medium that is not already prepared, and it will be up to you to put it together. It is for situations like this that the information in this exercise will be useful.

The first portion of this exercise pertains to the different types of media and how they relate to the needs of microorganisms. The last part of the exercise pertains to the actual mechanics of making up a batch of medium. Whether you will be provided an opportunity to prepare some media during a designated laboratory period will depend on the availability of time and classroom needs. Your instructor will indicate how this exercise is to be used.

Media Consistency

A microbiological **medium** (*media,* plural) is the food that we use for culturing bacteria, molds, and other microorganisms. It can exist in three consistencies: liquid, solid, and semisolid. If you have performed all of the exercises in Part 3, you are already familiar with all of them.

Liquid media include nutrient broth, citrate broth, glucose broth, litmus milk, etc. These media are used for the propagation of large numbers of organisms, fermentation studies, and various other tests.

Solid media are made by adding a solidifying agent, such as agar, gelatin, or silica gel, to a liquid medium. A good solidifying agent is one that is not utilized by microorganisms, does not inhibit bacterial growth, and does not liquefy at room temperature. Agar and silica gel do not liquefy at room temperature and are utilized by very few organisms. Gelatin, on the other hand, is hydrolyzed by quite a few organisms and liquefies at room temperature.

Nutrient agar, blood agar, and Sabouraud's agar are examples of solid media that are used for developing surface colony growth of bacteria and molds. As we will see in the next exercise, the development of colonies on the surface of a medium is essential when trying to isolate organisms from mixed cultures.

Semisolid media fall in between liquid and solid media. Although they are similar to solid media in that they contain solidifying agents such as agar and gelatin, they are more jellylike due to lower percentages of these solidifiers. These media are particularly useful in determining whether certain bacteria are motile (Exercise 19).

Nutritional Needs of Bacteria

Before one can construct a medium that will achieve a desired result in the growth of organisms, one must understand their basic needs. Any medium that is to be suitable for a specific group of organisms must take into account the following seven factors: water, carbon, energy, nitrogen, minerals, growth factors, and pH. The role of each one of these factors follows.

Water   Protoplasm consists of 70% to 85% water. The water in a single-celled organism is continuous with the water of its environment, and the molecules pass freely in and out of the cell, providing a vehicle for nutrients inward and secretions or excretions outward. All the enzymatically controlled chemical reactions that occur within the cell occur only in the presence of an adequate amount of water.

The quality of water used in preparing media is important. Hard tap water, high in calcium and magnesium ions, should not be used. Insoluble phosphates of calcium and magnesium may precipitate in the presence of peptones and beef extract. The best policy is to *always use distilled water.*

Carbon   Organisms are divided into two groups with respect to their sources of carbon. Those that can utilize the carbon in carbon dioxide for synthesis of all cell materials are called *autotrophs.* If they must have one or more organic compounds for their carbon source, they are called *heterotrophs.* In addition to organic sources of carbon, the heterotrophs are also dependent on carbon dioxide. If this gas is completely excluded from their environment, their growth is greatly retarded, particularly in the early stages of starting a culture.

Specific organic carbon needs are as diverse as the organisms themselves. Where one organism may require only a single simple compound such as acetic acid, another may require a dozen or more organic nutrients of various degrees of complexity.

Energy Organisms that have pigments that enable them to utilize solar energy are called *photoautotrophs* (photosynthetic autotrophs). Media for such organisms will not include components to provide energy.

Autotrophs that cannot utilize solar energy but are able to oxidize simple inorganic substances for energy are called *chemoautotrophs* (chemosynthetic autotrophs). The essential energy-yielding substance for these organisms may be as elemental as nitrite, nitrate, or sulfide.

Most bacteria fall into the category of *chemoheterotrophs* (chemosynthetic heterotrophs) that require an organic source of energy, such as glucose or amino acids. The amounts of energy-yielding ingredients in media for both chemosynthetic types is on the order of 0.5%.

A small number of bacteria are classified as *photoheterotrophs* (photosynthetic heterotrophs). These organisms have photosynthetic pigments that enable them to utilize sunlight for energy. Their carbon source must be an organic compound such as alcohol.

Nitrogen Although autotrophic organisms can utilize inorganic sources of nitrogen, the heterotrophs get their nitrogen from amino acids and intermediate protein compounds such as peptides, proteoses, and peptones. Beef extract and peptone, as used in nutrient broth, provide the nitrogen needs for the heterotrophs grown in this medium.

Minerals All organisms require several metallic elements such as sodium, potassium, calcium, magnesium, manganese, iron, zinc, copper, phosphorus, and cobalt for normal growth. Bacteria are no exception. The amounts required are very small.

Growth Factors Any essential component of cell material that an organism is unable to synthesize from its basic carbon and nitrogen sources is classified as being a *growth factor*. This may include certain amino acids or vitamins. Many heterotrophs are satisfied by the growth factors present in beef extract of nutrient broth. Most fastidious pathogens require enriched media such as blood agar for ample growth factors.

Hydrogen Ion Concentration The growth of organisms in a particular medium may be completely inhibited if the pH of the medium is not within certain limits. The enzymes of microorganisms are greatly affected by this factor. Since most bacteria grow best around pH 7 or slightly lower, the pH of nutrient broth should be adjusted to pH 6.8. Pathogens, on the other hand, usually prefer a more alkaline pH. Trypticase soy broth, a suitable medium for the more fastidious pathogens, should be adjusted to pH 7.3.

Exact Composition Media

Media can be prepared to exact specifications so that the exact composition is known. These media are generally made from chemical compounds that are highly purified and precisely defined. Such media are readily reproducible. They are known as **synthetic media.** Media such as nutrient broth that contain ingredients of imprecise composition are called **nonsynthetic media.** Both the beef extract and peptone in nutrient broth are inexact in composition.

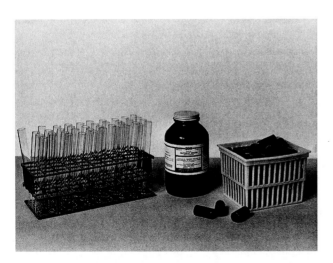

Figure 20.1 Basic supplies and equipment needed for making up a batch of medium.

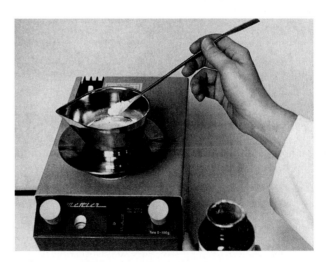

Figure 20.2 Correct amount of dehydrated medium is weighed on balance.

Special Media

Two kinds of special media that will be widely used in this manual are selective and differential media.

Selective media are media that allow only certain types of organisms to grow in or on them because of (1) the absence of certain critical nutrients that make it unfavorable for most, but not all, organisms, or (2) the presence of inhibitory substances that prevent certain types of organisms to grow on them. The inhibitory substance may be salt (NaCl), acid, a toxic chemical (crystal violet), an antibiotic (streptomycin), or some other substance.

Differential media are media that contain substances that cause some bacteria to take on a different appearance from other species, allowing one to differentiate one species from another.

In some cases media have been formulated that are both selective and differential. A good example is Levine EMB agar, which is used to determine the presence of coliforms in water analysis (Exercise 61).

Dehydrated Media

Until around 1930, the laboratory worker had to spend a good deal of time preparing laboratory media from various raw materials. If a medium contained five or six ingredients, it was not only necessary to measure the various materials, but, also, in many instances, to fabricate some of the components such as beef extract or veal infusion by long tedious cooking methods. Today, dehydrated media have revolutionized media preparation techniques in much the same way that commercial cake mixes have taken over in the kitchen. For most routine bacteriological work, media preparation has been

simplified to the extent that all that is necessary is to dissolve a measured amount of dehydrated medium in water, adjust the pH, dispense into tubes, and sterilize. In many cases pH adjustment is not even necessary.

Media Preparation Assignment

In this laboratory period you will work with your laboratory partner to prepare tubes of media that will be used in future laboratory experiments. Your instructor will indicate which media you are to prepare. Record in the space below the number of tubes of specific media that have been assigned to you and your partner.

nutrient broth   _____
nutrient agar pours   _____
nutrient agar slants   _____
other   _____

Several different sizes of test tubes are used for media, but the two sizes most generally used are either 16 mm or 20 mm diameter by 15 cm long. Select the correct size tubes first, according to these guidelines:

Large tubes (20 mm dia): Use these test tubes for *all pours:* i.e., nutrient agar, Sabouraud's agar, EMB agar, etc. Pours are used for filling Petri plates.

Small tubes (16 mm dia): Use these tubes for all *broths, deeps,* and *slants.*

If the tubes are clean and have been protected from dust or other contamination, they can be used without cleaning. If they need cleaning, scrub out the insides with warm water and detergent, using a test-tube brush. Rinse twice, first with tap water, and

Figure 20.3 Dehydrated medium is dissolved in a measured amount of distilled water.

Figure 20.4 If medium contains agar, it must be brought to a boil to bring agar into solution.

finally with distilled water to rid them of all traces of detergent. Place them in a wire basket or rack, inverted, so that they can drain. Do not dry with a towel.

Measurement and Mixing

The amount of medium you make for a batch should be determined as precisely as possible to avoid shortage or excess.

Materials:
>   graduate, beaker, glass stirring rod
>   bottles of dehydrated media
>   Bunsen burner and tripod, or hot plate

1. Measure the correct amount of water needed to make up your batch. The following volumes required per tube must be taken into consideration:

   pours .... 12 ml
   deeps .... 6 ml
   slants .... 4 ml
   broths .... 5 ml
   broths with fermentation tubes .... 5–7 ml

2. Consult the label on the bottle to determine how much powder is needed for 1000 ml and then determine by proportionate methods how much you need for the amount of water you are using. Weigh this amount on a balance and add it to the beaker of water. If the medium does not contain agar, the mixture usually goes into solution without heating.

3. **If the medium contains agar,** heat the mixture over a Bunsen burner (figure 20.4) or on an electric hot plate until it comes to a boil. To safeguard against water loss, *before heating, mark the level of the top of the medium on the side of the beaker with a china marking pencil.*

As soon as it "froths up," turn off the heat. If an electric hot plate is used, the medium must be removed from the hot plate or it will boil over the sides of the container.

Caution: Be sure to keep stirring from the bottom with a glass stirring rod so that the medium does not char on the bottom of the beaker.

4. Check the level of the medium with the mark on the beaker to note if any water has been lost. Add sufficient distilled water as indicated. Keep the temperature of the medium at about 60° C to avoid solidification. The medium will solidify at around 40° C.

Adjusting the pH

Although dehydrated media contain buffering agents to keep the pH of the medium in a desired range, the pH of a batch of medium may differ from that stated on the label of the bottle. Before the medium is tubed, therefore, one should check the pH and make any necessary adjustments.

If a pH meter (figure 20.5) is available and already standardized, use it to check the pH of your medium. If the medium needs adjustment use the bottles of HCl and NaOH to correct the pH. If no meter is available pH papers will work about as well. Make pH adjustment as follows:

Materials:
>   beaker of medium
>   acid and base kits (dropping bottles of 1N and 0.1N HCl and NaOH)
>   glass stirring rod
>   pH papers
>   pH meter (optional)

1. Dip a piece of pH test paper into the medium to determine the pH of the medium.

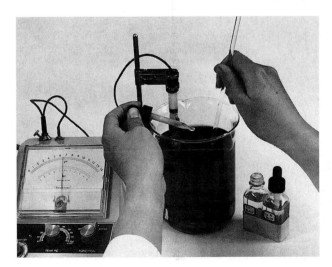

Figure 20.5 The hydrogen ion concentration of a medium must be adjusted to its recommended pH.

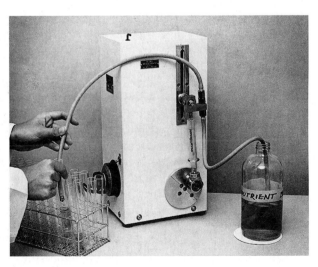

Figure 20.6 An automatic pipetting machine will deliver precise amounts of media at a controlled rate.

2. **If the pH is too high,** add a drop or two of HCl to lower the pH. For large batches use 1N HCl. If the pH difference is slight, use the 0.1N HCl. Use a glass stirring rod to mix the solution as the drops are added.

3. **If the pH is too low,** add NaOH, one drop at a time, to raise the pH. For slight pH differences, use 0.1N NaOH; for large differences use 1N NaOH. Use a glass stirring rod to mix the solution as the drops are added.

Filling the Test Tubes

Once the pH of the medium is adjusted it must be dispensed into test tubes. If an automatic pipetting machine is to be used, as shown in figure 20.6, it will have to be set up for you by your instructor. These machines can be adjusted to deliver any amount of medium at any desired speed. When large numbers of tubes are to be filled, the automatic pipetting machine should be used. For smaller batches, the funnel method shown in figure 20.7 is adequate. Use the following procedure when filling tubes with a funnel assembly.

Materials:
    ring stand assembly
    funnel assembly (glass funnel, rubber tubing, hose clamp, and glass tip)
    graduate (small size)

1. Fill one test tube with a measured amount of medium. This tube will be your guide for filling the other tubes.
2. Fill the funnel and proceed to fill the test tubes to the proper level, holding the guide tube alongside of each empty tube to help you to determine the amount to allow into each tube.
3. Keep the beaker of medium over heat if it contains agar.
4. If fermentation tubes are to be used, add one to each tube at this time *with the open end down.*

Capping the Tubes

The last step before sterilization is to provide a closure for each tube. Plastic (polypropylene) caps are suitable in most cases. All caps that slip over the tube end have inside ridges that grip the side of the tube and provide an air gap to allow steam to escape during sterilization. If you are using tubes with plastic screw-caps, *the caps should not be screwed tightly before sterilization; instead, each one must be left partly unscrewed.*

If no slip-on caps of the correct size are available, it may be necessary to make up some cotton plugs. A properly made cotton plug should hold firmly in the tube so that it is not easily dislodged.

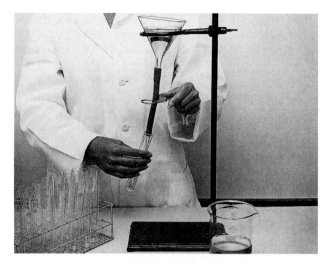
Figure 20.7 A glass funnel assembly and hose clamp are adequate for filling small batches of tubes.

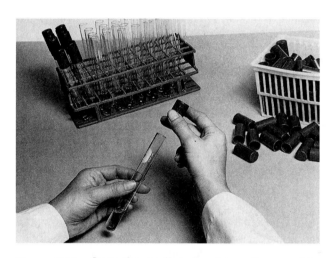
Figure 20.8 Once the medium has been dispensed to all the tubes, they are capped prior to sterilization.

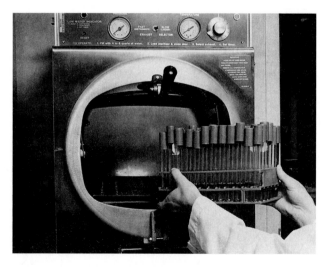
Figure 20.9 Tubes of media are sterilized in an autoclave for 20 to 30 minutes at 15 psi steam pressure.

Sterilization

As soon as the tubes of media have been stoppered they must be sterilized. Organisms on the walls of the tubes, in the distilled water, and in the dehydrated medium will begin to grow within a short period of time at room temperature, destroying the medium.

Prior to sterilization, the tubes of media should be placed in a wire basket with a label taped on the outside of the basket. The label should indicate the type of medium, the date, and your name.

Sterilization must be done in an autoclave. The following considerations are important in using an autoclave:

- Check with your instructor on the procedure to be used with your particular type of autoclave. Complete sterilization occurs at 250° F (121.6° C). To achieve this temperature the autoclave has to develop 15 pounds per square inch (psi) of steam pressure. To reach the correct temperature there must be some provision in the chamber for the escape of air. On some of the older units it is necessary to allow the steam to force air out through the door before closing it.

- *Don't overload the chamber.* One should not attempt to see how much media can be packed into it. Provide ample space between baskets of media to allow for circulation of steam.

- *Adjust the time of sterilization to the size of load.* Small loads may take only 10 to 15 minutes. An autoclave full of media may require 30 minutes for complete sterilization.

After Sterilization

Slants   If you have a basket of tubes that are to be converted to slants, it is necessary to lay the tubes down in a near-horizontal manner as soon as they are removed from the autoclave. The easiest way to do this is to use a piece of rubber tubing (½″ dia) to support the capped end of the tube as it rests on the countertop. Solidification should occur in about 30–60 minutes.

Other Media   Tubes of broth, agar deeps, nutrient gelatin, etc., should be allowed to cool to room temperature after removal from the autoclave. Once they have cooled down, place them in a refrigerator or cold storage room.

Storage   If tubes of media are not to be used immediately they should be stored in a cool place. When stored for long periods of time at room temperature media tend to lose moisture. At refrigerated temperatures media will keep for months.

Laboratory Report

Complete the Laboratory Report for this exercise.

Pure Culture Techniques

When we try to study the bacterial flora of the body, soil, water, food, or any other part of our environment, we soon discover that bacteria exist in mixed populations. It is only in very rare situations that they occur as a single species. To be able to study the cultural, morphological, and physiological characteristics of an individual species, it is essential, first of all, that the organism be separated from the other species that are normally found in its habitat; in other words, we must have a **pure culture** of the microorganism.

Several different methods of getting a pure culture from a mixed culture are available to us. The two most frequently used methods involve making a streak plate or a pour plate. Both plate techniques involve thinning the organisms so that the individual species can be selected from the others.

In this exercise you will have an opportunity to use both methods in an attempt to separate three distinct species from a tube that contains a mixture. The principal difference between the three organisms will be their colors: *Serratia marcescens* is red, *Micrococcus luteus* is yellow, and *Escherichia coli* is white. If *Chromobacterium violaceum* is used in place of *M. luteus,* the three colors will be red, white, and purple.

Streak Plate Method

For economy of materials and time, this method is best. It requires a certain amount of skill, however, which is forthcoming with experience. A properly executed streak plate will give as good an isolation as is desired for most work. Figure 21.1 illustrates how colonies of a mixed culture should be spread out on a properly made streak plate. The important thing is to produce good spacing between colonies.

Materials:
electric hot plate (or tripod and wire gauze)
Bunsen burner and beaker of water
wire loop, thermometer, and china marking pencil
1 nutrient agar pour and 1 sterile Petri plate
1 mixed culture of *Serratia marcescens,* *Escherichia coli,* and *Micrococcus luteus* (or *Chromobacterium violaceum*)

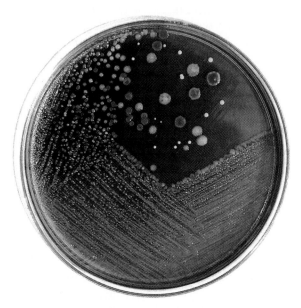

Figure 21.1 If your streak plate reveals well-isolated colonies of three colors (red, white, and yellow), you will have a plate suitable for subculturing.

1. Prepare your tabletop by disinfecting its surface with the disinfectant that is available in the laboratory (Roccal, Zephiran, Betadine, etc.). Use a sponge to scrub it clean.
2. Label the bottom surface of a sterile Petri plate with your name and date. Use a china marking pencil.
3. Liquefy a tube of nutrient agar, cool to 50° C, and pour the medium into the bottom of the plate, following the procedure illustrated in figure 21.2. Be sure to flame the neck of the tube prior to pouring to destroy any bacteria around the end of the tube.

   After pouring the medium into the plate, gently rotate the plate so that it becomes evenly distributed, but do not splash any medium up over the sides.

   Agar-agar, the solidifying agent in this medium becomes liquid when boiled and resolidifies at around 42° C. Failure to cool it prior to pouring into the plate will result in condensation of moisture on the cover. Any moisture on the cover is undesirable because if it drops down on the colonies, the organisms of one colony can spread to other colonies, defeating the entire isolation technique.

4. Streak the plate by one of the methods shown in figure 21.4. Your instructor will indicate which technique you should use.

Caution: Be sure to follow the routine in figure

21.3 for getting the organism out of culture.

5. Incubate the plate in an *inverted position* at 25° C for 24–48 hours. By incubating plates upside down, the problem of moisture on the cover is minimized.

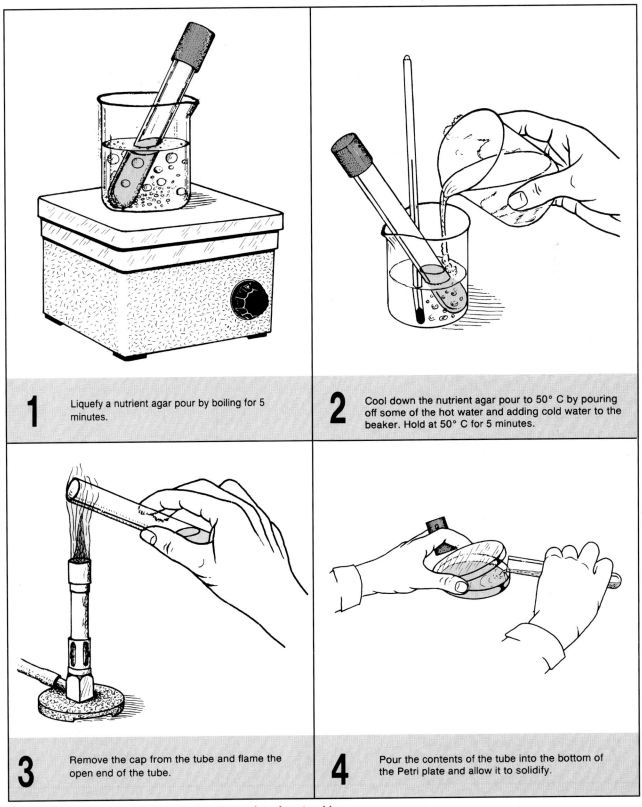

1 Liquefy a nutrient agar pour by boiling for 5 minutes.

2 Cool down the nutrient agar pour to 50° C by pouring off some of the hot water and adding cold water to the beaker. Hold at 50° C for 5 minutes.

3 Remove the cap from the tube and flame the open end of the tube.

4 Pour the contents of the tube into the bottom of the Petri plate and allow it to solidify.

Figure 21.2 Procedure for pouring an agar plate for streaking

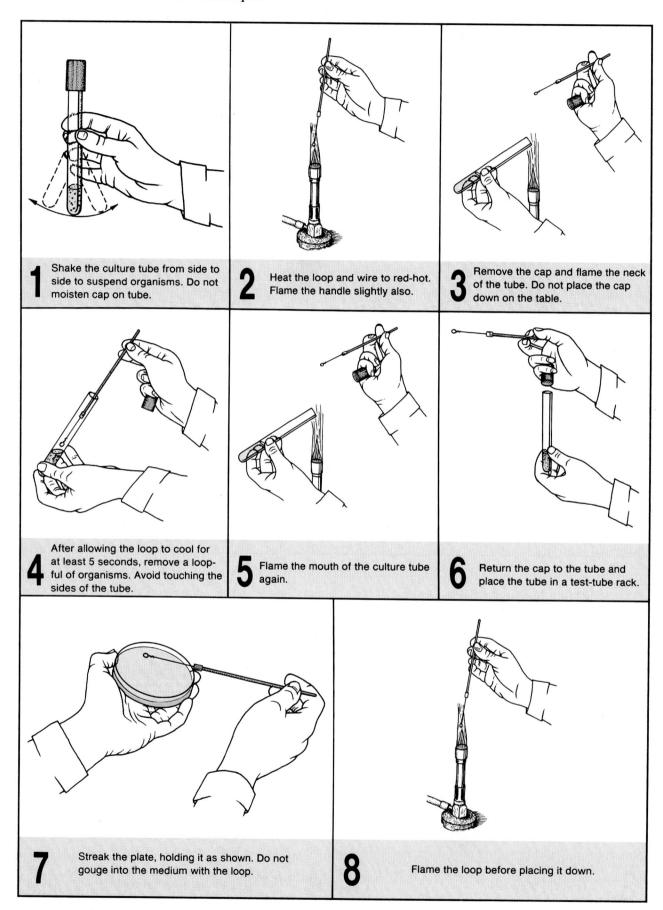

1 Shake the culture tube from side to side to suspend organisms. Do not moisten cap on tube.

2 Heat the loop and wire to red-hot. Flame the handle slightly also.

3 Remove the cap and flame the neck of the tube. Do not place the cap down on the table.

4 After allowing the loop to cool for at least 5 seconds, remove a loopful of organisms. Avoid touching the sides of the tube.

5 Flame the mouth of the culture tube again.

6 Return the cap to the tube and place the tube in a test-tube rack.

7 Streak the plate, holding it as shown. Do not gouge into the medium with the loop.

8 Flame the loop before placing it down.

Figure 21.3 Routine for inoculating a Petri plate

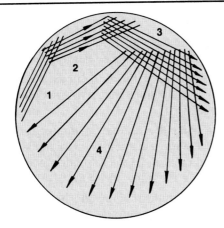

QUADRANT STREAK

1. Streak one loopful of organisms over Area 1 near edge of the plate. *Apply the loop lightly.* Don't gouge into the medium.

2. Flame the loop, cool 5 seconds, and make 5 or 6 streaks from Area 1 through Area 2.

3. Flame the loop again, cool it, and make 6 or 7 streaks from Area 2 through Area 3.

4. Flame the loop again and make as many streaks as possible from Area 3 into Area 4, using up the remainder of the plate surface.

5. Flame the loop before putting it aside.

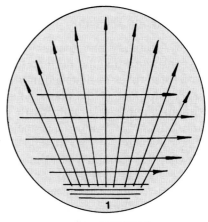

RADIANT STREAK

1. Spread a loopful of organisms in small area near edge of plate **(Area 1).**

2. Flame the loop and allow it to cool for 5 seconds.

3. **From the edge** of Area 1 make 7 or 8 straight streaks to the opposite side of the plate.

4. Flame the loop again and cross-streak over the last streaks, **starting near Area 1.**

5. Flame the loop again before putting it aside.

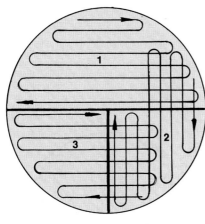

"T" STREAK

1. With marking pencil draw a "T" on bottom of plate, dividing the plate into one half and two quarters.

2. Inoculate the half portion with one loopful of organisms in a continuous line from edge of the plate to the midline of the plate. Don't gouge the medium.

3. After flaming the loop, cross-streak from Area 1 into Area 2 with one continuous streak, filling Area 2.

4. After flaming loop again, cross-streak from Area 2 to Area 3 with a single continuous streak.

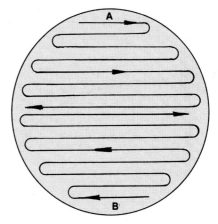

CONTINUOUS STREAK

1. Starting at the edge of the plate (Area A) with a loopful of organisms, spread the organisms in a single continuous movement to the center of the plate.

2. Rotate the plate 180 degrees so that the uninoculated portion of the plate is away from you.

3. Without flaming loop, and using the same face of the loop, continue streaking the other half of the plate by starting at Area B and working toward the center.

4. Flame your loop before setting it aside.

Figure 21.4   Streak plate methods

Pour Plate Method
(Loop Dilution)

This method of separating one species of bacteria from another consists of diluting out one loopful of organisms with three tubes of liquefied nutrient agar in such a manner that one of the plates poured will have an optimum number of organisms to provide good isolation. Figure 21.5 illustrates the general procedure. One advantage of this method is that it requires somewhat less skill than that required for a good streak plate; a disadvantage, however, is that it requires more media, tubes, and plates. Proceed as follows to make three dilution pour plates, using the same mixed culture you used for your streak plate.

Materials:
> mixed culture of bacteria
> 3 nutrient agar pours
> 3 sterile Petri plates
> electric hot plate
> beaker of water
> thermometer
> inoculating loop and china marking pencil

1. Label the three nutrient agar pours **I, II,** and **III** with a marking pencil and place them in a beaker of water on an electric hot plate to be liquefied.

To save time, start with hot tap water if it is available.

2. While the tubes of media are being heated, label the bottoms of the three Petri plates **I, II,** and **III**.

3. Cool down the tubes of media to 50° C, using the same method that was used for the streak plate.

4. Following the routine in figure 21.5, inoculate tube I with one loopful of organisms from the mixed culture. Note the sequence and manner of handling the tubes in figure 21.6.

5. Inoculate tube II with one loopful from tube I after thoroughly mixing the organisms in tube I by shaking the tube from side to side or by rolling the tube vigorously between the palms of both hands. ***Do not splash any of the medium up onto the tube closure.*** Return tube I to the water bath.

6. Agitate tube II to completely disperse the organisms and inoculate tube III with one loopful from tube II. Return tube II to the water bath.

7. Agitate tube III, flame its neck, and pour its contents into plate III.

8. Flame the necks of tubes I and II and pour their contents into their respective plates.

9. After the medium has completely solidified, incubate the *inverted* plates at 25° C for 24–48 hours.

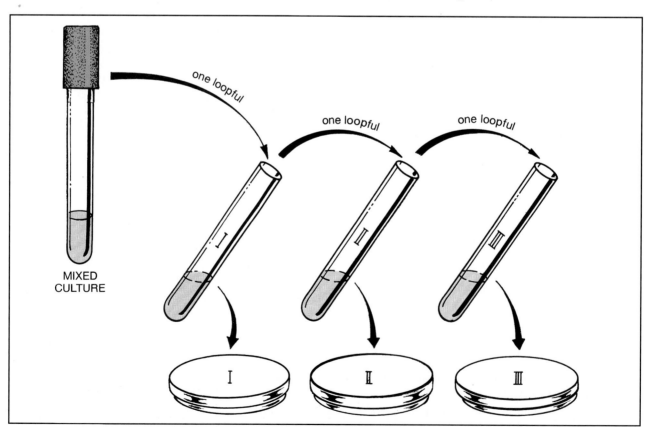

Figure 21.5 Three steps in the loop dilution technique for separating out organisms

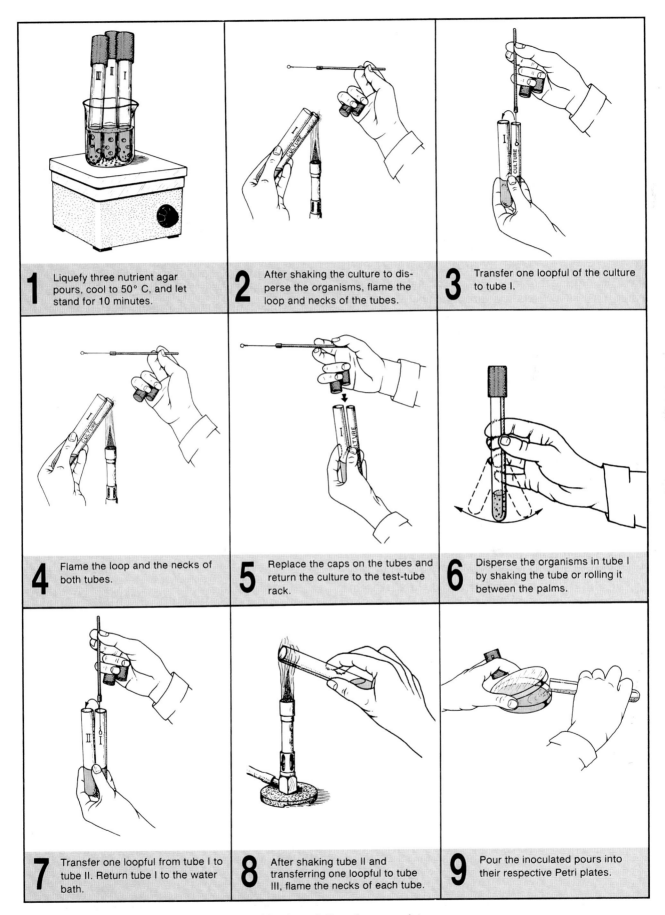

1 Liquefy three nutrient agar pours, cool to 50° C, and let stand for 10 minutes.

2 After shaking the culture to disperse the organisms, flame the loop and necks of the tubes.

3 Transfer one loopful of the culture to tube I.

4 Flame the loop and the necks of both tubes.

5 Replace the caps on the tubes and return the culture to the test-tube rack.

6 Disperse the organisms in tube I by shaking the tube or rolling it between the palms.

7 Transfer one loopful from tube I to tube II. Return tube I to the water bath.

8 After shaking tube II and transferring one loopful to tube III, flame the necks of each tube.

9 Pour the inoculated pours into their respective Petri plates.

Figure 21.6 Tube-handling procedure in making inoculations for pour plates

Evaluation of the Two Methods

After 24 to 48 hours of incubation examine all four Petri plates. Look for colonies that are well isolated from the others. Note how crowded the colonies appear on plate I as compared with plates II and III. Plate I will be unusable. Either plate II or III will have the most favorable isolation of colonies. Can you pick out three well-isolated colonies on your best pour plate that are white, yellow, and red?

Draw the appearance of your streak plate and pour plates on the Laboratory Report.

Subculturing Techniques

The next step in the development of a pure culture is to transfer the organisms from the Petri plate to a tube of nutrient broth or a slant of nutrient agar. After this subculture has been incubated for 24 hours, a stained slide of the culture can be made to determine if a pure culture has been achieved. When transferring the organisms from the plate, an inoculating needle (straight wire) is used instead of the wire loop. The needle is inserted into the center of the colony where there is a greater probability of getting only one species of organism. Use the following routine in subculturing out the three different organisms.

Materials:

   3 nutrient agar slants
   inoculating needle
   Bunsen burner

1. Label one tube *S. marcescens,* another *E. coli,* and the third *M. luteus* or *C. violaceum.*
2. Select a well-isolated red colony on either the streak plate or pour plate for your first transfer. Insert the inoculating needle into the center of the colony.

3. In the tube labeled *S. marcescens,* streak the slant by placing the needle near the bottom of the slant and drawing it up over its surface. One streak is sufficient.
4. Repeat this inoculating procedure on the other two slants for a white colony and a yellow (or purple) colony.
5. Incubate for 24 to 48 hours at 25° C.

Evaluation of Slants

After incubation, examine the slants. Is *S. marcescens* red? Is *E. coli* white? Is your third slant yellow or purple? If the incubation temperature has been too high, *S. marcescens* may appear white due to the fact that the red pigment forms only at a moderate temperature, such as 25° C. Draw the appearance of the slants with colored pencils on the Laboratory Report.

Although the colors of the growths on the slants may lead you to think that you have pure cultures, you cannot be absolutely certain until you have made a microscopic examination of each culture. For example, it is entirely possible that the yellow slant (*M. luteus*) may have some *E. coli* present that are masked by the yellow pigment.

To find out if you have a pure culture on each slant, make a gram-stained slide from each slant. Knowing that *S. marcescens, E. coli,* and *C. violaceum* are gram-negative rods, and that *M. luteus* is a gram-positive coccus, you should be able to evaluate your slants more precisely microscopically. Draw the organisms on the Laboratory Report.

Laboratory Report

Complete the Laboratory Report for this exercise.

Cultivation of Anaerobes

The procedures for culturing bacteria that were used in the last exercise work well only if the organisms will grow in the presence of oxygen. Unfortunately, there are many bacteria that find oxygen toxic or at least inhibitory to their existence. For these organisms we need to create an anaerobic environment by using special media deficient or lacking in oxygen and containers that are oxygen-free. In this laboratory period we will learn how to find out what the oxygen requirements are for specific organisms and how to grow them in liquid and solid media. In doing so we will be inoculating special media with several organisms of different cultural requirements to evaluate their oxygen needs.

The oxygen requirements of bacteria range from **strict** (obligate) **aerobes** that cannot exist without this gas to the **strict** (obligate) **anaerobes** that die in its presence. In between these extremes are the facultatives, indifferents, and microaerophilics. The **facultatives** are bacteria that have enzyme systems enabling them to utilize free oxygen or some alternative oxygen source such as nitrate. If oxygen is present, they tend to utilize it in preference to the alternative. The **indifferents,** however, show no preference for either condition, growing equally well in aerobic and anaerobic conditions. **Microaerophiles,** on the other hand, are organisms that require free oxygen, but only in limited amounts. Figure 22.1 illustrates where these various types tend to grow with respect to the degree of oxygen tension in a medium.

In this experiment we will inoculate one liquid medium and two solid media with several organisms that have different oxygen requirements. The media are fluid thioglycollate medium (FTM), tryptone glucose yeast agar (TGYA), and Brewer's anaerobic agar. Each medium will serve a different purpose. A discussion of the function of each medium follows:

TGYA Shake This solid medium will be used in what is called a "shake tube." The medium is not primarily an anaerobic medium; instead it is a rich general purpose medium that favors the growth of a broad spectrum of organisms. It will be inoculated in the liquefied state, shaken to mix the organisms throughout the medium, and allowed to solidify. After incubation one determines the oxygen require-

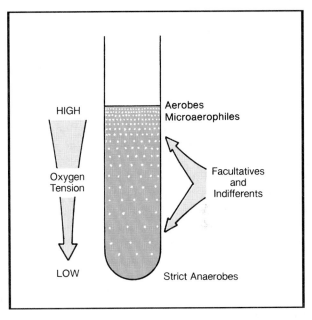

Figure 22.1 **Oxygen needs of microorganisms**

ments on the basis of where the growth occurs in the tube: top, middle, or bottom.

FTM Fluid thioglycollate medium is a rich liquid medium that supports the growth of both aerobic and anaerobic bacteria. It contains glucose, cystine, and sodium thioglycollate to reduce its oxidation-reduction (O/R) potential. It also contains the dye resazurin that is an indicator for the presence of oxygen. In the presence of oxygen the dye becomes pink. Since the oxygen tension is always higher near the surface of the medium, the medium will be pink at the top and colorless in the middle and bottom. The medium also contains a small amount of agar that helps to localize the organisms and favors anaerobiasis in the bottom of the tube.

Brewer's Anaerobic Agar This solid medium is an excellent medium for culturing anaerobic bacteria in Petri dishes. It contains thioglycollate as a reducing agent and resazurin as an O/R indicator. For strict anaerobic growth it is essential that plates be incubated in an oxygen-free environment.

To provide an oxygen-free incubation environment for the Petri plates of anaerobic agar we will use the **GasPak anaerobic jar.** Note in figure 22.2 that hydrogen is generated in the jar, which removes the oxygen by forming water. Palladium pellets catalyze the reaction at room temperature. The generation of hydrogen is achieved by adding water to a plastic envelope of chemicals. Note also that CO_2 is produced, which is a requirement for the growth of many fastidious bacteria. To make certain that anaerobic conditions actually exist in the jar, an indicator strip of methylene blue becomes colorless in the total absence of oxygen. If the strip is not reduced (decolorized) within 2 hours, the jar has not been sealed properly, or the chemical reaction has failed to occur.

In addition to doing a study of the oxygen requirements of six organisms in this experiment, an opportunity will be provided during the second period to do a microscopic study of the types of endospores formed by three spore-formers used in the inoculations. Proceed as follows:

First Period
(Inoculations and Incubation)

Since six microorganisms and three kinds of media are involved in this experiment, it will be necessary for economy of time and materials to have each student work with only three organisms. The materials list for this period indicates how the organisms will be distributed.

During this period each student will inoculate three tubes of medium and only one Petri plate of Brewer's anaerobic agar. The tubes and all of the plates will be placed in a GasPak jar to be incubated in a 37° C incubator. Students will share results.

Materials:
per student:

    3 tubes of fluid thioglycollate medium
    3 TGYA shake tubes (liquefied)
    1 Petri plate of Brewer's anaerobic agar

broth cultures for **odd-numbered students:**
    Staphylococcus aureus, Streptococcus faecalis, and *Clostridium sporogenes*

broth cultures for **even-numbered students:**
    Bacillus subtilis, Escherichia coli, and *Clostridium rubrum*

GasPak anaerobic jar, 3 GasPak generator envelopes, 1 GasPak anaerobic generator strip, scissors, and one 10 ml pipette
water baths at student stations (electric hot plate, beaker of water, and thermometer)

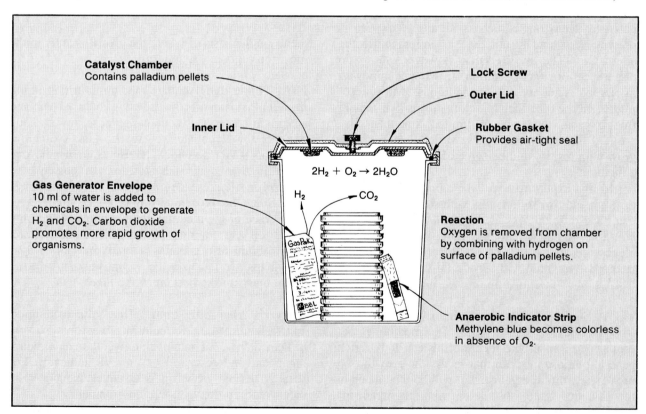

Figure 22.2 The GasPak anaerobic jar

1. Set up a 45° C water bath at your station in which you can keep your tubes of TGYA shakes from solidifying. One water bath for you and your laboratory partner will suffice. (Note in the materials list that the agar shakes have been liquefied for you prior to lab time.)

2. Label the six tubes with the organisms assigned to you (one organism per tube), your initials, and assignment number.

   Note: *Handle the tubes gently to avoid taking on any unwanted oxygen into the media. If the tubes of FTM are pink in the upper 30%, they must be boiled a few minutes to drive off the oxygen, then cooled to inoculate.*

3. Heavily inoculate each of the TGYA shake tubes with several loopfuls of the appropriate organism for that tube. To get good dispersion of the organisms in the medium, roll each tube gently between the palms as shown in figure 22.3. To prevent oxygen uptake do not overly agitate the medium. Allow these tubes to solidify at room temperature.

4. Inoculate each of the FTM tubes with the appropriate organisms.

5. Streak your three organisms on the plate of anaerobic agar in the manner shown in figure 22.4. Note that only three straight-line streaks, well separated, are made. Place the Petri plate (inverted) in a cannister with the plates of other students that is to go into the GasPak jar.

6. Once all the students' plates are in cannisters, place the cannisters and tubes into the jar.

7. To activate and seal the GasPak jar, proceed as follows:

   a. Peel apart the foil at one end of a GasPak indicator strip and pull it halfway down. The indicator will turn blue on exposure to the air. Place the indicator strip in the jar so that the wick is visible.

   b. Cut off the corner of each of three GasPak gas generator envelopes with a pair of scissors. Place them in the jar in an upright position.

   c. Pipette 10 ml of tap or distilled water into the open corner of each envelope. Avoid forcing the pipette into the envelope.

   d. Place the inner section of the lid on the jar, making certain it is centered on top of the jar. Do not use grease or other sealant on the rim of the jar since the O-ring gasket provides an effective seal when pressed down on a clean surface.

   e. Unscrew the thumbscrew of the outer lid until the exposed end is completely withdrawn into the threaded hole. Unless this is done, it will be impossible to engage the lugs of the jar with the outer lid.

   f. Place the outer lid on the jar directly over the inner lid and rotate the lid slightly to allow it to drop in place. Now rotate the lid firmly to engage the lugs. The lid may be rotated in either direction.

   g. Tighten the thumbscrew by turning clockwise. If the outer lid raises up, the lugs are not properly engaged.

8. Place the jar in a 37° C incubator. After 2 or 3 hours check the jar to note if the indicator strip has lost its blue color. If decolorization has not occurred, replace the palladium pellets and repeat the entire process.

9. Incubate the tubes and plates for 24 to 48 hours.

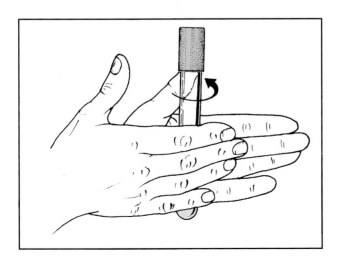

Figure 22.3 Organisms are dispersed in medium by rolling tube gently between palms.

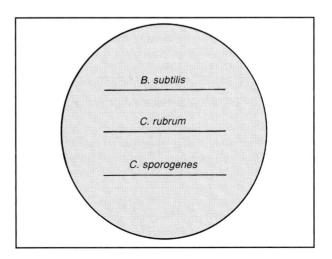

Figure 22.4 Three organisms are streaked on agar plate as straight-line streaks.

Second Period
(Culture Evaluations and Spore Staining)

Remove the lid from the GasPak jar. If vacuum holds the inner lid firmly in place, break the vacuum by sliding the lid to the edge. When transporting the plates and tubes to your desk *take care not to agitate the FTM tubes.* The position of growth in the medium can be easily changed if handled carelessly.

Materials:
> tubes of FTM
> shake tubes of TGYA
> 2 Brewer's anaerobic agar plates
> spore-staining kits and slides

1. Compare the six FTM and TGYA shake tubes that you and your laboratory partner share with figure 22.5 to evaluate the oxygen needs of the six organisms.
2. Compare the growths (or lack of growth) on your Petri plate and the plate of your laboratory partner.
3. Record your results on the Laboratory Report.
4. If time permits, make a combined slide with three separate smears of the three spore-formers, using either one of the two spore-staining methods in Exercise 15. Draw the organisms in the circles provided on the Laboratory Report.

Laboratory Report

Complete the Laboratory Report for this exercise.

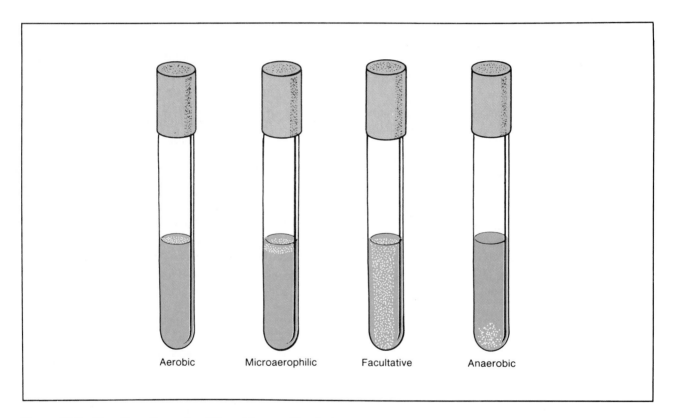

Figure 22.5 Growth patterns for different types of bacteria.

Bacterial Population Counts

23

Many bacteriological studies require that we be able to determine the number of organisms that are present in a given unit of volume. Several different methods are available to us for such population counts. The method one uses is determined by the purpose of the study.

To get by with a minimum of equipment, it is possible to do a population count by diluting out the organisms and counting the organisms in a number of microscopic fields on a slide. Direct examination of milk samples with this technique can be performed very quickly, and the results obtained are quite reliable. A technique similar to this can be performed on a Petrof-Hauser counting chamber.

Bacterial counts of gas-forming bacteria can be made by inoculating a series of tubes of lactose broth and using statistical probability tables to estimate bacterial numbers. This method, which we will use in Exercise 61 to estimate numbers of coliform bacteria in water samples, is easy to use, works well in water testing, but is limited to water, milk, and food testing.

In this exercise we will use **quantitative plating** (Standard Plate Count, or SPC) and **turbidity measurements** to determine the number of bacteria in a culture sample. Although the two methods are somewhat parallel in the results they yield, there are distinct differences. For one thing, the SPC reveals information only as related to viable organisms; that is, colonies that are seen on the plates after incubation represent only living organisms, not dead ones. Turbidimetry results, on the other hand, reflect the presence of all organisms in a culture, dead and living.

Quantitative Plating Method
(Standard Plate Count)

In determining the number of organisms present in water, milk, and food, the **standard plate count** (SPC) is universally used. It is relatively easy to perform and gives excellent results. We can also use this basic technique to calculate the number of organisms in a bacterial culture. It is in this respect that this assignment is set up.

The procedure consists of diluting the organisms with a series of sterile water blanks as illustrated in

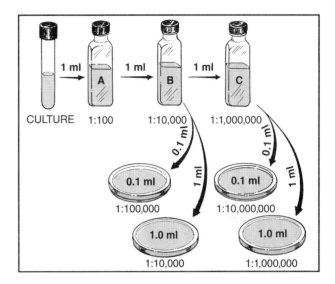

Figure 23.1 Quantitative plating procedure

figure 23.1. Generally, only three bottles are needed, but more could be used if necessary. By using the dilution procedure indicated here, a final dilution of 1:1,000,000 occurs in blank C. From blanks B and C, measured amounts of the diluted organisms are transferred into empty Petri plates. Nutrient agar, cooled to 50° C, is then poured into each plate. After the nutrient agar has solidified, the plates are incubated for 24 to 48 hours and examined. A plate that has between 30 and 300 colonies is selected for counting. From the count it is a simple matter to calculate the number of organisms per milliliter of the original culture. It should be pointed out that greater accuracy can be achieved by pouring two plates for each dilution and averaging the counts. Duplicate plating, however, has been avoided for obvious economic reasons.

Pipette Handling

Success in this experiment depends considerably on proper pipetting techniques. Pipettes may be available to you in metal cannisters or in individual envelopes; they may be disposable or reusable. In the distant past pipetting by mouth was routine practice. However, the hazards are obvious, and today it must be avoided. Your instructor will

89

indicate the techniques that will prevail in this laboratory. If this is the first time that you have used sterile pipettes, consult figure 23.2, keeping the following points in mind:

- When removing a sterile pipette from a cannister, do so without contaminating the ends of the other pipettes with your fingers. This can be accomplished by *gently* moving the cannister from side to side in an attempt to isolate one pipette from the rest.

- After removing your pipette, replace the cover on the cannister to maintain sterility of the remaining pipettes.

- Don't touch the body of the pipette with your fingers or lay the pipette down on the table before or after you use it. **Keep that pipette sterile** until you have used it, and don't contaminate the table or yourself with it after you have used it.

- Always use a mechanical pipetting device such as the one in illustration 3, figure 23.2. For safety reasons, deliveries by mouth are not acceptable in this laboratory.

- Remove and use only one pipette at a time; if you need 3 pipettes for the whole experiment and remove all 3 of them at once, there is no way that you will be able to keep 2 of them sterile while you are using the first one.

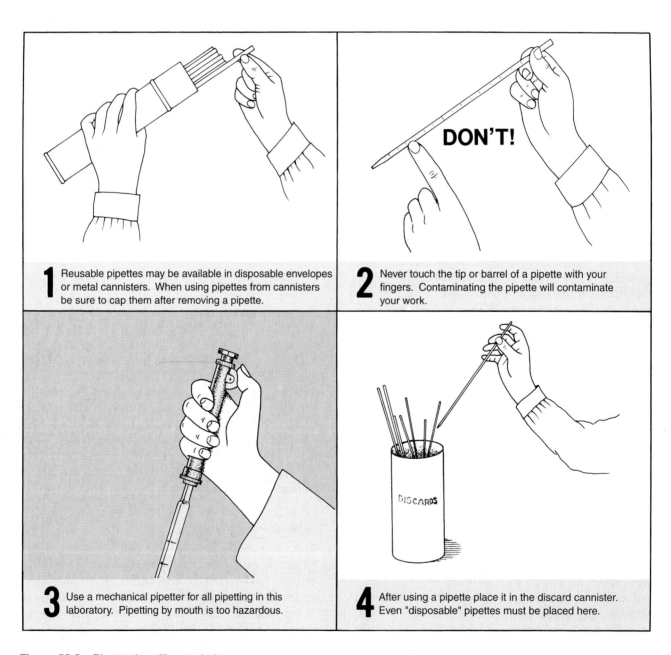

1 Reusable pipettes may be available in disposable envelopes or metal cannisters. When using pipettes from cannisters be sure to cap them after removing a pipette.

2 Never touch the tip or barrel of a pipette with your fingers. Contaminating the pipette will contaminate your work.

3 Use a mechanical pipetter for all pipetting in this laboratory. Pipetting by mouth is too hazardous.

4 After using a pipette place it in the discard cannister. Even "disposable" pipettes must be placed here.

Figure 23.2 Pipette-handling techniques

• When finished with a pipette, place it in the *discard cannister*. The discard cannister will have a disinfectant in it. At the end of the period reusable pipettes will be washed and sterilized by the laboratory assistant. Disposable pipettes will be discarded. Students have been known to absent-mindedly return used pipettes to the original sterile cannister, and, occasionally, even toss them into the wastebasket. We are certain that no one in this laboratory would *ever* do that!

Diluting and Plating Procedure

Proceed as follows to dilute out a culture of *E. coli* and pour four plates, as illustrated in figure 23.1.

Materials:
per 4 students:
    1 bottle (40 ml) broth culture of *E. coli per*
       student:
    1 bottle (80 ml) nutrient agar
    4 Petri plates
    1.1 ml pipettes
    3 sterile 99 ml water blanks
    cannister for discarded pipettes

1. Liquefy a bottle of nutrient agar. While it is being heated, label 3 99 ml sterile water blanks **A, B,** and **C.** Also, label the 4 Petri plates **1:10,000, 1:100,000, 1:1,000,000,** and **1:10,000,000.** In addition, indicate with labels the amount to be pipetted into each plate (**0.1 ml** or **1.0 ml**).

2. Shake the culture of *E. coli* and transfer 1 ml of the organisms to blank A, using a sterile 1.1 ml pipette. After using the pipette, place it in the discard cannister.

3. Shake blank A 25 times in an arc of 1 foot for 7 seconds with your elbow on the table as shown in figure 23.3. Forceful shaking not only brings about good distribution, but it also breaks up clumps of bacteria.

4. With a different 1.1 ml pipette, transfer 1 ml from blank A to blank B.

5. Shake water blank B 25 times in same manner.

6. With another sterile pipette, transfer 0.1 ml from blank B to the 1:100,000 plate and 1.0 ml to the 1:10,000 plate. With the same pipette, transfer 1.0 ml to blank C.

7. Shake blank C 25 times.

8. With another sterile pipette, transfer from blank C 0.1 ml to the 1:10,000,000 plate and 1.0 ml to the 1:1,000,000 plate.

9. After the bottle of nutrient agar has boiled for 8 minutes, cool it down in a water bath at 50° C for **at least 10 minutes.**

10. Pour one-fourth of the nutrient agar (20 ml) into each of 4 plates. Rotate the plates **gently** to get adequate mixing of medium and organisms. **This step is critical!** Too little action will result in poor dispersion and too much action may slop inoculated medium over the edge.

11. After the medium has cooled completely, incubate at 35° C for 48 hours, inverted.

Counting and Calculations
Materials:
    4 culture plates
    Quebec colony counter
    mechanical hand counter
    felt pen (optional)

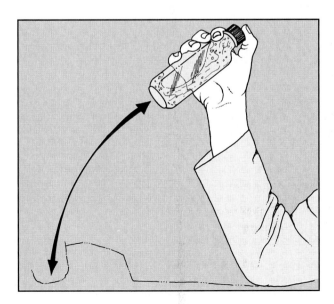

Figure 23.3 Standard procedure for shaking water blanks requires elbow to remain fixed on table.

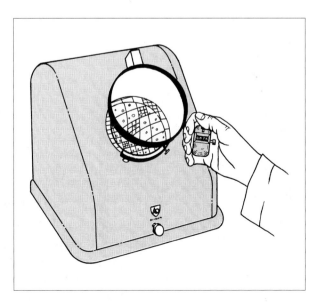

Figure 23.4 Colony counts are made on a Quebec counter, using a mechanical hand tally.

1. Lay out the plates on the table in order of dilution and compare them. *Select the plates that have no fewer than 30 nor more than 300 colonies for your count.* Plates with less than 30 or more than 300 colonies are statistically unreliable.

2. Place the plate on the Quebec colony counter with the lid removed. See figure 23.4. Start counting at the top of the plate, using the grid lines to prevent counting the same colony twice. Use a mechanical hand counter. Count every colony, regardless of how small or insignificant. Record counts on the table in section A of the Laboratory Report.

   Alternative Counting Method: Another way to do the count is to remove the lid and place the plate upside down on the colony counter. Instead of using the grid to keep track, use a felt pen to mark off each colony as you do the count.

3. Calculate the number of bacteria per ml of undiluted culture using the data recorded in section A of the Laboratory Report. Multiply the number of colonies counted by the dilution factor (the reciprocal of the dilution).

   Example: If you counted 220 colonies on the plate that received 1.0 ml of the 1:1,000,000 dilution: $220 \times 1,000,000$ (or 2.2×10^8) bacteria per ml. If 220 colonies were counted on the plate that received 0.1 ml of the 1:1,000,000 dilution, then the above results would be multiplied by 10 to convert from number of bacteria per 0.1 ml to number of bacteria per 1.0 ml (2,200,000,000, or 2.2×10^9).

Use only two significant figures. If the number of bacteria per ml was calculated to be 227,000,000, it should be recorded as 230,000,000, or 2.3×10^8.

Turbidimetry Determinations

When it is necessary to make bacteriological counts on large numbers of cultures, the quantitative plate count method becomes a rather cumbersome tool. It not only takes a considerable amount of glassware and media, but it is also time-consuming. A much faster method is to measure the turbidity of the culture with a spectrophotometer and translate this into the number of organisms. To accomplish this, however, the plate count must be used to establish the count for one culture of known turbidity.

To understand how a spectrophotometer works, it is necessary, first, to recognize the fact that a culture of bacteria acts as a colloidal suspension, which will intercept the light as it passes through. Within certain limits the amount of light that is absorbed is directly proportional to the concentration of cells.

Figure 23.5 illustrates the path of light through a spectrophotometer. Note that a beam of white light passes through two lenses and an entrance slit into a diffraction grating that disperses the light into horizontal beams of all colors of the spectrum. Short wavelengths (violet and ultraviolet) are at one end and long wavelengths (red and infrared) are at the other end. The spectrum of light falls on a dark screen with a slit (exit slit) cut in it. Only that portion

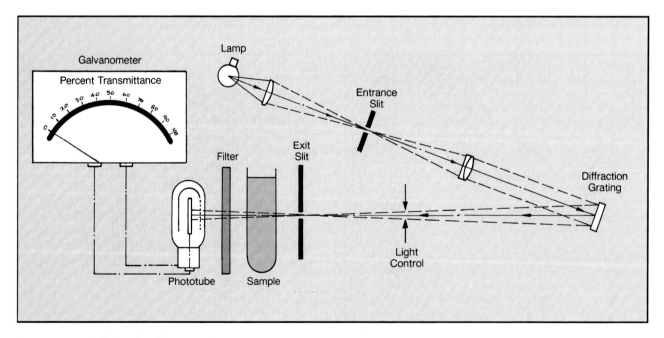

Figure 23.5 Schematic of a spectrophotometer

of the spectrum that happens to fall on the slit goes through into the sample. It will be a monochromatic beam of light. By turning a wavelength control knob on the instrument, the diffraction grating can be reoriented to allow different wavelengths to pass through the slit. The light that passes through the culture activates a phototube, which, in turn, registers **percent transmittance (% T)** on a galvanometer. The higher the percent transmittance, the fewer are the cells in suspension.

There should be a direct proportional relationship between the concentration of bacterial cells and the absorbance (optical density, O.D.) of the culture. To demonstrate this principle, you will measure the %T of various dilutions of the culture provided to you. These values will be converted to O.D. and plotted on a graph as a function of culture dilution. You may find that there is a linear relationship between concentration of cells and O.D. only up to a certain O.D. At higher O.D. values the relationship may not be linear. That is, for a doubling in cell concentration, there may be less than a doubling in O.D.

Materials:
    broth culture of *E. coli* (same one as used for
      plate count)
    spectrophotometer cuvettes
      (2 per student)
    4 small test tubes and test-tube rack
    5 ml pipettes
    bottle of sterile nutrient broth
      (20 ml per student)

1. Calibrate the spectrophotometer, using the procedure described in figure 23.7. These instructions are specifically for the Bausch and Lomb Spectonic 20. In handling the cuvettes, keep the following points in mind:
   a. Rinse the cuvette several times with distilled water to get it clean before using.
   b. Keep the lower part of the cuvette spotlessly clean by keeping it free of liquids, smudges, and fingerprints. Wipe it clean with Kimwipes or some other lint-free tissue. Don't wipe the cuvettes with towels or handkerchiefs.
   c. Insert the cuvette into the sample holder with its index line registered with the index line on the holder.
   d. After the cuvette is seated, line up the index lines exactly.
   e. Handle these tubes with great care. They are expensive.
2. Label a cuvette 1:1 (near top of tube) and four test tubes 1:2, 1:4, 1:8, and 1:16. These tubes will be used for the serial dilutions shown in figure 23.6.

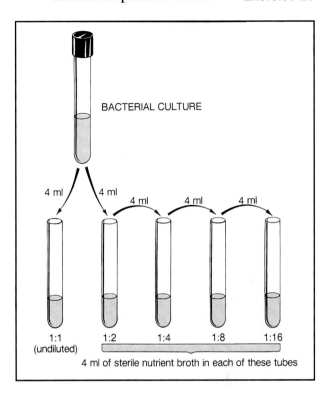

Figure 23.6 Dilution procedure for cuvettes

3. With a 5 ml pipette, dispense 4 ml of sterile nutrient broth into tubes 1:2, 1:4, 1:8, and 1:16.
4. Shake the culture of *E. coli* vigorously to suspend the organisms, and with the same 5 ml pipette, transfer 4 ml to the 1:1 cuvette and 4 ml to the 1:2 test tube.
5. Mix the contents in the 1:2 tube by drawing the mixture up into the pipette and discharging it into the tube three times.
6. Transfer 4 ml from the 1:2 tube to the 1:4 tube, mix three times, and go on to the other tubes in a similar manner. Tube 1:16 will have 8 ml of diluted organisms.
7. Measure the percent transmittance of each of the five tubes, starting with the 1:16 tube first. The contents of each of the test tubes must be transferred to a cuvette for measurement. Be sure to close the lid on the sample holder when making measurements. A single cuvette can be used for all the measurements.
8. Convert the percent transmittance values to optical density (O.D.) using the following formula:

O.D. = 2 − log of percent transmittance

Example: If the percent transmittance of one of your dilutions is 53.5, you would solve the problem in this way:

O.D. = 2 − log of 53.5
    = 2 − 1.7284
    = 0.272

Table II of Appendix A is a log table. Of course, if you have a calculator, all this is much simpler.

Logarithm Refresher In case you have forgotten how to use logarithms, recall these facts:

Mantissa: The value you find in the log table (0.7284 in the above example) is the mantissa.

Characteristic: The number to the left of the decimal (1 in the example) is the characteristic.

This figure (the characteristic) is always one number less than the number of digits of the figure you are looking up.

Examples:

number	characteristic	mantissa
5.31	0	.7251
531	2	.7251

Although the galvanometer may show absorbance (O.D.) values, greater accuracy will result from calculating them from percent transmittance.

9. Record the O.D. values in the table of the Laboratory Report.
10. Plot the O.D. values on the graph of the Laboratory Report.

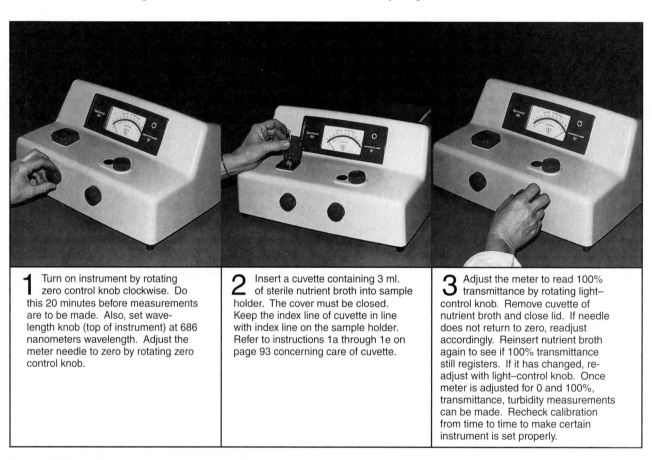

1 Turn on instrument by rotating zero control knob clockwise. Do this 20 minutes before measurements are to be made. Also, set wavelength knob (top of instrument) at 686 nanometers wavelength. Adjust the meter needle to zero by rotating zero control knob.

2 Insert a cuvette containing 3 ml. of sterile nutrient broth into sample holder. The cover must be closed. Keep the index line of cuvette in line with index line on the sample holder. Refer to instructions 1a through 1e on page 93 concerning care of cuvette.

3 Adjust the meter to read 100% transmittance by rotating light–control knob. Remove cuvette of nutrient broth and close lid. If needle does not return to zero, readjust accordingly. Reinsert nutrient broth again to see if 100% transmittance still registers. If it has changed, re-adjust with light–control knob. Once meter is adjusted for 0 and 100%, transmittance, turbidity measurements can be made. Recheck calibration from time to time to make certain instrument is set properly.

Figure 23.7   Calibration procedure for the B & L Spectronic 20

Slide Culture: Molds

The isolation, culture, and microscopic examination of molds require the use of suitable selective media and special microscopic slide techniques. If simple wet mount slides of molds were attempted in Exercise 8, it became apparent that wet mount slides made from mold colonies usually don't reveal the arrangement of spores that is so necessary in identification. The process of merely transferring hyphae to a slide breaks up the hyphae and sporangiophores in such a way that identification becomes very difficult. In this exercise a slide culture method will be used to prepare stained slides of molds. The method is superior to wet mounts in that the hyphae, sporangiophores, and spores remain more or less intact when stained.

When molds are collected from the environment, as in Exercise 8, Sabouraud's agar is most frequently used. It is a simple medium consisting of 1% peptone, 4% glucose, and 2% agar-agar. The pH of the medium is adjusted to 5.6 to inhibit bacterial growth.

Unfortunately, for some molds the pH of Sabouraud's agar is too low and the glucose content is too high. A better medium for these organisms is one suggested by C. W. Emmons that contains only 2% glucose, with 1% neopeptone, and an adjusted pH of 6.8–7.0. To inhibit bacterial growth, 40 mg of chloramphenicol is added to one liter of the medium.

In addition to the above two media, cornmeal agar, Czapek solution agar, and others are available for special applications in culturing molds.

Figure 24.2 illustrates the procedure that will be used to produce a mold culture on a slide that can be stained directly on the slide. Note that a sterile cube of Sabouraud's agar is inoculated on two sides with spores from a mold colony. Figure 24.1 illustrates how the cube is held with a scalpel blade as inoculation takes place. The cube is placed in the center of a microscope slide with one of the inoculated surfaces placed against the slide. On the other inoculated surface of the cube is placed a cover glass. The assembled slide is incubated at room temperature for 48 hours in a moist chamber (Petri dish with a small amount of water). After incubation the cube of medium is carefully separated from the slide and discarded.

During incubation the mold will grow over the glass surfaces of the slide and cover glass. By adding

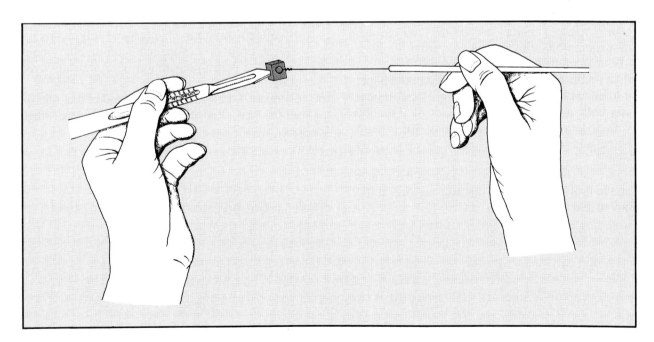

Figure 24.1   Inoculation technique

a little stain to the slide a semipermanent slide can be made by placing a cover glass over it. The cover glass can also be used to make another slide by placing it on another clean slide with a drop of stain on it. Before the stain (lactophenol cotton blue) is used, it is desirable to add to the hyphae a drop of alcohol, which acts as a wetting agent.

First Period
(Slide Culture Preparation)

Proceed as follows to make slide cultures of one or more mold colonies.

Materials:

> Petri dishes, glass, sterile
> filter paper (9 cm dia, sterile)
> glass U-shaped rods
> mold culture plate (mixture)
> 1 Petri plate of Sabouraud's agar or Emmons'
>   medium per 4 students
> scalpels
> inoculating loop
> sterile water
> microscope slides and cover glasses (sterile)
> forceps

1. Aseptically, with a pair of forceps, place a sheet of sterile filter paper in a Petri dish.
2. Place a sterile U-shaped glass rod on the filter paper. (Rod can be sterilized by flaming, if held by forceps.)
3. Pour enough sterile water (about 4 ml) on filter paper to completely moisten it.
4. With forceps, place a sterile slide on the U-shaped rod.
5. *Gently* flame a scalpel to sterilize, and cut a 5 mm square block of the medium from the plate of Sabouraud's agar or Emmons' medium.
6. Pick up the block of agar by inserting the scalpel into one side as illustrated in figure 24.1. Inoculate both top and bottom surfaces of the cube with spores from the mold colony. Be sure to flame and cool the loop prior to picking up spores.
7. Place the inoculated block of agar in the center of a microscope slide. Be sure to place one of the inoculated surfaces down.

8. Aseptically, place a sterile cover glass on the upper inoculated surface of the agar cube.
9. Place the cover on the Petri dish and incubate at room temperature for 48 hours.
10. After 48 hours examine the slide under low power. If growth has occurred you should see hyphae and spores. If growth is inadequate and spores are not evident, allow the mold to grow another 24–48 hours before making the stained slides.

Second Period
(Application of Stain)

As soon as there is evidence of spores on the slide, prepare two stained slides from the slide culture, using the following procedure:

Materials:

> microscope slides and cover glasses
> 95% ethanol
> lactophenol cotton blue stain
> forceps

1. Place a drop of lactophenol cotton blue stain on a clean microscope slide.
2. Remove the cover glass from the slide culture and discard the block of agar.
3. Add a drop of 95% ethanol to the hyphae on the cover glass. As soon as most of the alcohol has evaporated place the cover glass, mold side down, on the drop of lactophenol cotton blue stain on the slide. This slide is ready for examination.
4. Remove the slide from the Petri dish, add a drop of 95% ethanol to the hyphae and follow this up with a drop of lactophenol cotton blue stain. Cover the entire preparation with a clean cover glass.
5. Compare both stained slides under the microscope; one slide may be better than the other one.

Laboratory Report

There is no Laboratory Report for this exercise.

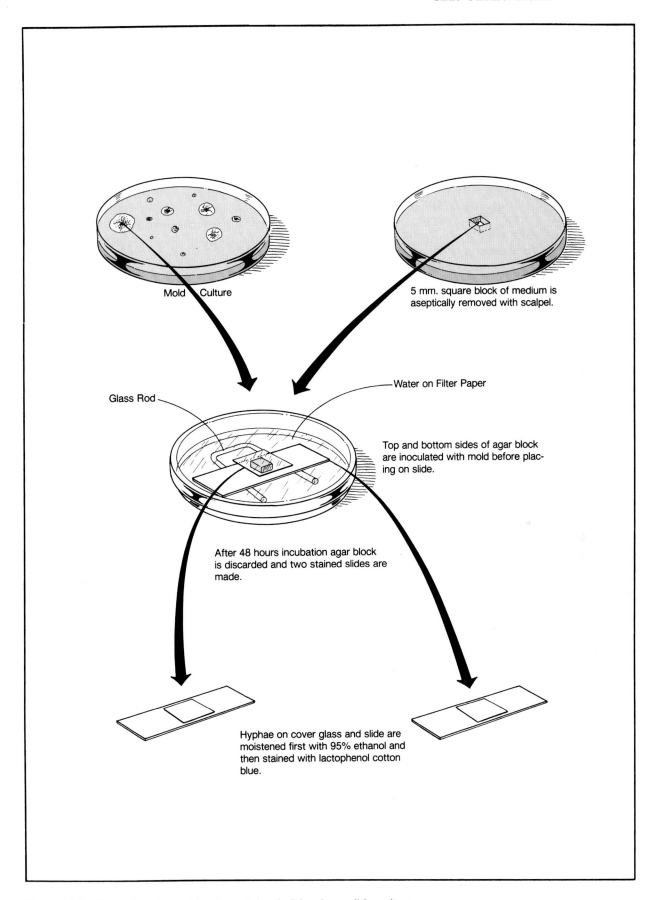

Mold   Culture

5 mm. square block of medium is
aseptically removed with scalpel.

Water on Filter Paper

Glass Rod

Top and bottom sides of agar block
are inoculated with mold before placing on slide.

After 48 hours incubation agar block
is discarded and two stained slides are
made.

Hyphae on cover glass and slide are
moistened first with 95% ethanol and
then stained with lactophenol cotton
blue.

Figure 24.2   Procedure for making two stained slides from slide culture

25

Slide Culture: Autotrophs

There is probably no single medium or method that one can use to do a comprehensive population count of all living microorganisms in a specific biosphere. Those media that we categorize as being "general purpose" will, for various reasons, inhibit the growth of many organisms. To make comparative studies of free-living organisms in freshwater lakes, A. T. Henrici, in 1932, devised an immersed slide technique that revealed the presence of many organisms that did not show up by other methods. Although his original concern was with algal populations, the technique worked as well for bacteria and other microorganisms.

His method consists of suspending glass microscope slides in the body of water for a specified period of time. Microorganisms in the water adhere to the glass and multiply to form small colonies that are observable under the microscope.

Materials:
>   adhesive tape (½″ width)
>   2 microscope slides
>   copper wire
>   gummed labels
>   acid-alcohol

1. Clean 2 microscope slides as follows:
    a. Scrub with green soap or Bon Ami.
    b. Dip them in acid-alcohol for 1 minute and dry with tissue.
    c. Place them in a beaker of distilled water for 5 minutes to allow any residual solvent to dissipate.
2. Tape a piece of copper wire to one edge as illustrated in figure 25.1. Hold the slides back to back by their edges. Do not touch the flat surfaces with your fingers. Wrap all four edges with tape. For identification, attach a gummed label with your name to the wire.
3. Suspend the slide in an aquarium or container of water that is known to have a stabilized natural flora of bacteria.

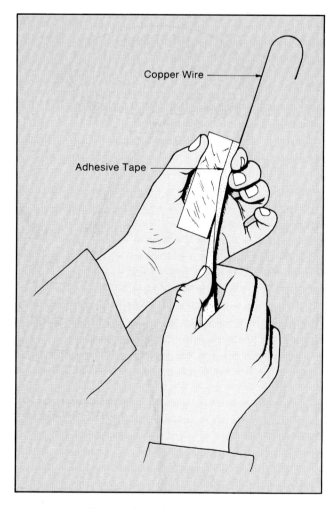

Figure 25.1 Preparation of slide for immersion

4. After 1 week remove the binding from the slides. Prepare one slide with Gram's stain and place a drop of water and cover glass on the other one.
5. Examine both slides under oil immersion and record your observations on the first portion of Laboratory Report 25, 26.

Reference: Henrici, A. T. 1933. Studies of fresh water bacteria. *J. Bact.* 25 (3): 277–286.

Slime Mold Culture

26

The classification system proposed by Alexopoulos and Mims places the slime molds in Division Gymnomycota of the Kingdom Myceteae. These heterotrophic microorganisms exist in cool, shady, moist places in the woods—on decaying logs, dead leaves, and other organic matter. Unlike the holophytic bacteria and other Myceteae, they ingest their food in a manner similar to the amoebas; that is, they are phagotrophic. In the vegetative stages, these microorganisms are unlike the other Myceteae in that the cells lack cell walls; when fruiting bodies are formed, however, cell walls are present.

The categorization of slime molds as protozoans or as fungi has always been problematical. Certainly, they are intermediate in that they have characteristics of both groups.

Figure 26.1 illustrates the life cycle of one type of slime mold, the plasmodial type. The genus *Physarum* is the one to be studied in this experiment. The assimilative stage of this organism is the **plasmodium.** This multinucleate structure is slimy in appearance and moves slowly by flowing its cytoplasm in amoeboidlike fashion over surfaces on which it feeds. Most species feed on bacteria and possibly on other small organisms that they encounter.

Plasmodial growth continues as long as adequate food supply and moisture are available. Eventually, however, environmental changes may result in the formation of sclerotia or sporangia. A **sclerotium** is a hardened mass of irregular shape that forms from the plasmodium when moisture and temperature conditions become less than ideal. When conditions improve, the sclerotium reverts back to a plasmodium. Figure 26.2 is a photograph of two sclerotia that formed on a laboratory culture. **Sporangia** are fructifications that form under conditions similar to those required for sclerotia. Exactly why sporangia form instead of sclerotia is still not clearly understood. Sporangia form by the separation of the plasmodium into many rounded mounds of protoplasm that extend upward on stalks. The nuclei within the sporangia undergo meiosis to become haploid spores with tough cell walls. The sclerotia and sporangia of figures 26.2 and 26.3 were

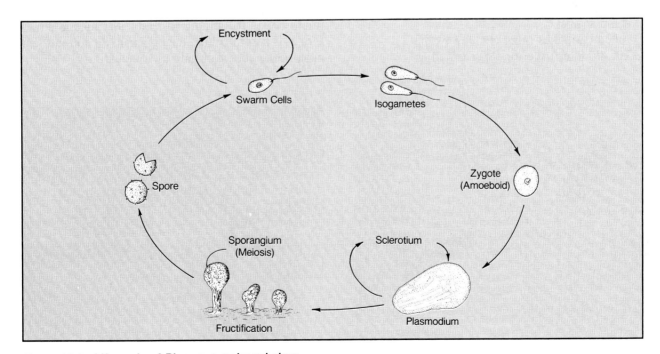

Figure 26.1   Life cycle of *Physarum polycephalum*

99

photographed on the same culture of laboratory-grown *Physarum.*

Both sclerotia and spores may survive adverse environmental conditions for long periods of time. Once environmental conditions improve, the spores germinate to produce flagellated pear-shaped **swarm cells.** These swarm cells may do one of three things: (1) they may encyst if conditions suddenly become adverse, (2) they may divide one or more times to form isogametes, or (3) they may act as isogametes and unite directly to form a **zygote.** Once a zygote is formed, it takes on an amoeboid form and undergoes a series of mitotic divisions to produce a plasmodium. This completes the life cycle.

Three procedures will be described here for the study of *Physarum polycephalum:* (1) moist chamber culture, (2) agar culture method, and (3) spore germination technique. The techniques used will be determined by the availability of time and materials.

Moist Chamber Culture

To grow large numbers of plasmodia, sclerotia, and sporangia that can be used for an entire class, one needs to create a rather large moisture chamber. Any covered glass or plastic container that is 10 to 12 inches square or round is suitable.

Materials:

    sclerotia of *Physarum polycephalum*
    container for culture (10½″ dia Pyrex casserole
        dish with cover or 10–12″ square plastic
        box with cover)
    glass Petri dish cover
    sharp scalpel
    rolled oat flakes (long-cooking type)
    10″ dia filter paper or paper toweling

1. In the center of the container place a Petri dish cover, open end down. Lay a large piece of filter paper or paper toweling over the Petri dish and saturate with distilled water. The Petri dish provides a raised area above any excess water that may make the paper too wet.

2. With a sharp scalpel transfer a small fragment of sclerotium from the *Physarum* culture to the filter paper. A sclerotium may vary from dark orange to brown in color. See figure 26.2. Moisten the sclerotium with a drop of distilled water.

3. After a few hours the organism will be awakened to activity and begin to seek food. At this point, place a flake of rolled oats near the edge of the spreading growth for it to feed on.

4. Incubate the moist chamber in a dark place at room temperature. Add moisture (distilled water) and oat flakes periodically as needed. It is better to add a few fresh flakes daily than to overfeed by applying all flakes at once. Such a culture should keep for several weeks. To promote the formation of sclerotia, allow some of the water to evaporate away by leaving the lid partially open for a while. To bring about sporangia formation, withhold food while keeping the culture moist.

Agar Culture Method
(Plasmodial Study)

An actively metabolizing plasmodium is dark yellow and streaked with vessels. The streaming of protoplasm in these vessels is best observed under the microscope. To be able to study this unique structure, it is best to culture the organism on non-nutrient agar. Make such a culture as follows:

Figure 26.2 Sclerotia of *Physarum polycephalum* (3X)

Figure 26.3 Sporangia of *Physarum polycephalum* (20X)

Materials:
    rolled oat flakes
    scalpel
    Petri plate with 15 ml of nonsterile,
        non-nutrient agar

1. Lift some occupied oat flakes from the filter paper in the moist chamber and transfer to a plate of nonsterile, non-nutrient 1.5% agar. Maintain this culture by adding fresh oat flakes periodically, but don't add water.
2. After a well-developed plasmodium has formed, study the streaming protoplasm under low power of the microscope. Observation is made by transmitted light through the agar on the microscope stage. Look for periodical reversal of direction of flow.
3. Cut one of the vessels through in which the flow is active and observe the effect.
4. Transfer a piece of plasmodium to another part of the medium and watch it reconstitute itself.
5. Leave the cover slightly open on the Petri dish for several days and note any changes that might occur as time goes by.

Spore Germination

The observation of spore germination can be achieved with a hanging drop slide. Once sporangia are in abundance, one can make such a slide as follows:

Materials:
    depression slides (sterile)
    plain microscope slides (sterile)
    cover glasses (sterile)
    Vaseline
    toothpicks
    sporangia of *Physarum polycephalum*
    Bunsen burner
    70% alcohol

1. With a toothpick, place a small amount of Vaseline near each corner of the cover glass. (See figure 19.1, page 69.)
2. Saturate a sporangium with a drop of 70% alcohol on the center of a sterile plain microscope slide.
3. As soon as the alcohol has evaporated, add a drop of distilled water and place another sterile slide over the wet sporangium.
4. Crush the sporangium with thumb pressure on the upper slide. Separate the two slides to expose the crushed sporangium.
5. Transfer a few loopfuls of crushed sporangial material to a drop of distilled water on a sterile cover glass.
6. Place the depression slide over the cover glass, make contact, and quickly invert to produce a completed hanging drop slide.
7. Examine under low and high power.

Laboratory Report

Complete all the answers on Laboratory Report 25, 26 that pertain to this exercise.

A considerable number of pigmented bacteria exist that derive their energy by photosynthesis. At present three families are recognized that contain around 20 genera. These phototrophic bacteria are a diverse group of cocci, rods, vibrio-, and spiral-shaped forms. All of them are gram-negative. Although cell suspensions of various species occur as purple, red, orange-brown, brown, or green, the genera are subdivided into two main groups: those that are purple and those that are green. All of them are aquatic, being abundant in the ooze of ponds.

The purple photosynthetic bacteria (suborder Rhodospirillineae) include the families Rhodospirillaceae and Chromatiaceae. The green species (suborder Chlorobiineae) belong to one family, the Chlorobiaceae.

The photosynthetic bacteria differ from the algae and green plants in that the bacteria are anaerobic and utilize bacteriochlorophyll instead of chlorophyll a for photosynthesis. While the algae produce oxygen as a by-product of photosynthesis, the Chromatiaceae and Chlorobiaceae produce sulfur instead. The following chemical reaction of photosynthesis in these organisms indicates how hydrogen sulfide is the hydrogen donor.

$$CO_2 + 2H_2S \xrightarrow[\text{sunlight}]{\text{bacteriochlorophyll}} C(H_2O) + 2S + H_2O$$

In addition to differing in color, Chromatiaceae and Chlorobiaceae also differ in that the former produce intracellular sulfur granules (one exception) and the Chlorobiaceae produce extracellular sulfur granules. The Rhodospirillaceae, incidentally, are nonsulfur phototrophs and were formerly called the Athiorhodaceae.

In this exercise an attempt will be made to isolate species of *Chromatium* (family Chromatiaceae) and *Chlorobium* (family Chlorobiaceae) from the ooze of stagnant ponds. Bacterial decomposition in the pond ooze provides the hydrogen sulfide essential to their photosynthesis. Figure 27.1 illustrates the procedure. Note that special enrichment media are first inoculated with ooze collected from various ponds by students. These bottles of inoculated media are then incubated at room temperature

for 7 days while being exposed to incandescent lamps 24 hours a day. To get isolated colonies of phototrophic bacteria, agar shake deeps will be made from bottles that appear green, red, brown, or purple after 7 days.

Enrichment

The enrichment media for these two genera contain essentially the same ingredients with differences being only in amounts. The chromatium medium will turn red in 4 to 7 days if any chromatia are present. The chlorobium medium should turn green in about the same length of time, but it may become overgrown by red chromatia in a short period of time; thus, subculture timing is more critical with the latter.

Materials:
    samples of pond mud
    chromatium enrichment medium (nonsterile, in 3 oz prescription bottle)
    chlorobium enrichment medium (nonsterile, in 3 oz prescription bottle)
    desk lamp with 40 watt bulb

1. Into bottles of each type of medium place enough pond mud so that a depth of 0.5 cm exists in the bottom. Fill the bottles to maximum capacity with medium and twist the lids securely. Ideally, there should be no air bubble in the bottle. Invert each bottle several times to mix the mud and medium.
2. Place the bottles about 12 inches away from a desk lamp and incubate the bottles at room temperature for 7 days. The lamp should remain on 24 hours per day. Examine the bottles each day to look for color changes. The chromatium medium will become reddish if species of this genus are present in the pond mud. The chlorobium medium will become green if the mud contains species of *Chlorobium*.
3. Make a hanging drop slide of organisms from both bottles and examine it with phase optics or brightfield oil immersion. Record your observations on the Laboratory Report.

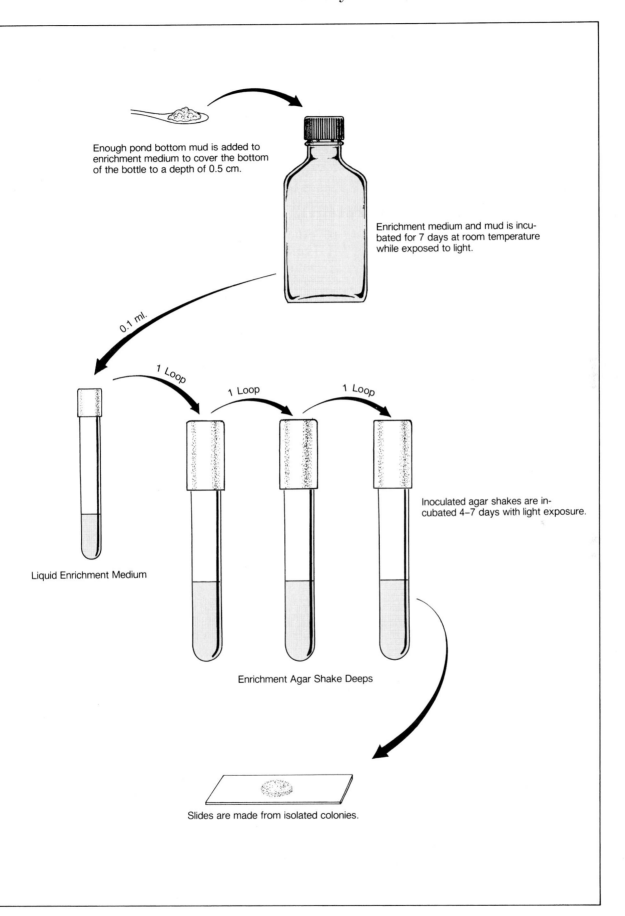

Enough pond bottom mud is added to enrichment medium to cover the bottom of the bottle to a depth of 0.5 cm.

Enrichment medium and mud is incubated for 7 days at room temperature while exposed to light.

0.1 ml.

1 Loop

1 Loop

1 Loop

Liquid Enrichment Medium

Inoculated agar shakes are incubated 4–7 days with light exposure.

Enrichment Agar Shake Deeps

Slides are made from isolated colonies.

Figure 27.1 Enrichment and isolation procedure for photosynthetic bacteria

Pure Cultures

The use of enrichment media never produces a pure culture. To get pure isolates it is necessary to serial dilute the mixed cultures with sterile agar shake tubes. Tubes of sterile chromatium and chlorobium agar must be maintained in a 50° C water bath until inoculated. If the medium is allowed to solidify before inoculation it must be discarded since it cannot be reliquefied without its chemical composition being affected. After the agar shakes have been inoculated and allowed to solidify, they must be sealed with paraffin to exclude air for the anaerobic chromatia and chlorobia.

Materials:

enrichment cultures of *Chromatium* and *Chlorobium*
1 tube of nonsterile chromatium enrichment medium (5 ml)
1 tube of nonsterile chlorobium enrichment medium (5 ml)
3 chromatium agar shake deeps (in 50° C water bath)
3 chlorobium agar shake deeps (in 50° C water bath)
1 ml serological pipettes
paraffin

1. Set up a 50° C water bath at your desk and place 3 agar shake deeps of each medium in the bath.
2. Mark the three tubes of each medium **I, II,** and **III** with a china marking pencil. Be sure to keep the tubes at 50° C until all inoculations are complete. Remember that these tubes cannot be reliquefied once they solidify.

3. Pipette 0.1 ml from the *Chromatium* culture to the tube of nonsterile chromatium enrichment medium.
4. Pipette 0.1 ml from the *Chlorobium* culture to the tube of nonsterile chlorobium enrichment medium.
5. After mixing the bacterial suspensions in these two tubes of nonsterile media, make serial dilutions from each tube to three agar shakes according to the routine shown in figure 27.1. *Be sure to rotate each agar deep between the palms before removing loopful of organisms for next tube.*
6. After the agar media have solidified, seal the surface of the agar by pouring liquefied paraffin into each tube. One centimeter of paraffin should be sufficient.
7. Incubate the tubes in a test-tube rack in front of a desk lamp for 4 to 7 days.
8. Make hanging drop slides from colonies removed from the agar deeps.
The ideal way to get the organisms out of the agar is to sterilize the outer surface of the tube, cut the glass to expose the agar, and aseptically dissect out a colony. This should be the procedure if one plans to maintain a pure culture for some time.
A cruder method that may be used to economize on tubes is to dissect out the paraffin with a sharp-pointed scalpel and apply a little heat to the surface of the tube to cause the agar to slide out. Once the agar is extracted from the tube, deep colonies can be fished out for examination.
9. Record your results on the Laboratory Report.

PART

Bacterial Viruses: Isolation and Propagation

The viruses differ from bacteria in being much smaller, noncellular and intracellular parasites. In addition, they cannot be grown on ordinary media. Despite these seemingly difficult obstacles to laboratory study, we are readily able to detect their presence by observing their effects upon the cells they parasitize.

Specific viruses are associated with all types of cells, eukaryotic and prokaryotic. Their dependence on other cells is due to their inability to synthesize enzymes needed for their own metabolism. By existing within cells, however, they are able to utilize the

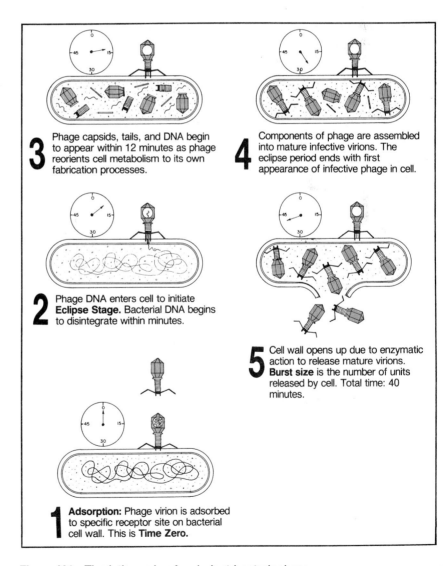

3 Phage capsids, tails, and DNA begin to appear within 12 minutes as phage reorients cell metabolism to its own fabrication processes.

4 Components of phage are assembled into mature infective virions. The eclipse period ends with first appearance of infective phage in cell.

2 Phage DNA enters cell to initiate **Eclipse Stage.** Bacterial DNA begins to disintegrate within minutes.

5 Cell wall opens up due to enzymatic action to release mature virions. **Burst size** is the number of units released by cell. Total time: 40 minutes.

1 **Adsorption:** Phage virion is adsorbed to specific receptor site on bacterial cell wall. This is **Time Zero.**

Figure V.1 The lytic cycle of a virulent bacteriophage

enzymes of their host. They may contain DNA or RNA, but never both of these nucleic acids.

The study of viruses that parasitize plant and animal cells is time-consuming and requires special tissue culture techniques. Viruses that parasitize bacteria, however, are relatively easy to study, utilizing ordinary bacteriological techniques. It is for this reason that bacterial viruses will be studied here. Principles learned from studying the viruses of bacteria apply to viruses of eukaryotic cells.

Viruses that parasitize bacteria are called **bacteriophage,** or **phage.** These viruses exist in many shapes and sizes. Some of the simplest ones exist as a single-stranded DNA virion. Most of them are tadpole-like, with "heads" and "tails" as seen in figure V.1 on the previous page. The head, or **capsid,** may be round, oval, or polyhedral and is composed of protein. It forms a protective envelope for the DNA of the organism. The tail structure is hollow and provides an exit for the DNA from the capsid into the cytoplasm of the bacterial cell. The extreme end of the tail has the ability to become attached to specific receptor sites on the surface of phage-sensitive bacteria. Once the tail of the virus attaches itself to a cell, it literally digests its way through the wall of the host cell.

With the invasion of a bacterial cell by the DNA, one of two things will occur: lysis or lysogeny. In the event that **lysis** occurs, as illustrated on the previous page, the metabolism of the bacterial cell becomes reoriented to the synthesis of new viral DNA and protein to produce mature phage particles. Once all the cellular material is used up, the cell bursts to release phage virions that, in turn, are prepared to invade other cells.

Phage that cause lysis are said to be *virulent.* If the phage does not cause lysis, however, it is termed *temperate* and establishes a relationship with the bacterial cell known as **lysogeny.** In these cells, the DNA of the phage becomes an integral part of the bacterial chromosome. Lysogenic bacteria grow normally, but their cultures always contain some phage. Periodically, however, phage virions are released by lysogenized cells in lytic bursts similar to that seen in the lytic cycle.

Visual evidence of lysis is demonstrated by mixing a culture of bacteria with phage and growing the mixture on nutrient agar. Areas where the phage are active will show up as clear spots called *plaques.*

The most thoroughly studied bacterial viruses are those that parasitize *Escherichia coli.* They are collectively referred to as the *coliphages.* They are readily isolated from raw sewage and co-prophagous (dung-eating) insects. Exercises 28 and 29 pertain to these techniques. Exercise 30 provides a method for determining the burst size of a phage. Before attempting any of these experiments, be certain that you thoroughly understand the various stages in the phage lytic cycle as depicted here.

Isolation of Phage from Sewage

Establishing the presence of phage virions in sewage involves three steps. First, it is necessary to increase the phage numbers by *enrichment* with special media. Second, it is necessary to separate the phage from the bacteria by *filtration*. The final step is to produce plaque evidence by *seeding* a "lawn" of bacteria with phage in the filtrate. Figures 28.1 and 28.2 illustrate the three steps we will go through. Before beginning this experiment, however, here are a few comments about safety procedures that must be followed in handling sewage:

- When collecting sewage samples always wear latex gloves. Raw sewage is a potent source of bacterial, viral, and fungal pathogens.

- Raw sewage rich in bacteriophage is best collected at municipal sewage treatment plants. Usually, collection is made through manhole access.
- As emphasized in previous pages of this manual, no mouth pipetting permitted!

Enrichment

To increase the number of phage virions in a raw sewage sample, it is necessary to add 5 ml of deca-strength phage broth (DSPB) and 5 ml of *E. coli* to 45 ml of raw sewage as in illustration 1 of figure 28.1. The DSPB medium is 10 times as strong as ordinary broth to accommodate dilution with 45 ml of sewage. This mixture is incubated at 37° C for 24 hours.

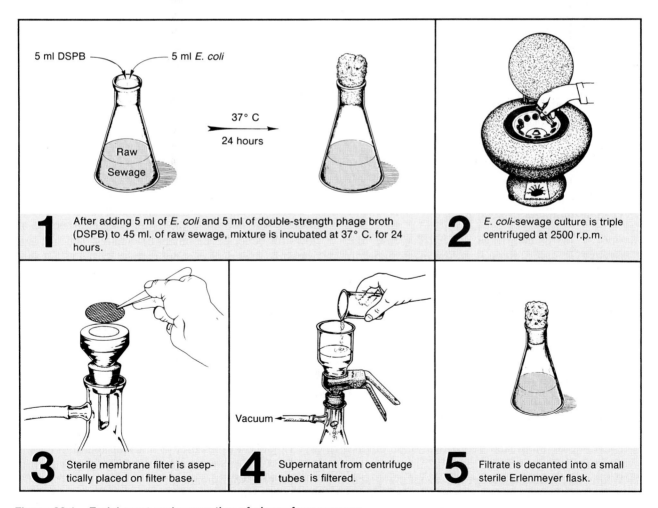

1 After adding 5 ml of *E. coli* and 5 ml of double-strength phage broth (DSPB) to 45 ml. of raw sewage, mixture is incubated at 37° C. for 24 hours.

2 *E. coli*-sewage culture is triple centrifuged at 2500 r.p.m.

3 Sterile membrane filter is aseptically placed on filter base.

4 Supernatant from centrifuge tubes is filtered.

5 Filtrate is decanted into a small sterile Erlenmeyer flask.

Figure 28.1 Enrichment and separation of phage from sewage

Materials:
    flask of raw sewage
    1 Erlenmeyer flask (125 ml size)
    5 ml of DSPB medium
    nutrient broth culture of *E. coli* (strain B)
    graduate and 5 ml pipettes

1. With a graduate, measure out 45 ml of raw sewage and decant into an Erlenmeyer flask.
2. Pour 5 ml of DSPB medium and 5 ml of *E. coli* into the flask of sewage. If these constituents are not premeasured, use a 5 ml pipette. If the medium is pipetted first, the same pipette can be used for pipetting the *E. coli.*
3. Place the flask in the 37° C incubator for 24 hours.

Filtration

Rapid filtration to separate the phage from *E. coli* in the enrichment mixture requires adequate centrifugation first. If centrifugation is inadequate, the membrane filter will clog quickly and impair the rate of filtration. To minimize filter clogging, a **triple centrifugation** procedure will be used. To save time in the event that filter clogging does occur, an extra filter assembly and an adequate supply of membrane filters should be available. These membrane filters have a maximum pore size of 0.45 μm, which holds back all bacteria, allowing only the phage virions to pass through.

Materials:
    centrifuge and centrifuge tubes (6–12)
    2 sterile membrane filter assemblies (funnel,
        glass base, clamp, and vacuum flask)
    package of sterile membrane filters
    sterile Erlenmeyer flask with cotton plug
        (125 ml size)
    forceps and Bunsen burner
    vacuum pump and rubber hose

1. Into 6 or 8 centrifuge tubes, dispense the sewage–*E. coli* mixture, filling each tube to within ½″ of the top. Place the tubes in the centrifuge so that the load is balanced. Centrifuge the tubes at 2500 rpm for 10 minutes.
2. Without disturbing the material in the bottom of the tubes, decant all material from the tubes to within 1″ of the bottom into another set of tubes.
3. Centrifuge this second set of tubes at 2500 rpm for another 10 minutes. While centrifugation is taking place, rinse out the first set of tubes.
4. When the second centrifugation is complete, pour off the top two-thirds of each tube into the clean set of tubes and centrifuge again in the same manner.
5. While the third centrifugation is taking place, aseptically place a membrane filter on the glass base of a sterile filter assembly (illustration 3,

figure 28.1). Use flamed forceps. Note that the filter is a thin sheet with grid lines on it.
6. Place the glass funnel over the filter and fix the clamp in place.
7. Hook up a rubber hose between the vacuum flask and pump.
8. Carefully decant the top three-fourths of each tube into the filter funnel. **Do not disturb the material in the bottom of the tube.**
9. Turn on the vacuum pump. If centrifugation has removed all bacteria, filtration will occur almost instantly. If the filter becomes clogged and you have enough filtrate to complete the experiment, go on to step 10. (If this filtrate is to be used by the entire class, you will need 25–50 ml.)

    If the filter clogs before you have enough filtrate, pour the unfiltered material from the funnel back into another set of centrifuge tubes and recentrifuge for 10 minutes at 2500 rpm.

    While centrifugation is taking place, set up the other filter assembly and pour whatever filtrate you have from the first flask into the funnel of the new setup. After centrifugation, decant the top three-fourths of material from each tube into the funnel and turn on the vacuum pump. Filtration should take place rapidly now.
10. Aseptically transfer the final filtrate from the vacuum flask to a sterile 125 ml Erlenmeyer flask that has a sterile cotton plug. Putting the filtrate in a small flask is necessary to facilitate pipetting. *Be sure to flame the necks of both flasks while pouring from one to the other.*

Seeding

Evidence of phage in the filtrate is produced by providing a "lawn" of *E. coli* and phage. The medium used is soft nutrient agar. Its jellylike consistency allows for better development of plaques. The soft agar is poured over the top of prewarmed hard nutrient agar. Prewarmed plates result in a smoother top agar surface. Figure 28.2 illustrates the general procedure.

Materials:
    nutrient broth culture of *E. coli* (strain B)
    flask of enriched sewage filtrate
    4 metal-capped tubes of soft nutrient agar (5 ml
        per tube)
    4 Petri plates of nutrient agar (15 ml per plate,
        preferably prewarmed at 37° C)
    1 ml serological pipettes

1. Liquefy 4 tubes of soft nutrient agar and cool to 50° C. Keep the tubes in a 50° C water bath to prevent solidification.
2. Label the tubes 1, 2, 3, and 4. Label the plates 1, 2, 3, and control.

3. With a 1 ml pipette, transfer one drop of filtrate to tube 1, three drops to tube 2, and six drops to tube 3. Don't put any filtrate into tube 4.
4. With a fresh 1 ml pipette, transfer 0.3 ml of *E. coli* to each of the four tubes of soft agar.
5. After flaming the necks of each of the soft agar tubes, pour the contents of each tube over the hard agar of similarly numbered agar plates. Note that tube 4 is poured over the control plate.
6. Once the agar is cooled completely, put the

plates, inverted, into a 37° C incubator. If possible, examine the plates **3 hours later** to look for plaque formation. If some plaques are visible, measure them and record their diameters on the Laboratory Report. Plaque size should be checked **every hour** for changes.

Laboratory Report

Record all results on Laboratory Report 28, 29.

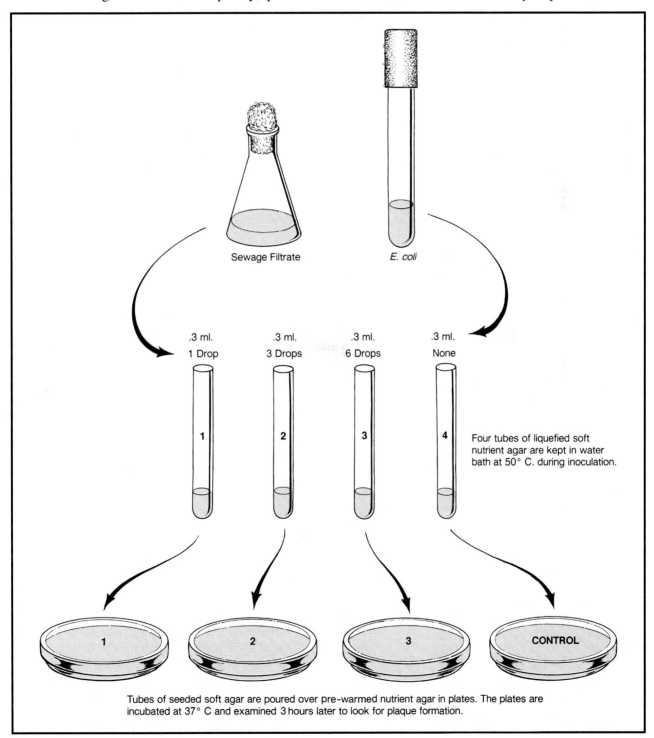

Figure 28.2 Overlay method of seeding *Escherichia coli* cultures with phage

As stated earlier, coprophagous insects, as well as raw sewage, contain various kinds of bacterial viruses. Houseflies fall into the coprophagous category because they deposit their eggs in fecal material where the young larvae feed, grow, pupate, and emerge as adult flies. This type of environment is heavily populated by *E. coli* and its inseparable parasitic phages.

In this experiment we will follow a procedure that is quite similar to the one used in working with raw sewage. An enrichment medium, utilizing cyanide, will be substituted for the DSPB, however. Figures 29.1 and 29.2 illustrate the procedure.

Fly Collection

To increase the probability of success in isolating phage, it is desirable that one use 20 to 24 houseflies. A smaller number might be sufficient; the larger number, however, increases the probability of initial success. Houseflies should not be confused with the smaller blackfly or the larger blowfly. An ideal spot for collecting these insects is a barnyard or riding stable. One should not use a cyanide killing bottle or any other chemical means. Flies should be kept alive until just prior to crushing and placing them in the growth medium. There are many ways that one might use to capture them—use your ingenuity!

Enrichment

Within the flies' digestive tracts are several different strains of *E. coli* and bacteriophage. Our first concern is to enhance the growth of both organisms to ensure an adequate supply of phage. To accomplish this the flies must be ground up with a mortar and pestle and then incubated in a special growth medium for a total of 48 hours. During the last 6 hours of incubation, a lysing agent, sodium cyanide, is included in the growth medium to augment the lysing properties of the phage.

Materials:
    bottle of phage growth medium* (50 ml)
    bottle of phage lysing medium* (50 ml)
    Erlenmeyer flask (125 ml capacity) with cotton
      plug
    mortar and pestle (glass)
    *see Appendix C for composition

1. Into a clean nonsterile mortar place 24 freshly killed houseflies. Pour half of the growth medium into the mortar and grind the flies to a fine pulp with the pestle.
2. Transfer this fly-broth mixture to an empty flask. Use the remainder of the growth medium to rinse out the mortar and pestle, pouring all the medium into the flask.
3. Wash the mortar and pestle with soap and hot water before returning them to the cabinet.
4. Incubate the fly-broth mixture for 42 hours at 37° C.
5. At the end of the 42-hour incubation period add 50 ml of lysing medium to the fly-broth mixture. Incubate this mixture for another 6 hours.

Centrifugation

Before attempting filtration, you will find it necessary to separate the fly fragments and miscellaneous bacteria from the culture medium. If centrifugation is incomplete, the membrane filter will clog quickly and filtration will progress slowly. To minimize filter clogging, a triple centrifugation procedure will be used. To save time in the event filter clogging does occur, an extra filter assembly and an adequate supply of membrane filters should be available. These filters have a maximum pore size of 0.45 μm, which holds back all bacteria, allowing only the phage virions to pass through.

Materials:
    centrifuge
    6–12 centrifuge tubes
    2 sterile membrane filter assemblies (funnel,
      glass base, clamp, and vacuum flask)
    package of sterile membrane filters
    sterile Erlenmeyer flask with cotton plug (125
      ml size)
    vacuum pump and rubber hose

1. Into 6 or 8 centrifuge tubes, dispense the enrichment mixture, filling each tube to within in ½″ of the top. Place the tubes in the centrifuge so that the load is balanced. Centrifuge the tubes at 2500 rpm for 10 minutes.
2. Without disturbing the material in the bottom of the tubes, decant all material from the tubes to within 1″ of the bottom into another set of tubes.

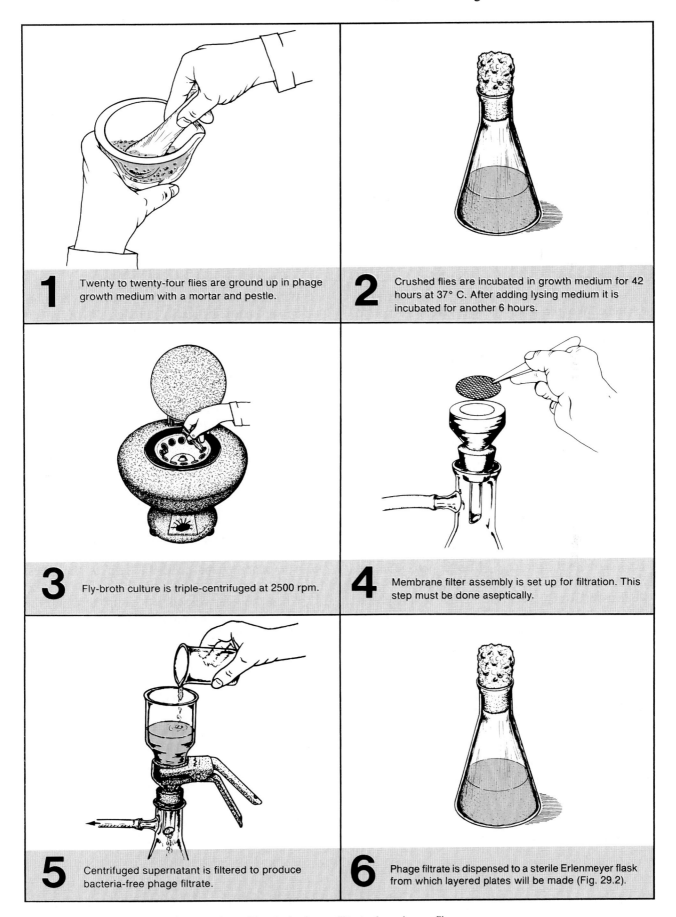

1 Twenty to twenty-four flies are ground up in phage growth medium with a mortar and pestle.

2 Crushed flies are incubated in growth medium for 42 hours at 37° C. After adding lysing medium it is incubated for another 6 hours.

3 Fly-broth culture is triple-centrifuged at 2500 rpm.

4 Membrane filter assembly is set up for filtration. This step must be done aseptically.

5 Centrifuged supernatant is filtered to produce bacteria-free phage filtrate.

6 Phage filtrate is dispensed to a sterile Erlenmeyer flask from which layered plates will be made (Fig. 29.2).

Figure 29.1 Procedure for preparation of bacteriophage filtrate from houseflies

3. Centrifuge this second set of tubes at 2500 rpm for another 10 minutes. While centrifugation is taking place, rinse out the first set of tubes.

4. When the second centrifugation is complete, pour off the top two-thirds of each tube into the clean set of tubes and centrifuge again in the same manner.

Filtration

While the third centrifugation is taking place, aseptically place a membrane filter on the glass base of a sterile filter assembly (illustration 4, figure 29.1). Use flamed forceps. Note that the filter is a thin sheet with grid lines on it. Place the glass funnel over the filter and fix the clamp in place. Hook up a rubber hose between the vacuum flask and pump.

Now, carefully decant the top three-fourths of each tube into the filter funnel. Take care not to disturb the material in the bottom of the tube. Turn on the vacuum pump. If centrifugation and decanting have been performed properly, filtration will occur almost instantly. If the filter clogs before you have enough filtrate, recentrifuge all material and pass it through the spare filter assembly.

Aseptically, transfer the final filtrate from the vacuum flask to a sterile 125 ml Erlenmeyer flask that has a sterile cotton plug. Putting the filtrate in a small flask is necessary to facilitate pipetting. Be sure to flame the necks of both flasks while pouring from one to the other.

Inoculation and Incubation

To demonstrate the presence of bacteriophage in the fly-broth filtrate, a strain of phage-susceptible *E. coli* will be used. To achieve an ideal proportion of phage to bacteria, a proportional dilution method will be used. The phage and bacteria will be added to tubes of soft nutrient agar that will be layered over plates of hard nutrient agar. Soft nutrient agar contains only half as much agar as ordinary nutrient agar. This medium and *E. coli* provide an ideal "lawn" for phage growth. Its jelly-like consistency allows for better diffusion of phage particles; thus, more even development of plaques occurs.

Figure 29.2 illustrates the overall procedure. It is best to perform this inoculation procedure in the morning so that the plates can be examined in late afternoon. As plaques develop, one can watch them increase in size with the multiplication of phage and simultaneous destruction of *E. coli*.

Materials:

nutrient broth cultures of *Escherchia coli* (ATCC #8677 phage host)

flask of fly-broth filtrate

10 tubes of soft nutrient agar (5 ml per tube) with metal caps

10 plates of nutrient agar (15 ml per plate, and prewarmed at 37° C)

1 ml serological pipettes, sterile

1. Liquefy 10 tubes of soft nutrient agar and cool to 50° C. Keep tubes in water bath to prevent solidification.

2. With a china marking pencil, number the tubes of soft nutrient agar 1 through 10. Keep the tubes sequentially arranged in the test-tube rack.

3. Label 10 plates of prewarmed nutrient agar 1 through 10. Also, label plate 10 **negative control.** Prewarming these plates will allow the soft agar to solidify more evenly.

4. With a 1 ml serological pipette, deliver 0.1 ml of fly-broth filtrate to tube 1, 0.2 ml to tube 2, etc., until 0.9 ml has been delivered to tube 9. Refer to figure 29.2 for sequence. **Note that no fly-broth filtrate is added to tube 10.** This tube will be your negative control.

5. With a fresh 1 ml pipette, deliver 0.9 ml of *E. coli* to tube 1, 0.8 ml to tube 2, etc., as shown in figure 29.2. **Note that tube 10 receives 1.0 ml of *E. coli*.**

6. After flaming the necks of each of the tubes, pour them into similarly numbered plates.

7. When the agar has cooled completely, put the plates, inverted, into a 37° C incubator.

8. **After about 3 hours** incubation, examine the plates, looking for plaques. If some are visible, measure them and record their diameters on the Laboratory Report.

9. If no plaques are visible, check the plates again in another **2 hours.**

10. Check the plaque size again at **12 hours,** if possible, recording your results. Incubate a total of 24 hours.

11. Complete Laboratory Report 28, 29.

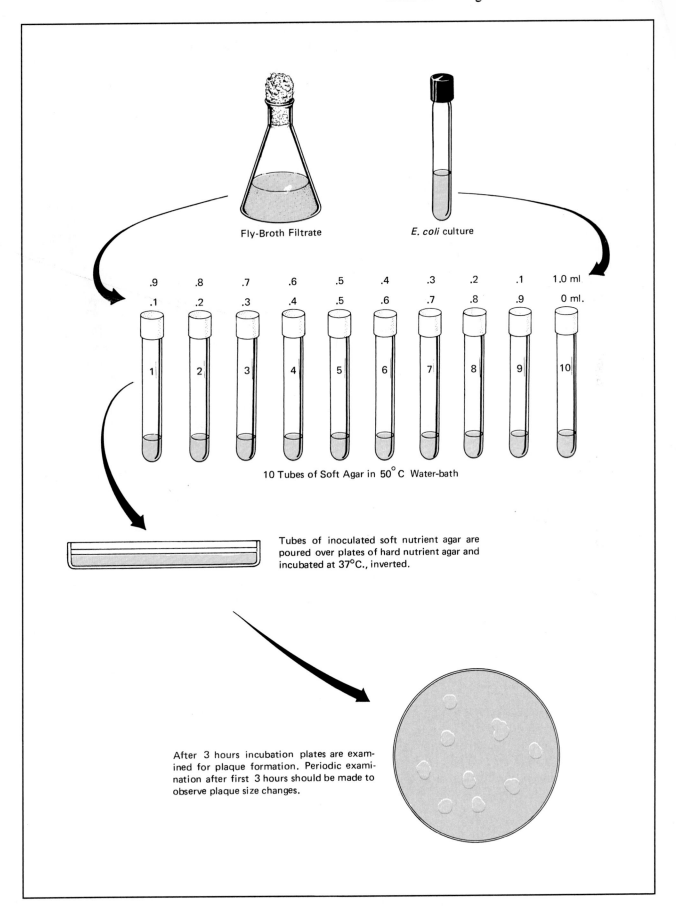

Fly-Broth Filtrate

E. coli culture

.9   .8   .7   .6   .5   .4   .3   .2   .1   1.0 ml
.1   .2   .3   .4   .5   .6   .7   .8   .9   0 ml.

1   2   3   4   5   6   7   8   9   10

10 Tubes of Soft Agar in 50°C Water-bath

Tubes of inoculated soft nutrient agar are poured over plates of hard nutrient agar and incubated at 37°C., inverted.

After 3 hours incubation plates are examined for plaque formation. Periodic examination after first 3 hours should be made to observe plaque size changes.

Figure 29.2   Inoculation of *Escherichia coli* with bacteriophage from fly-broth filtrate

Burst Size Determination:
A One-Step Growth Curve

The average number of mature phage virions released by the lysis of a single bacterial cell is between 20 and 200. This number is called the **burst size.** It can be determined by adding a small amount of phage to a known quantity of bacteria and then lysing the cells at 5-minute intervals with chloroform. The chloroform-lysed cells, in turn, are mixed with bacteria, plated out, and incubated. By counting the plaques, it is possible to determine the burst size. In this experiment we will determine the burst size of coliphage T4 on host cells *E. coli,* strain B.

Adsorption

Figure 30.1 illustrates the procedure of this experiment. The first step is to add the phage to the susceptible bacteria. As soon as the two are mixed, adsorption begins. The phage collide in random fashion with the bacterial cells and attach their tails to specific receptor sites on the surfaces of host cells. The adsorption process can be stopped at any time by dilution. **Time zero** of adsorption is the time of mixture of phage and bacteria.

Note in figure 30.1 that 0.1 ml of coliphage T4 (2×10^8/ml) and 2 ml of *E. coli* (5×10^8/ml) are mixed in the first tube, which is labeled ADS. The ratio of phage to bacteria in this case is 0.02, which calculates out in this manner:

$$\frac{0.1 \times 2 \times 10^8}{2 \times 5 \times 10^8} = \frac{0.2 \times 10^8}{10 \times 10^8} = 0.02, \text{ or } 1/50$$

This ratio is called the **multiplicity of infection,** or **m.o.i.**

By referring back to figure V.1 on page 105 we can see what is occurring in this experiment. Note that during the adsorption stage, DNA in the capsid passes down through the tail into the host through a hole produced in the cell wall by enzymatic action at the tip of the phage tail.

Eclipse Stage

As soon as the phage DNA gets inside the bacterial cell, the phage enters the *eclipse stage.* During this stage, which lasts approximately 12 minutes, the entire physiology of the host cell is reoriented toward the production of phage components: capsids, tails, and DNA. If the cell is experimentally lysed with chloroform during this period of time, it will be seen that the incomplete components of phage are unable to infect new cells (no plaques are formed).

Maturation Stage

As phage components begin to assemble late in the eclipse stage to form mature infective virions, the phage enters the *maturation stage.* The lysing of cultures with chloroform beyond 12 minutes of time zero will reveal the presence of these mature units by producing plaques on poured plates. Lysis of a population of infected cells does not occur instantaneously, but instead follows a normal distribution curve, or **rise period.** The rise period, which lasts for several minutes, represents the growth in numbers of mature phage present. *The peak of the curve is the burst size.* It is this value that will be determined here.

Two Methods

To accommodate the availability of time and materials, there are two options for performing this experiment. The first option is for students to work in pairs to perform the entire experiment. Figure 30.1 illustrates the procedure for this method. The other option, which requires much less media and time, utilizes a team approach in which students, working in pairs, do just a portion of the experiment; in this case, data are pooled to complete the experiment. Figure 30.3 illustrates the procedure for this method. Your instructor will indicate which method will be used.

The Entire Experiment

To perform the experiment in its entirety, follow the procedures that are shown in figure 30.1.

Materials:
    1 sterile serological tube (for ADS tube)
    15 tubes tryptone broth (9.9 ml in each one)
    8 tubes of nutrient soft agar (3 ml per tube)
    8 Petri plates of tryptone agar
    16 pipettes (1 ml size)

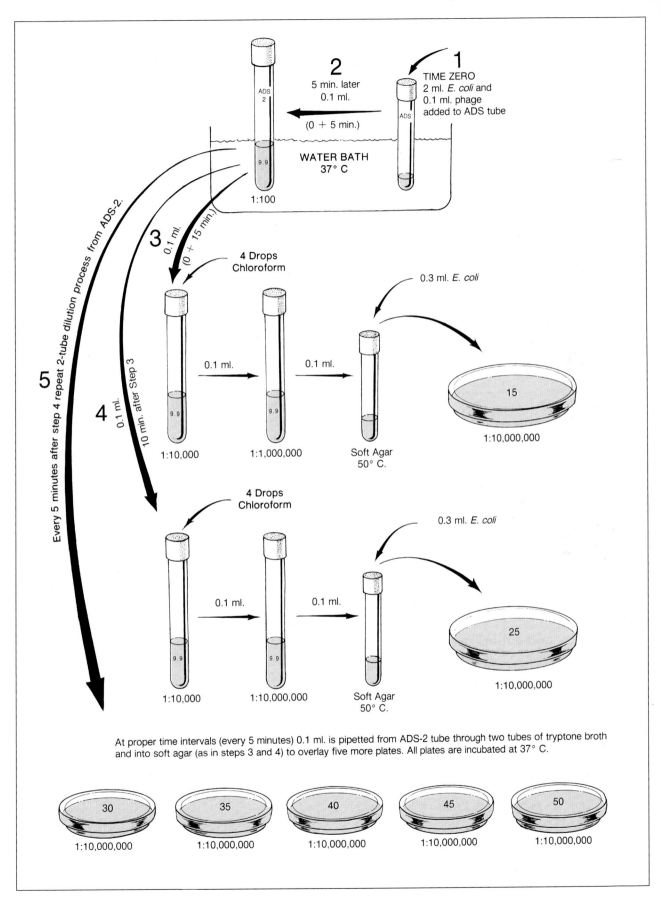

At proper time intervals (every 5 minutes) 0.1 ml. is pipetted from ADS-2 tube through two tubes of tryptone broth and into soft agar (as in steps 3 and 4) to overlay five more plates. All plates are incubated at 37° C.

Figure 30.1 Procedure for entire experiment

1 dropping bottle of chloroform
1 small wire basket to hold 7 tubes of soft agar in water bath
1 wire test-tube rack
2 water baths (37° C and 50° C)
1 culture of *E. coli,* strain B (5 ml) with concentration of 5×10^8 per ml
1 tube of T4 phage (2×10^8 per ml)

Preliminaries

1. Liquefy 8 tubes of soft nutrient agar by boiling in a beaker of water. Cool to 50° C and place in a wire basket or rack in 50° C water bath.
2. Label a sterile serological tube "ADS" to signify the adsorption tube.
3. Label 1 tube of tryptone broth "ADS-2."
4. Label 8 tryptone agar plates: control, 15, 25, 30, 35, 40, 45, and 50.
5. Arrange the ADS, ADS-2, and the 14 tryptone broth tubes in a rack as shown in figure 30.2. Place the rack in a 37° C water bath.

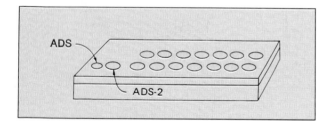

Figure 30.2 Tube arrangement

6. Dispense 3 to 4 drops of chloroform in each of the 7 tubes of tryptone broth that are in the front row.

Inoculations and Dilutions

1. Pipette 0.1 ml of *E. coli,* strain B, into a tube of liquefied soft nutrient agar and pour into the control plate. Swirl the plate gently to spread evenly. This plate will indicate whether any phage was in the original bacterial culture. Set this plate aside to harden.
2. With the same pipette as above, transfer 0.3 ml of *E. coli* into each of the tubes of soft nutrient agar. Keep the tubes in the 50° C water bath.
3. Still using the same pipette, transfer 2.0 ml of *E. coli* into the ADS tube.
4. With a fresh pipette, deliver 0.1 ml of T4 phage into the ADS tube and *immediately* record the time (*time zero*) of this mixing with *E. coli* in the following table. Mix gently and allow to remain in the 37° C water bath for 5 minutes.

	TIME	PLATE
Time zero	_____	none
Step 6 time (5 min later)	_____	none
Step 7 time (10 min later)	_____	15
Step 8 time (10 min later)	_____	25
Five minutes later	_____	30
Five minutes later	_____	35
Five minutes later	_____	40
Five minutes later	_____	45
Five minutes later	_____	50

5. While the mixture is incubating, fill in the table, recording all the projected times so that you will know when each step is to begin. *
6. After the 5-minute incubation time, transfer 0.1 ml of the mixture to the ADS-2 tube, gently mix, and incubate at 37° C for another 10 minutes.
7. After 10 minutes, transfer 0.1 ml from the ADS-2 tube to the first front row tube of tryptone broth. Keep the ADS-2 tube in the water bath. Mix this dilution tube gently and transfer 0.1 ml to the adjacent tryptone broth tube in the second row. Mix this tube gently, also.
8. Transfer 0.1 ml from the second tube of tryptone broth to a tube of soft nutrient agar, mix gently, flame the tube neck, and pour the soft agar over the tryptone agar plate that is labeled "15."

    Swirl the plate carefully to disperse the soft agar mixture evenly.
9. Follow the above procedure 10 minutes later to produce a soft agar overlay plate on the plate labeled "25."
10. Repeat at the allotted times for 30-, 35-, 40-, 45-, and 50-minute plates.
11. Invert and incubate all plates for 24–48 hours at 37° C.

Examination of the Plates

Once the plates have been incubated, count the plaques on all the plates, using a Quebec colony counter and hand tally counter. Record all counts on the Laboratory Report and determine burst size.

Abbreviated Procedure
(Team Method)

Performance of this experiment in teams will require a minimum of seven pairs of students. Each pair of students (team) will follow the procedure shown in figure 30.3 to produce one soft agar overlay plate for a designated time.

Materials:

per team:

1 sterile serological tube
3 tubes of tryptone broth (9.9 ml per tube)
2 tubes of soft nutrient agar (3 ml per tube)
2 Petri plates of tryptone agar
4 1 ml pipettes
1 dropping bottle of chloroform
1 wire test-tube rack (small size)
1 small beaker (150 ml size)
1 tube of T4 phage (2×10^8 per ml)
1 culture of *E. coli,* strain B (5×10^8 per ml)
water bath at 37° C (a small pan that will hold a test-tube rack)

Preliminaries

1. Liquefy two tubes of soft nutrient agar in boiling water. Use a small beaker. Cool the water to 50° C and keep the tubes of media at this temperature.

2. Label a sterile serological tube "ADS" to signify the adsorption tube.
3. Label one tube of tryptone broth "ADS-2."
4. Label the other tryptone tubes "I" and "II."
5. Label one tryptone agar plate "control" and the other your designated time (15, 25, 30, 35, 40, 45, or 50). *Your instructor will assign you a specific time.* Put your names on both plates.
6. Arrange the ADS, ADS-2, and two tryptone tubes in a small test-tube rack in same order as shown in figure 30.3.
7. Place the rack of tubes in a pan of 37° C water. Although it is only necessary to incubate the ADS and ADS-2 tubes, it will be more convenient if they are all together.

Inoculations and Dilutions

1. Pipette 0.1 ml of *E. coli,* strain B, into a tube of liquefied soft nutrient agar and pour it into the control plate. Swirl the plate gently to spread evenly.

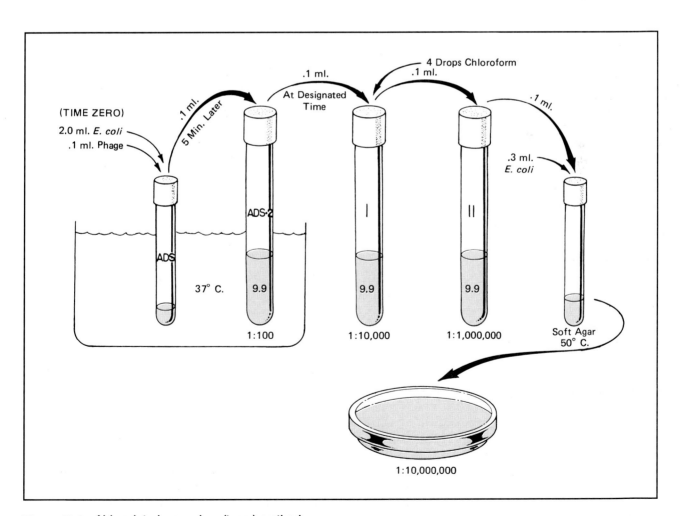

Figure 30.3 Abbreviated procedure (team) method

This plate will indicate whether any phage was in the original bacterial culture. Set this plate aside to harden.

2. With the same pipette, transfer 0.3 ml of *E. coli* to the other tube of soft nutrient agar. Keep this tube in the beaker of water at 50° C.

3. Still using the same pipette, transfer 2.0 ml of *E. coli* into the ADS tube.

4. With a *fresh pipette,* deliver 0.1 ml of T4 phage into the ADS tube.

    Record this time (time zero):_____

5. After 5 minutes, transfer 0.1 ml of the *E. coli*–phage mixture from the ADS tube to ADS-2 tube. Mix the ADS-2 tube gently.

6. After the designated time (time zero plus designated time), transfer 0.1 ml from ADS-2 tube to tryptone broth tube I. Mix gently.

7. Add 3 or 4 drops of chloroform to tube I.

8. With a *fresh pipette,* transfer 0.1 ml from tube I to tube II. Mix tube II gently.

9. With *another fresh pipette,* transfer 0.1 ml from tube II to the tube of soft agar.

10. After mixing the soft agar tube, pour it over the tryptone agar plate. Swirl the plate carefully to disperse the soft agar. Set aside to cool for a few minutes.

11. Incubate both plates at 37° C for 24–48 hours.

Examination of Plates

Once the plates have been incubated, examine both of them on a Quebec colony counter. The control plate should be free of plaques. Count the plaques on the other plate, using a hand tally counter if the number is great. Record your count on the Laboratory Report and on the chalkboard.

PART 6 Microbial Interrelationships

Populations within the microbial world relate to each other in various ways. Although many of them will be neutralistic toward each other by not interacting in any way, others will establish relationships that are quite different.

The three exercises in this unit reveal how certain organisms have developed relationships that are commensalistic, synergistic, and antagonistic. While most of these relationships are between bacteria, some are between bacteria and molds.

31

Bacterial Commensalism

There are many commensalistic relationships that exist between organisms in a mixed microbial population. The excretory products of one organism often become the nutrients of another. The oxygen usage of one species may produce the desired oxidation-reduction potential for another organism. In all cases of commensalism, the beneficiary contributes nothing in the way of benefit or injury to the other.

In this exercise we will culture two organisms separately and together to observe an example of commensalism. One of the organisms is *Staphylococcus aureus* and the other is *Clostridium sporogenes.* From your observations of the results, you are to determine which organism profits from the association and what controlling factor is changed when the two are grown together.

Materials:
   3 tubes of nutrient broth
   1 ml pipette
   nutrient broth culture of *S. aureus*

fluid thioglycollate medium culture of *C. sporogenes*

First Period

1. Label one tube of nutrient broth *S. aureus,* a second tube *C. sporogenes,* and the third tube *S. aureus* and *C. sporogenes.*
2. Inoculate the first and third tubes with one loopful each of *S. aureus.*
3. With a 1 ml pipette, transfer 0.1 ml of *C. sporogenes* to tubes 2 and 3.
4. Incubate the three tubes at 37° C for 48 hours.

Second Period

1. Compare the turbidity in the three tubes, noting which ones are most turbid. Record these results on the Laboratory Report.
2. After shaking the tubes for good dispersion, make a gram-stained slide of the organisms in each tube and record your observations on combined Laboratory Report 31–33.

Bacterial Synergism

Two or more organisms acting together to produce a substance that none can produce separately is a synergistic relationship. Such relationships are not uncommon among microorganisms. This phenomenon is readily demonstrated in the ability of some bacteria acting, synergistically, to produce gas by fermenting certain disaccharides.

In this exercise we will observe the fermentation capabilities of three organisms on two disaccharides. The two sugars, lactose and sucrose, will be inoculated with the individual organisms as well as with various combinations of the organisms to detect which organisms can act synergistically on which sugars. To conserve on media, the class will be divided into three groups (A, B, and C). Results of inoculations will be shared.

Materials:
per pair of students:
    3 Durham* tubes of lactose broth with
        bromthymol blue indicator
    3 Durham* tubes of sucrose broth with
        bromthymol blue indicator
    1 nutrient broth culture of *S. aureus*
    1 nutrient broth culture of *P. vulgaris*
    1 nutrient broth culture of *E. coli*

    *A *Durham tube* is a fermentation tube of
        sugar broth that has a small inverted vial in
        it. See figure 48.2, page 158.

First Period

Group A
1. Label one tube of each kind of broth *E. coli*.
2. Label one tube of each kind of broth *P. vulgaris*.
3. Label one tube of each kind of broth *E. coli* and *P. vulgaris*.
4. Inoculate each tube with one loopful of the appropriate organisms.
5. Incubate the six tubes at 37° C for 48 hours.

Group B
1. Label one tube of each kind of broth *E. coli*.
2. Label one tube of each kind of broth *S. aureus*.
3. Label one tube of each kind of broth *E. coli* and *S. aureus*.
4. Inoculate each tube with one loopful of the appropriate organisms.
5. Incubate the six tubes at 37° C for 48 hours.

Group C
1. Label one tube of each kind of broth *S. aureus*.
2. Label one tube of each kind of broth *P. vulgaris*.
3. Label one tube of each kind of broth *S. aureus* and *P. vulgaris*.
4. Inoculate each tube with one loopful of the appropriate organisms.
5. Incubate the six tubes at 37° C for 48 hours.

Second Period
1. Look for acid and gas production in each tube, recording your results on the Laboratory Report.
2. Determine which organisms acted synergistically on which disaccharides.
3. Answer the questions for this exercise on combined Laboratory Report 31–33.

Microbial antagonisms, in which one organism is inhibited and the other is unaffected, are easily demonstrated. Usually, the inhibitor produces a substance that inhibits or kills one or more organisms. The substance may be specific in its action, affecting only a few species, or it may be nonspecific, affecting a large number of organisms.

In this exercise we will attempt to evaluate the antagonistic capabilities of three organisms on two test organisms. The antagonists are *Bacillus cereus* var. *mycoides, Pseudomonas fluorescens,* and *Penicillium notatum.* The test organisms are *Escherichia coli* (gram-negative) and *Staphylococcus aureus* (gram-positive).

Materials:
    6 nutrient agar pours
    6 sterile Petri plates
    nutrient broth cultures of *E. coli, S. aureus,*
        B. cereus var. *mycoides,* and
        P. fluorescens
    flask culture of *Penicillium notatum*
        (8–12 day old culture)

First Period

1. Liquefy six nutrient agar pours and cool to 50° C. Hold in 50° C water bath.

2. While the pours are being liquefied, label six plates as follows:

Test Organism	Antagonist
I *S. aureus*	*B. cereus* var. *mycoides*
II *S. aureus*	*P. fluorescens*
III *S. aureus*	*Penicillium notatum*
IV *E. coli*	*B. mycoides*
V *E. coli*	*P. fluorescens*
VI *E. coli*	*Penicillium notatum*

3. Label three liquefied pours *S. aureus,* and label the other three *E. coli.*
4. Inoculate each of the pours with a loopful of the appropriate organisms, flame their necks, and pour into their respective plates.
5. After the nutrient agar in the plates has hardened, streak each plate with the appropriate antagonist. Use a good isolation technique.
6. Invert and incubate the plates for 24 hours at 37° C.

Second Period

1. Examine each plate carefully, looking for evidence of inhibition.
2. Record your results on combined Laboratory Report 31–33 and answer all the questions.

PART 7

Environmental Influences and Control of Microbial Growth

The 11 exercises of this unit are concerned with two aspects of microbial growth: promotion and control. On the one hand, the microbiologist is concerned with providing optimum growth conditions to favor maximization of growth. The physician, nurse, and other members of the medical arts profession, on the other hand, are concerned with the limitation of microbial populations in disease prevention and treatment. An understanding of one of these facets of microbial existence enhances the other.

In Part 4 we were primarily concerned with providing media for microbial growth that contain all the essential nutritional needs. Very little emphasis was placed on other limiting factors such as temperature, oxygen, or hydrogen ion concentration. An organism provided with all its nutritional needs may fail to grow if one or more of these essentials are not provided. The total environment must be sustained to achieve the desired growth of microorganisms.

Microbial control by chemical and physical means involves the use of antiseptics, disinfectants, antibiotics, ultraviolet light, and many other agents. The exercises of this unit that are related to these aspects are intended, primarily, to demonstrate methods of measurement; no attempt has been made to make in-depth evaluation.

34

Temperature:
Effects on Growth

Temperature is one of the most important factors influencing the activity of bacterial enzymes. Unlike warm-blooded animals, the bacteria lack mechanisms that conserve or dissipate heat generated by metabolism, and consequently their enzyme systems are directly affected by ambient temperatures. Enzymes have minimal, optimal, and maximal temperatures. At the **optimum** temperature the enzymatic reactions progress at maximum speed. Below the **minimum** and above the **maximum** temperatures the enzymes become inactive. At some point above the maximum temperature, destruction of a specific enzyme will occur. Low temperatures are less deleterious in most cases.

Microorganisms grow in a broad temperature range that extends from approximately 0° C to above 90° C. They are divided into three groups: **mesophiles** that grow between 10° C and 47° C, **psychrophiles** that are able to grow between 0° C and 5° C, and **thermophiles** that grow at high temperatures (above 50° C).

The psychrophiles and thermophiles are further subdivided into obligate and facultative groups. Obligate psychrophiles seldom grow above 22° C and facultative psychrophiles (psychrotrophs) grow very well above 25° C. Thermophiles that thrive *only* at high temperatures (above 50° C and not below 40° C) are considered to be obligate thermophiles; those that will grow below 40° C are considered to be facultative thermophiles.

In this experiment we will attempt to measure the effects of various temperatures on two physiological reactions: pigment production and growth rate. Nutrient broth and nutrient agar slants will be inoculated with three different organisms that have

different optimum growth temperatures. One organism, *Serratia marcescens,* produces a red pigment called *prodigiosin* that is produced only in a certain temperature range. It is our goal here to determine the optimum temperature for prodigiosin production and the approximate optimum growth temperatures for all three microorganisms. To determine optimum growth temperatures we will be incubating cultures at five different temperatures. A spectrophotometer will be used to measure turbidity densities in the broth cultures after incubation.

First Period
(Inoculations)

To economize on time and media it will be necessary for each student to work with only two organisms and seven tubes of media. Refer to table 34.1 to determine your assignment. Figure 34.1 illustrates the procedure.

Materials:
>   nutrient broth cultures of *Serratia marcescens,*
>   *Bacillus stearothermophilus,* and
>   *Escherichia coli*

per student:
>   2 nutrient agar slants
>   5 tubes of nutrient broth

1.   Label the tubes as follows:
     Slants: Label both of them *S. marcescens;* label one tube 25° C and the other tube 38° C.
     Broths: Label each tube of nutrient broth with your other organism and one of the following five temperatures: 5° C, 25° C, 38° C, 42° C, or 55° C.

Table 34.1 Inoculation Assignments

Student Number	*S. marcescens*	*B. stearothermophilus*	*E. coli*
1, 4, 7, 10, 13, 16, 19, 22, 25	2 slants and 5 broths		
2, 5, 8, 11, 14, 17, 20, 23, 26	2 slants	5 broths	
3, 6, 9, 12, 15, 18, 21, 24, 27	2 slants		5 broths

2. Inoculate each of the tubes with the appropriate organisms. Use a wire loop.

3. Place each tube in one of the five baskets that is labeled according to incubation temperature.
Note: The instructor will see that the 5° C basket is placed in the refrigerator and the other four are placed in incubators that are set at the proper temperatures.

Second Period
(Tabulation of Results)

Materials:
> slants and broth cultures that have been incubated at various temperatures
> spectrophotometer and cuvettes
> tube of sterile nutrient broth

1. Compare the nutrient agar slants of *S. marcescens*. Using colored pencils, draw the appearance of the growths on the Laboratory Report.

2. Shake the broth cultures and compare them, noting the differences in turbidity. Those tubes that appear to have no growth should be compared with a tube of sterile nutrient broth.

3. If a spectrophotometer is available, determine the turbidity of each tube following the instructions on the Laboratory Report.

4. If no spectrophotometer is available, record turbidity by visual observation. The Laboratory Report indicates how to do this.

5. Exchange results with other students to complete data collection for experiment.

Laboratory Report

After recording all data, answer the questions on the Laboratory Report for this exercise.

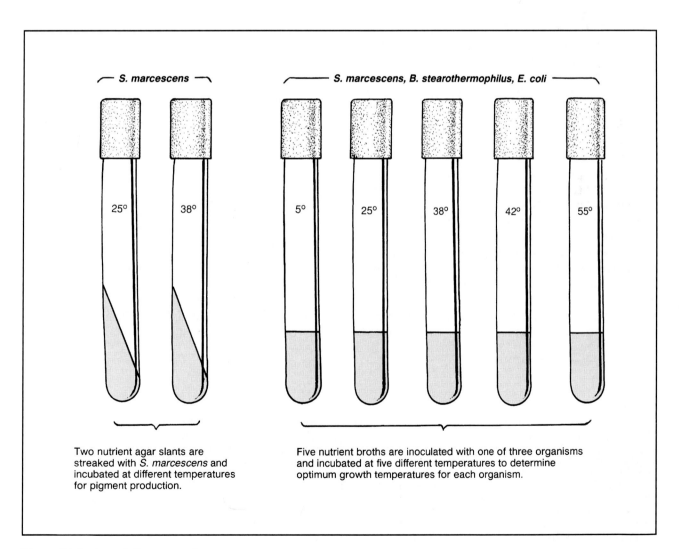

Figure 34.1 Inoculation procedure

125

Temperature:
Lethal Effects

In attempting to compare the susceptibility of different organisms to elevated temperatures, it is necessary to use some yardstick of measure. Two methods of comparison are used: the thermal death point and the thermal death time. The **thermal death point** (TDP) is the temperature at which an organism is killed in 10 minutes. The **thermal death time** (TDT) is the time required to kill a suspension of cells or spores at a given temperature. Since various factors such as pH, moisture, composition of medium, and age of cells will greatly influence results, these variables must be clearly stated.

In this exercise we will subject cultures of three different organisms to temperatures of 60°, 70°, 80°, 90°, and 100° C. At intervals of 10 minutes organisms will be removed and plated out to test their viability. The spore-former *Bacillus megaterium* will be compared with the non-spore-formers *Staphylococcus aureus* and *Escherichia coli*. The overall procedure is illustrated in figure 35.1.

Note in figure 35.1 that *before* the culture is heated a **control plate** is inoculated with 0.1 ml of the organism. When the culture is placed in the water bath, a tube of nutrient broth with a thermometer inserted into it is placed in the bath at the same time. Timing of the experiment starts when the thermometer reaches the test temperature.

Due to the large number of plates that have to be inoculated to perform the entire experiment, it will be necessary for each member of the class to be assigned a specific temperature and organism to work with. Table 35.1 provides assignments by student number. After the plates have been incubated, each student's results will be tabulated on a Laboratory Report chart at the demonstration table. The instructor will have copies made of it to give each student so that everyone will have all the pertinent data needed to draw the essential conclusions.

Although this experiment is not difficult, it often fails to turn out the way it should because of student error. Common errors are (1) omission of the control plate inoculation, (2) putting the thermometer in the culture tube instead of in a tube of sterile broth, and (3) not using fresh sterile pipettes when instructed to do so.

Materials:
per student:
    5 Petri plates
    5 pipettes (1 ml size)
    1 tube of nutrient broth
    1 bottle of nutrient agar (60 ml)
    1 culture of organisms

class equipment:
    water baths set up at 60°, 70°, 80°, 90°, and
        100° C

broth cultures:
    Staphylococcus aureus, Escherichia coli, and
        Bacillus megaterium (minimum of 5
        cultures of each species per lab section)

1. Consult table 35.1 to determine what organism and temperature has been assigned to you. If several thermostatically controlled water baths have been provided in the lab, locate the one that you will use. If a bath is not available for your

Table 35.1 Inoculation Assignments

Organism	Student Number				
	60° C	70° C	80° C	90° C	100° C
Staphylococcus aureus	1, 16	4, 19	7, 22	10, 25	13, 28
Escherichia coli	2, 17	5, 20	8, 23	11, 26	14, 29
Bacillus megaterium	3, 18	6, 21	9, 24	12, 27	15, 30f

temperature, set up a bath on an electric hot plate or over a tripod and Bunsen burner.

If your temperature is 100° C, a hot plate and beaker of water are the only way to go. When setting up a water bath use hot tap water to start with to save heating time.

2. Liquefy a bottle of 60 ml of nutrient agar and cool to 50° C. This can be done while the rest of the experiment is in progress.

3. Label five Petri plates: **control, 10 min, 20 min, 30 min,** and **40 min.**

4. Shake the culture of organisms and transfer 0.1 ml of organisms with a 1 ml pipette to the control plate.

5. Place the culture and a tube of sterile nutrient broth into the water bath. Remove the cap from the tube of nutrient broth and insert a thermometer into the tube. *Don't make the mistake of inserting the thermometer into the culture of organisms!*

6. As soon as the temperature of the nutrient broth reaches the desired temperature, record the time here: _____.

Watch the temperature carefully to make sure it does not vary appreciably.

7. After 10 minutes have elapsed, transfer 0.1 ml from the culture to the 10-minute plate with a fresh 1 ml pipette. Repeat this operation at 10-minute intervals until all the plates have been inoculated. *Use fresh pipettes each time and be sure to shake the culture before each delivery.*

8. Pour liquefied nutrient agar (50° C) into each plate, rotate, and cool.

9. Incubate at 37° C for 24 to 48 hours. After evaluating your plates, record your results on the chart on the Laboratory Report and on the chart on the demonstration table.

Laboratory Report

Complete the Laboratory Report once you have a copy of the class results.

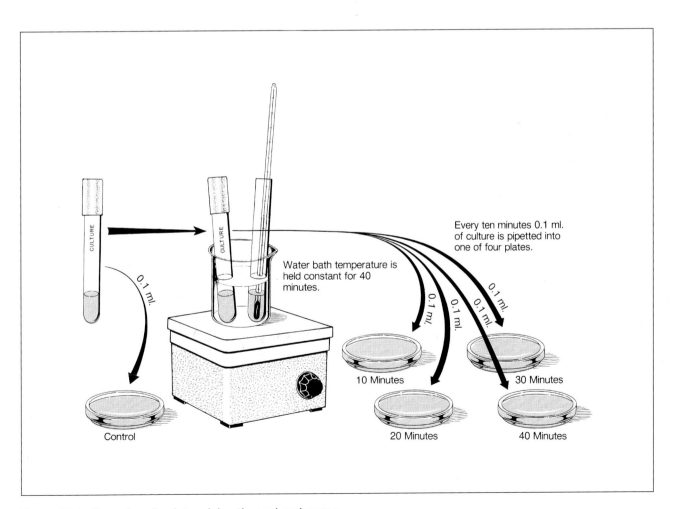

Figure 35.1 Procedure for determining thermal endurance

pH and Microbial Growth

Aside from temperature, the hydrogen ion concentration of an organism's environment exerts the greatest influence on its growth. The concentration of hydrogen ions, which is customarily designated by the term **pH** ($-\log 1/H^+$), limits the activity of enzymes with which an organism is able to synthesize new protoplasm. As in the case of temperature, there exists for each organism an optimum concentration of hydrogen ions in which it grows best. The pH values above and below which an organism fails to grow are, respectively, referred to as the minimum and maximum hydrogen ion concentrations. These values hold only when other environmental factors remain constant. If the composition of the medium, incubation temperature, or osmotic pressure is varied, the hydrogen ion requirements become different.

In this exercise we will test the degree of inhibition of microorganisms that results from media containing different pH concentrations. Note in the materials list that tubes of six different hydrogen concentrations are listed. Your instructor will indicate which ones, if not all, will be tested.

First Period

Materials:

per student:

    1 tube of nutrient broth of pH 3.0
    1 tube of nutrient broth of pH 5.0
    1 tube of nutrient broth of pH 7.0
    1 tube of nutrient broth of pH 8.0
    1 tube of nutrient broth of pH 9.0
    1 tube of nutrient broth of pH 10.0

class materials:

    broth cultures of *Escherichia coli*
    broth cultures of *Staphylococcus aureus*
    broth cultures of *Alcaligenes faecalis**
    broth cultures of *Saccharomyces cerevisiae***

---

*Sporosarcina ureae can be used as a substitute for *Alcaligenes faecalis.*

**Candida glabrata is a good substitute for *Saccharomyces cerevisiae.*

1. Inoculate a tube of each of these broths with one organism. Use the organism following your assigned number from the table below:

Student Number	Organism
1,5,9,13,17,21,25	*Escherichia coli*
2,6,10,14,18,22,26	*Staphylcoccus aureus*
3,7,11,15,19,23,27	*Alcaligenes faecalis*
4,8,12,16,20,24,28	*Saccharomyces cerevisiae**

2. Incubate the tubes of *E. coli, S. aureus,* and *A. faecalis* at 37° C for 48 hours. Incubate the tubes of *S. ureae, C. glabrata,* and *S. cervisiae* at 20° C for 48 to 72 hours.

Second Period

Materials:

    spectrophotometer
    1 tube of sterile nutrient broth
    tubes of incubated cultures at various pHs

1. Use the tube of sterile broth to calibrate the spectrophotometer and measure the %T of each culture (page 93, Ex. 23). Record your results in the tables on the Laboratory Report.
2. Plot the O.D. values in the graph on the Laboratory Report and answer all the questions.

Growth of bacteria can be profoundly affected by the amount of water entering or leaving the cell. When the medium surrounding an organism is **hypotonic** (low solute content), a resultant higher osmotic pressure occurs in the cell. Except for some marine forms, this situation is not harmful to most bacteria. The cell wall structure of most bacteria is so strong and rigid that even slight cellular swelling is generally inapparent.

In the reverse situation, however, when bacteria are placed in a **hypertonic** solution (high solute content), their growth may be considerably inhibited. The degree of inhibition will depend on the type of solute and the nature of the organism. In media of growth-inhibiting osmotic pressure, the cytoplasm becomes dehydrated and shrinks away from the cell wall. Such **plasmolyzed** cells are often simply inhibited in the absence of sufficient cellular water and return to normal when placed in an **isotonic** solution. In other instances, the organisms are irreversibly affected due to permanent inactivation of enzyme systems.

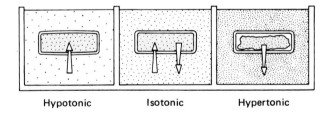

| Hypotonic | Isotonic | Hypertonic |

Figure 37.1 Osmotic variabilities

Organisms that thrive in hypertonic solutions are designated as halophiles or osmophiles. If they require minimum concentrations of salt (NaCl and other cations and anions) they are called **halophiles.** Obligate halophiles require a minimum of 13% sodium chloride. **Osmophiles,** on the other hand, require high concentrations of an organic solute, such as sugar.

In this exercise we will test the degree of inhibition of organisms that results with media containing different concentrations of sodium chloride. To accomplish this, you will streak three different organisms on four plates of media. The specific organisms used differ in their tolerance of salt

concentrations. The salt concentrations will be 0.5, 5, 10, and 15%. After incubation for 48 hours and several more days, comparisons will be made of growth differences to determine their degrees of salt tolerances.

Materials:
per student:
    1 Petri plate of nutrient agar (0.5% NaCl)
    1 Petri plate of nutrient agar (5% NaCl)
    1 Petri plate of nutrient agar (10% NaCl)
    1 Petri plate of milk salt agar (15% NaCl)
cultures:
    Escherichia coli (nutrient broth)
    Staphylococcus aureus (nutrient broth)
    Halobacterium salinarium (slant culture)

1. Mark the bottoms of the four Petri plates as indicated in figure 37.2.
2. Streak each organism in a straight line on the agar, using a wire loop.
3. Incubate all the plates for 48 hours at room temperature with exposure to light (the pigmentation of *H. salinarium* requires light to develop). Record your results on the Laboratory Report.
4. Continue the incubation of the milk salt agar plate for several more days in the same manner, and record your results again on the first portion of Laboratory Report 37, 38.

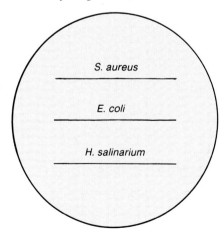

Streak pattern.

Figure 37.2 Streak pattern

The ability of small amounts of heavy metals to exert a lethal effect on bacteria is designated as **oligodynamic action** (Greek: *oligos,* small; *dynamis,* power). The effectiveness of these small amounts of metal is probably due to the high affinity of cellular proteins for the metallic ions. Although the concentration of ions in solution may be miniscule (a few parts per million), cells die due to the cumulative effects of ions within the cell.

The success of silver amalgam fillings to prevent secondary dental decay in teeth over long periods of time is due to the small amounts of silver and mercury ions that diffuse into adjacent tooth dentin. Its success in this respect has led to much debated concern that its toxicity may cause long-term injury to patients. In addition to its value (or harm) as a dental restoration material, oligodynamic action of certain other heavy metals has been applied to water purification, ointment manufacture, and the treatment of bandages and fabrics.

In this exercise we will compare the oligodynamic action of three metals (copper, silver, and aluminum) to note the differences.

Materials:
    1 Petri plate
    1 nutrient agar pour
    forceps and Bunsen burner
    acid-alcohol
    broth culture of *E. coli* and *S. aureus*
    3 metallic disks (copper, silver, aluminum)
    water bath at student station (beaker of water
        and electric hot plate)

1. Liquefy a tube of nutrient agar, cool to 50° C, and inoculate with either *E. coli* or *S. aureus* (odd: *E. coli;* even: *S. aureus*).
2. Pour half of the medium from each tube into a sterile Petri plate and leave the other half in a water bath (50° C). Allow agar to solidify in the plate.
3. Clean three metallic disks, one at a time, and place them on the agar, evenly spaced, as soon as they are cleaned. Use this routine:
    • Wash first with soap and water; then rinse with water.
    • With flamed forceps dip in acid-alcohol and rinse with distilled water.
4. Pour the remaining seeded agar from the tube over the metal disks. Incubate for 48 hours at 37° C.

Laboratory Report

After incubation compare the zones of inhibition and record your results on the last portion of Laboratory Report 37, 38.

Ultraviolet Light: Lethal Effects

39

Except for the photosynthetic bacteria, most bacteria are harmed by ultraviolet radiation. Those that contain photosynthetic pigments require exposure to sunlight in order to synthesize substances needed in their metabolism. Although sunlight contains the complete spectrum of short to long wavelengths of light, it is only the short, invisible ultraviolet wavelengths that are injurious to the nonphotosynthetic bacteria.

Wavelengths of light may be expressed in nanometers (nm) or angstrom units (Å). The angstrom unit is equal to 10^{-8} cm. In terms of nanometers, 10Å equal one nanometer. Thus, a wavelength of 4500×10^{-8} cm would be expressed as 4500Å, 450 nm, or 0.45 μm.

Figure 39.1 illustrates the relationship of ultraviolet to other types of radiations. By definition, ultraviolet light includes electromagnetic radiations that fall in the wavelength band between 40 and 4000Å. It bridges the gap between the X rays and the shortest wavelengths of light visible to the human eye. The visible range is approximately between 4000 and 7800Å. Actually, the practical range of ultraviolet, as far as we are concerned, lies between 2000 and 4000Å. The "extreme" range (40–2000Å) includes radiations that are absorbed by air and consequently function only in a vacuum. This region is also referred to as *vacuum ultraviolet.*

Ultraviolet is not a single entity, but is a very wide band of wavelengths. This fact is often not realized. Extending from 40 to 4000Å, it encom-

passes a span of 1:100; visible wavelengths (4000–7800Å), on the other hand, represent only a twofold spread.

The germicidal effects of the ultraviolet are limited to only a specific region of the ultraviolet spectrum. As indicated in figure 39.1, the most effective wavelength is 2650Å. Low-pressure mercury vapor lamps, which have a high output (90%) of 2437Å, make very effective bactericidal lamps.

In this exercise organisms that have been spread on nutrient agar will be exposed to ultraviolet radiation for various lengths of time to determine the minimum amount of exposure required to effect a 100% kill. One-half of each plate will be shielded from the radiation to provide a control comparison. *Bacillus megaterium,* a spore-former, and *Staphylococcus aureus,* a non-spore-former, will be used to provide a comparison of the relative resistance of vegetative and spore types.

Exposure to ultraviolet light may be accomplished with a lamp as shown in figure 39.2 or with a UV box that has built-in ultraviolet lamps. The UV exposure effectiveness varies with the type of setup used. The exposure times given in table 39.1 work well for a specific type of mercury arc lamp. Note in the table that space is provided under the times for adding in different timing. Your instructor will inform you as to whether you should write in new times that will be more suited to the equipment in your lab. Proceed as follows to do this experiment.

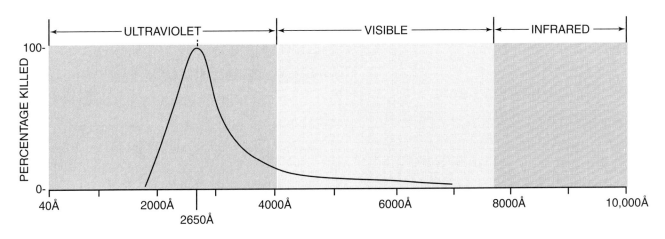

Figure 39.1 Lethal effectiveness of ultraviolet light

Materials:

Petri plates of nutrient agar (one or more per student)
ultraviolet lamp or UV exposure box
timers (bell type)
cards (3″ × 5″)
nutrient broth cultures of *S. aureus* with swabs
saline suspensions of *B. megaterium* with swabs

1. Refer to table 39.1 to determine which organism you will work with. You may be assigned more than one plate to inoculate. If different times are to be used, your instructor will inform you what times to write in. Since there are only 16 assignment numbers in the table, more student assignment numbers can be written in as designated by your instructor.
2. Label the bottoms of the plates with your assignment number and your initials.
3. Using a cotton-tipped swab that is in the culture tube, swab the entire surface of the agar in each plate. Before swabbing, express the excess culture from the swab against the inner wall of the tube.
4. Place the plates under the ultraviolet lamp *with the lids removed.* Cover half of each plate with a 3″ × 5″ card as shown in figure 39.2. Note that if your number is 8 or 16, you will not remove the lid from your plate. The purpose of this exposure is to see to what extent, if any, UV light can penetrate plastic.
5. After exposing the plates for the correct time durations, re-cover them with their lids, and incubate them inverted at 37° C for 48 hours.

Caution: Avoid looking directly into the ultraviolet lamp. These rays can cause cataracts and other eye injury.

Laboratory Report

Record your observations on the Laboratory Report and answer all the questions.

Table 39.1 Student Inoculation Assignments

	Exposure Times (Student Assignments)							
S. aureus	1	2	3	4	5	6	7	8
	10 sec	20 sec	40 sec	80 sec	2.5 min	5 min	10 min	20 min*
B. megaterium	9	10	11	12	13	14	15	16
	1 min	2 min	4 min	8 min	15 min	30 min	60 min	60 min*

*These Petri plates will be covered with dish covers during exposure.

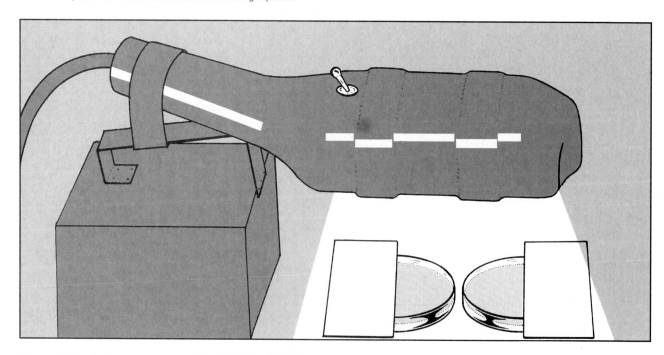

Figure 39.2 Plates are exposed to UV light with 50% coverage

Evaluation of Disinfectants: The Use-Dilution Method

40

When considering the relative effectiveness of different chemical agents against bacteria, some yardstick of comparison is necessary. Many different methods have been developed over the years since Robert Koch, in 1881, worked out the first scientific procedure by measuring the killing power of various germicides on silk threads that were impregnated with spores of *Bacillus anthracis.* Koch's method and many that followed proved unreliable for various reasons.

Finally, in 1931, the United States Food and Drug Administration adopted a method that was a modification of a test developed in England in 1903 by Rideal and Walker. In 1950 the Association of Official Agricultural Chemists adopted it as the official method of testing disinfectants. This method compares the effectiveness of various agents with phenol. A value called the **phenol coefficient** is arrived at that has significant meaning with certain limitations. The restrictions are that the test should be used only for phenol-like compounds that do not exert bacteriostatic effects and are not neutralized by the subculture media used. Many excellent disinfectants cannot be evaluated with this test. Disinfectants such as bichloride of mercury, iodine, metaphen, and quaternary detergents are unlike phenol in their germicidal properties and should not be evaluated in terms of phenol coefficients. Notwithstanding, however, many pharmaceutical companies have applied this test to such disinfectants with misleading results. A more suitable test for these nonphenolic disinfectants is the use-dilution method.

The **use-dilution method** makes use of small glass rods on which test organisms are dried for 30 minutes. The seeded rods are then exposed to the test solutions at 20° C for 1, 5, 10, and 30 minutes, rinsed with water or neutralizing solution, and transferred to the tubes of media. After incubation at 37° C for 48 hours, the tubes are examined for growth. When the results of this test are applied to practical conditions of use, they are found to be completely reliable.

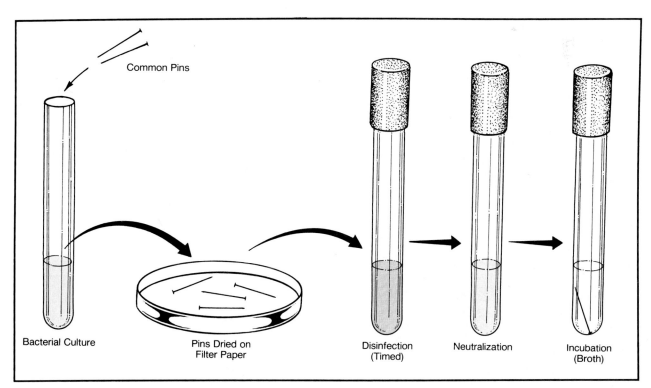

Common Pins

Bacterial Culture

Pins Dried on Filter Paper

Disinfection (Timed)

Neutralization

Incubation (Broth)

Figure 40.1 **Procedure for use-dilution evaluation of a disinfectant**

In this experiment you will follow a modified procedure of the use-dilution method to compare the relative merits of three different disinfectants on two kinds of bacteria: a spore-former, *Bacillus megaterium,* and a non-spore-former, *Staphylococcus aureus.* Instead of glass rods, we will use rustproof common pins.

Since each student will be performing only a small portion of the entire experiment it will be necessary to make student assignments with respect to agents used and timing. Table 40.1 indicates, according to student numbers, which agent each student will be working with, and the length of time to apply the agent to the pin. Note that a blank column is provided for write-in substitutions. Proceed as follows:

Materials:

per student:
> 2 tubes of one of the following agents:
>> 1:750 Zephiran for students 1, 4, 7, 10, 13, 16, 19, 22, 25, 28
>>
>> 5% phenol for students 2, 5, 8, 11, 14, 17, 20, 23, 26, 29
>>
>> 8% formaldehyde for students 3, 6, 9, 12, 15, 18, 21, 24, 27, 30
>
> 2 tubes of sterile water (about 7 ml each)
> 2 tubes of nutrient broth (about 7 ml each)
> forceps

on demonstration table:
> 1 nutrient broth culture of *S. aureus*
> 1 physiological saline suspension of a 48 hour nutrient agar slant culture of *B. megaterium*
> 2 sterile Petri plates with filter paper in bottom several forceps and Bunsen burner
> 2 test tubes containing 36 sterile common pins in each one (pins must be plated brass, which are rustproof)

1. Consult table 40.1 or the materials list to determine which disinfectant you are to use.
2. Get two tubes of the disinfectant, two tubes of sterile water, and two tubes of nutrient broth from the table. Label one of each pair *B. megaterium* and the other *S. aureus.*

Instructor: While the students are getting their supplies together, you can start the experiment by pouring the broth culture of *S. aureus* into one of the tubes of pins and the saline suspension of *B. megaterium* into the other tube of pins. After decanting the organisms into a beaker of disinfectant, the pins are deposited onto filter paper in separate Petri plates to dry. Plates should be clearly labeled as to contents. Allow a few minutes for the pins to dry before allowing students to take them. Make certain, also, that a Bunsen burner and forceps are set up near the two dishes of pins.

3. Gently flame a pair of forceps, let cool, and transfer one pin from each Petri plate to the separate tubes of disinfectant. Be sure to put them into the right tubes.
4. Leave the pins in the disinfectant for the length of time indicated in table 40.1. Find your number under the time indicated for your disinfectant.
5. At the end of the assigned time, flame the mouths of the tubes of disinfectant and *carefully pour the disinfectant into the sink without discarding the pins.* Then, transfer the pins into separate tubes of sterile water. Avoid transferring any of the disinfectant to the water tubes with the pins.
6. After 1 minute in the tubes of water, flame the mouths of the water and broth tubes, pour off the water, and shake the pins out of the emptied tubes into separate, labeled tubes of nutrient broth.

Instructor: At this point the instructor, or a designated class member, should put one pin from each of the Petri plates into separate labeled tubes of nutrient broth to be used as positive **controls** for each organism.

7. Incubate all nutrient broth tubes with pins for 48 hours at 37° C. Examine them and record your results on the Laboratory Report.

Table 40.1 Student Assignments for Agents and Timing

Disinfectant		Time in Minutes				
	Substitution	1	5	10	30	60
1:750 Zephiran		1, 16	4, 19	7, 22	10, 25	13, 28
5% Phenol		2, 17	5, 20	8, 23	11, 26	14, 29
8% Formaldehyde		3, 18	6, 21	9, 24	12, 27	15, 30

Evaluation of Alcohol:
Its Effectiveness as a Skin Degerming Agent

41

As a skin disinfectant, 70% alcohol is undoubtedly the most widely used agent. The ubiquitous prepackaged alcohol swabs used by nurses and technicians are evidence that these items are indispensible. The question that often arises is: How really effective is alcohol in routine use? When the skin is swabbed prior to penetration, are all, or mostly all, of the surface bacteria killed? To determine alcohol effectiveness, as it might be used in routine skin disinfection, we are going to perform a very simple experiment here that utilizes four thumbprints and a plate of enriched agar. Class results will be pooled to arrive at a statistical analysis.

Figure 41.1 illustrates the various steps in this test. Note that the Petri plate is divided into four parts. On the left side of the plate an unwashed left thumb is first pressed down on the agar in the lower quadrant of the plate. Next the left thumb is pressed down on the upper left quadrant. With the left thumb we are trying to establish the percentage of bacteria that are removed by simple contact with the agar.

On the right side of the plate an unwashed right thumb is pressed down on the lower right quadrant of the plate. The next step is to either dip the right thumb into alcohol or to scrub it with an alcohol swab and dry it. Half of the class will use the dipping method and the other half will use alcohol swabs. Your instructor will indicate what your assignment will be. The last step is to press the dried right thumb on the upper right quadrant of the plate.

After inoculating the plate it is incubated at 37° C for 24–48 hours. Colony counts will establish the effectiveness of the alcohol.

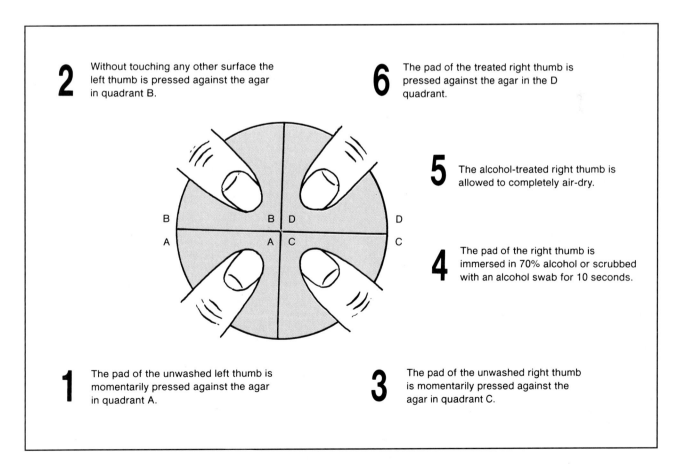

2 Without touching any other surface the left thumb is pressed against the agar in quadrant B.

6 The pad of the treated right thumb is pressed against the agar in the D quadrant.

5 The alcohol-treated right thumb is allowed to completely air-dry.

4 The pad of the right thumb is immersed in 70% alcohol or scrubbed with an alcohol swab for 10 seconds.

1 The pad of the unwashed left thumb is momentarily pressed against the agar in quadrant A.

3 The pad of the unwashed right thumb is momentarily pressed against the agar in quadrant C.

Figure 41.1   Procedure for testing the effectiveness of alcohol on the skin

Materials:
    1 Petri plate of veal infusion agar
    small beaker
    70% ethanol
    alcohol swab

1. Perform this experiment with unwashed hands.
2. With a china marking pencil, mark the bottom of the Petri plate with two perpendicular lines that divide it into four quadrants. Label the left quadrants **A** and **B** and the right quadrants **C** and **D** as shown in figure 41.1. (*Keep in mind that when you turn the plates over to label them, the A and B quadrants will be on the right and C and D will be on the left.*)
3. Press the pad of your left thumb against the agar surface in the A quadrant.

4. Without touching any other surface, press the left thumb into the B quadrant.
5. Press the pad of your right thumb against the agar surface of the C quadrant.
6. Disinfect the right thumb by one of the two following methods:
   - dip the thumb into a beaker of 70% ethanol for 5 seconds, or
   - scrub the entire pad surface of the right thumb with an alcohol swab.
7. Allow the alcohol to completely evaporate from the skin.
8. Press the right thumb against the agar in the D quadrant.
9. Incubate the plate at 37° C for 24–48 hours.
10. Follow the instructions on the Laboratory Report for evaluating the plate and answer all of the questions.

42

Evaluation of Antiseptics:
The Filter Paper Disk Method

The term *antiseptic* has, unfortunately, been somewhat ill-defined. Originally, the term was applied to any agent that prevents sepsis, or putrefaction. Since sepsis is caused by growing microorganisms, it follows that an antiseptic inhibits microbial multiplication without necessarily killing them. By this definition, we can assume that antiseptics are essentially bacteriostatic agents. Part of the confusion that has resulted in its definition is that the United States Food and Drug Administration rates antiseptics essentially the same as disinfectants. Only when an agent is to be used in contact with the body for a long period of time do they rate its bacteriostatic properties instead of its bactericidal properties.

If we are to compare antiseptics on the basis of their bacteriostatic properties, the filter paper disk method (figure 42.1) is a simple, satisfactory method to use. In this method a disk of filter paper (½″

diameter) is impregnated with the chemical agent and placed on a seeded nutrient agar plate. The plate is incubated for 48 hours. If the substance is inhibitory, a clear zone of inhibition will surround the disk. The size of this zone is an expression of the agent's effectiveness and can be compared quantitatively against other substances.

In this exercise we will measure the relative effectiveness of three agents (phenol, formaldehyde, and iodine) against two organisms: *Staphylococcus aureus* (gram-positive) and *Pseudomonas aeruginosa* (gram-negative). Table 42.1 will be used to assign each student one chemical agent to be tested against one organism. Note that space has been provided in the table for different agents to be written in as substitutes for the three agents listed. Your instructor may wish to make substitutions. Proceed as follows:

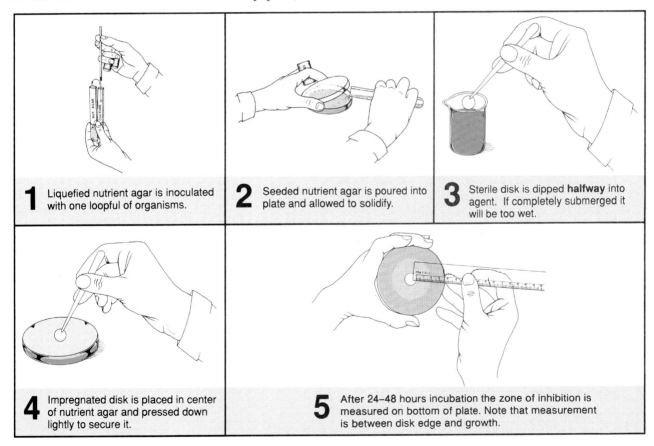

1 Liquefied nutrient agar is inoculated with one loopful of organisms.

2 Seeded nutrient agar is poured into plate and allowed to solidify.

3 Sterile disk is dipped **halfway** into agent. If completely submerged it will be too wet.

4 Impregnated disk is placed in center of nutrient agar and pressed down lightly to secure it.

5 After 24–48 hours incubation the zone of inhibition is measured on bottom of plate. Note that measurement is between disk edge and growth.

Figure 42.1 Filter paper disk method of evaluating an antiseptic

First Period
(Disk Application)

Materials:
per student:
    1 nutrient agar pour and 1 Petri plate
    broth culture of *S. aureus* or *P. aeruginosa*
on demonstration table:
    Petri dish containing sterile disks of filter paper
        (½" dia)
    forceps and Bunsen burner
    chemical agents in small beakers (5% phenol,
        5% formaldehyde, 5% aqueous iodine)

1. Consult table 42.1 to determine your assignment.
2. Liquefy a nutrient agar pour in a water bath and cool to 50° C.
3. Label the bottom of a Petri plate with the names of the organism and chemical agent.
4. Inoculate the agar pour with one loopful of the organism and pour into the plate.

5. After the medium has solidified in the plate, pick up a sterile disk with *lightly flamed* forceps, dip the disk *halfway* into a beaker of the chemical agent, and place the disk in the center of the medium.
    To secure the disk to the medium, press lightly on it with the forceps.
6. Incubate the plate at 37° C for 48 hours.

Second Period
(Evaluation)

1. Measure the zone of inhibition from the edge of the disk to the edge of the growth (see illustration 5, figure 42.1).
2. Exchange plates with other members of the class so that you will have an opportunity to complete the table on the Laboratory Report.

Table 42.1 Student Assignments

Chemical Agent		Student Number	
	Substitution	*S. aureus*	*P. aeruginosa*
5% Phenol		1, 7, 13, 19, 25	2, 8, 14, 20, 26
5% Formaldehyde		3, 9, 15, 21, 27	4, 10, 16, 22, 28
5% Iodine		5, 11, 17, 23, 29	6, 12, 18, 24, 30

Antimicrobic Sensitivity Testing: The Kirby-Bauer Method

43

Once the causative organism of a specific disease in a patient has been isolated, it is up to the attending physician to administer a chemotherapeutic agent that will inhibit or kill the pathogen without causing serious harm to the individual. The method must be relatively simple to use, be very reliable, and yield results in as short a time as possible. The Kirby-Bauer method of sensitivity testing is such a method. It is used for testing both antibiotics and drugs. **Antibiotics** are chemotherapeutic agents of low molecular weight produced by microorganisms that inhibit or kill other microorganisms. **Drugs,** on the other hand, are antimicrobic agents that are man-made. Both types of agents will be tested in this laboratory session according to the procedure shown in figure 43.1.

The effectiveness of an antimicrobic in sensitivity testing is based on the size of the zone of inhibition that surrounds a disk that has been impregnated with a specific concentration of the agent. The zone of inhibition, however, varies with the diffusibility of the agent, the size of the inoculum, the type of medium, and many other factors. Only by taking all these variables into consideration can a reliable method be worked out.

The **Kirby-Bauer method** is a standardized system that takes all variables into consideration. It is sanctioned by the U.S. FDA and the Subcommittee on Antimicrobial Susceptibility Testing of the National Committee for Clinical Laboratory Standards. Although time is insufficient here to consider all facets of this test, its basic procedure will be followed.

The recommended medium in this test is Mueller Hinton II agar. Its pH should be between 7.2 and 7.4, and it should be poured to a uniform thickness of 4 mm in the Petri plate. This requires 60 ml in a 150 mm plate and 25 ml in a 100 mm plate. For certain fastidious microorganisms, 5% defibrinated sheep blood is added to the medium.

Inoculation of the surface of the medium is made with a cotton swab from a broth culture. In clinical applications, the broth turbidity has to match a defined standard. Care must also be taken to express excess broth from the swab prior to inoculation.

High-potency disks are used that may be placed on the agar with a mechanical dispenser or sterile forceps. To secure the disks to the medium, it is necessary to press them down onto the agar.

After 16 to 18 hours incubation the plates are examined and the diameters of the zones are measured to the nearest millimeter. To determine the significance of the zone diameters, one must consult a table (Appendix A).

In this exercise we will work with four microorganisms: *Staphylococcus aureus, Escherichia coli, Proteus vulgaris,* and *Pseudomonas aeruginosa.* Each student will inoculate one plate with one of the four organisms and place the disks on the medium by whichever method is available. Since each student will be doing only a portion of the total experiment, student assignments will be made. Proceed as follows:

First Period
(Plate Preparation)

Materials:
> 1 Petri plate of Mueller-Hinton II agar
>   nutrient broth cultures (with swabs) of
>   *S. aureus, E. coli, P. vulgaris,* and *P.*
>   *aeurginosa*
> disk dispenser (BBL or Difco)
> cartridges of disks (BBL or Difco)
> forceps and Bunsen Burner
> zone interpretation charts (Difco or BBL)

1. Select the organisms you are going to work with from the following table.

Organism	Student Number
S. aureus	1, 5, 9, 13, 17, 21, 25
E. coli	2, 6, 10, 14, 18, 22, 26
P. vulgaris	3, 7, 11, 15, 19, 23, 27
P. aeruginosa	4, 8, 12, 16, 20, 24, 28

2. Label your plate with the name of your organism.
3. Inoculate the surface of the medium with the swab after expressing excess fluid from the swab by pressing and rotating the swab against the inside walls of the tube above the fluid level. Cover the surface of the agar evenly by swabbing in three directions. A final sweep should be made of the agar rim with the swab.
4. Allow **3 to 5 minutes** for the agar surface to dry before applying disks.

5. Dispense disks as follows:
   a. If an automatic dispenser is used, remove the lid, place the dispenser over the plate, and push down firmly on the plunger. With the sterile tip of forceps, tap each disk lightly to secure it to medium.
   b. If forceps are used, sterilize them first by flaming before picking up the disks. Keep each disk at least 15 mm from the edge of the plate. Place no more than 13 on a 150 mm plate, nor more than 5 on a 100 mm plate. Apply light pressure to each disk on the agar with the tip of a sterile forceps or inoculating loop to secure it to medium.
6. Invert and incubate the plate for 16 to 18 hours at 37° C.

Second Period
(Interpretation)

After incubation, measure the zone diameters with a metric ruler to the nearest whole millimeter. The zone of complete inhibition is determined without magnification. Ignore faint growth or tiny colonies that can be detected by very close scrutiny. Large colonies growing within the clear zone might represent resistant variants or a mixed inoculum and may require reidentification and retesting in clinical situations. Ignore the "swarming" characteristics of *Proteus,* measuring only to the margin of heavy growth.

Record the zone measurements on the table of the Laboratory Report and on the chart on the demonstration table, which has been provided by the instructor.

Use table 43.1 or 43.2 for identifying the various disks. Although BBL and Difco use essentially the same code numbers, there are slight differences in the two charts. Careful comparison of the charts will reveal that each company has certain antibiotics that are not listed by the other company.

To determine which antibiotics your organism is sensitive to (S), or resistant to (R), or intermediate (I), consult Table VII in Appendix A. It is important to note that the significance of a zone of inhibition varies with the type of organism. If you cannot find your antibiotic on the chart, consult a chart that is supplied by BBL or Difco that is on the demonstration table or bulletin board. Table VII is incomplete.

Table 43.1   Code for BBL Disks

AMD-10	Amdinocillin	E-15	Erythromycin
AN-30	Amikacin	GM-120	Gentamicin
AmC-30	Amoxicillin/	IPM-10	Imipenem
	Clavulanic Acid	K-30	Kanamycin
AM-10	Ampicillin	LOM-10	Lomefloxacin
SAM-20	Ampicillin/	LOR-30	Loracarbef
	Sulbactam	DP-5	Methicillin
AZM-15	Azithromycin	MZ-75	Meziocillin
AZ-75	Azlocillin	MI-30	Minocycline
ATM-30	Aztreonam	MOX-30	Moxalactam
B-10	Bacitracin	NF-1	Nafcillin
CB-100	Carbenicillin	NA-30	Nalidixic Acid
CEC-30	Cefactor	N-30	Neomycin
MA-30	Cefamandole	NET-30	Netilmicin
CZ-30	Cefazolin	NOR-10	Norfloxacin
CFM-5	Cefixime	NB-30	Novobiocin
CMZ-30	Cefmetrazole	OFX-5	Ofloxacin
CID-30	Cefonicid	OX-1	Oxacillin
CFP-75	Cefoperazone	OA-2	Oxolinic Acid
CTX-30	Cefotaxime	P-10	Penicillin
CTT-30	Cefotetan	PIP-100	Piperacillin
FOX-30	Cefoxitin	PB-300	Polymyxin B
CPD-10	Cefpodoxime	RA-5	Rifampin
CPR-30	Cefprozil	SPT-100	Spectinomycin
CAZ-30	Ceftazidime	S-300	Streptomycin
ZOX-30	Ceftizoxime	G-25	Sulfisoxazole
CRO-30	Ceftriaxone	Te-30	Tetracycline
CXM-30	Cefuroxime	TIC-75	Ticarcillin
CF-30	Cephalothin	TIM-85	Ticarcillin/
C-30	Chloramphenicol		Clavulanic Acid
CIN-100	Cinoxacin	NN-10	Tobramycin
CIP-5	Ciprofloxacin	TMP-5	Trimethoprim
CLR-15	Clarithromycin	SXT	Trimethoprim/
CC-2	Clindamycin		Sulfamethoxazole
CL-10	Colistin	Va-30	Vancomycin

Table 43.2   Code for Difco Disks

AN 30	Amikacin	E 15	Erythromycin
AMC 30	Amoxicillin/	FLX 5	Fleroxacin
	Clavulanic Acid	GM 10	Gentamycin
AM 10	Ampicillin	IPM 10	Imipenem
SAM 20	Ampicillin/	K 30	Kanamycin
	Sulbactam	LOM 10	Lomefloxacin
AZM 15	Azithromycin	LOR 30	Loracarbef
AZ 75	Azlocillin	MZ 75	Meziocillin
ATM 30	Aztreonam	Mi 30	Minocycline
CB 100	Carbenicillin	MOX 30	Moxalactam
CEC 30	Cefactor	NF 1	Nafcillin
MA 30	Cefamandole	NA 30	Nalidixic Acid
CZ 30	Cefazolin	NET 30	Netilmicin
FEP 30	Cefepime	FD 300	Nitrofurantoin
CAT 10	Cefetamet	NOR 10	Norfloxacin
CFM 5	Cefixime	OFX 5	Ofloxacin
CMZ 30	Cefmetrazole	P-10	Penicillin G
CID 30	Cefonicid	PTZ 110	Piperacillin/
CFP 75	Cefoperazone		Tazobactam
CTX 30	Cefotaxime	RA 5	Rifampin
CTT 30	Cefotetan	S 10	Streptomycin
FOX 30	Cefoxitin	G 300	Sulfisoxazole
CPD 10	Cefpodoxime	TEC 30	Telcoplanin
CPR 30	Cefprozil	TE 30	Tetracycline
CAZ 30	Ceftazidime	TIC 75	Ticarcillin
OX 30	Ceftizoxime	TIM 85	Ticarcillin/
CRO 30	Ceftriaxone		Clavulanic Acid
CXM 30	Cefuroxime	TN 10	Tobramycin
CF 30	Cephalothin	TMP 5	Trimethoprim
C 30	Chloramphenicol	SxT	Trimethorprim/
CIN 100	Cinoxacin		Sulfamethoxazole
CLR 15	Clarithromycin	VA 30	Vancomycin
CC 2	Clindamycin		
D 30	Doxycycline		
ENX 10	Enoxacin		

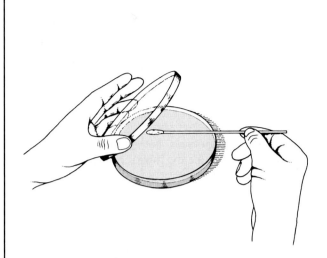

1 The entire surface of a plate of nutrient medium is swabbed with organism to be tested.

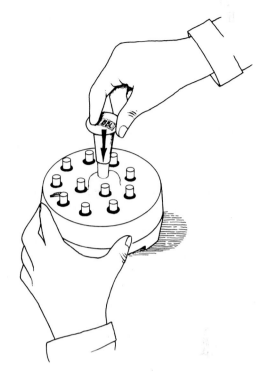

2 Handle of dispenser is pushed down to place 12 disks on the medium. In addition to dispensing disks, this dispenser also tamps disks onto medium.

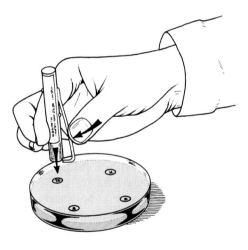

3 Cartridges (Difco) can be used to dispense individual disks. Only 4 or 5 disks should be placed on small (100 mm) plates.

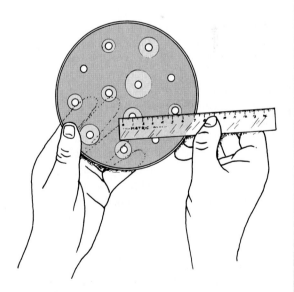

4 After 18 hours incubation, the zones of inhibition (diameters) are measured in millimeters. Significance of zones is determined from Kirby-Bauer chart.

Figure 43.1 Antimicrobic sensitivity testing

Effectiveness of Hand Scrubbing

The importance of hand disinfection in preventing the spread of disease is accredited to the observations of Semmelweis at the Lying-In Hospital in Vienna in 1846 and 1847. He noted that the number of cases of puerperal fever was closely related to the practice of sanitary methods. Until he took over his assignment in this hospital, it was customary for medical students to go directly from the autopsy room to a patient's bedside and assist in deliveries without scrubbing and disinfecting their hands. When the medical students were on vacation, only the nurses, who were not permitted in the autopsy room, attended the patients. Semmelweis noted that during this time, deaths due to puerperal fever fell off markedly.

As a result of his observations, he established a policy that no medical students would be allowed to examine obstetric patients or assist in deliveries until they had cleansed their hands with a solution of chloride of lime. This ruling caused the death rate from puerperal infections to drop from 12% to 1.27% in one year.

Today it is routine practice to wash hands prior to the examination of any patient and to do a complete surgical scrub prior to surgery. Scrubbing the hands involves the removal of **transient** (contaminant) and **resident** microorganisms. Depending on the condition of the skin and the numbers of bacteria present, it takes from 7 to 8 minutes of washing with soap and water to remove all transients, and they can be killed with relative ease using suitable antiseptics. Residents, on the other hand, are firmly entrenched and are removed slowly by washing. These organisms, which consist primarily of staphylococci of low pathogenicity, are less susceptible than the transients to the action of antiseptics.

In this exercise, an attempt will be made to evaluate the effectiveness of the length of time in removal of organisms from the hands using a surgical scrub technique. One member of the class will be selected to perform the scrub. Another student will assist by supplying the soap, brushes, and basins, as needed. During the scrub, at 2-minute intervals, the hands will be scrubbed into a basin of sterile water. Bacterial counts will be made of these basins to determine the effectiveness of the previous 2-minute scrub in reducing the bacterial flora of the hands.

Members of the class not involved in the scrub procedure will make the inoculations from the basins for the plate counts.

Scrub Procedure

The two members of the class who are chosen to perform the surgical scrub will set up their materials near a sink for convenience. As one student performs the scrub, the other will assist in reading the instructions and providing materials as needed. The basic steps, which are illustrated in figure 44.1, are also described in detail below. Before beginning the scrub, both students should read all the steps carefully.

Materials:
>   5 sterile surgical scrub brushes, individually
>   wrapped
>   5 basins (or 2000 ml beakers), containing 1000
>   ml each of sterile water. These basins
>   should be covered to prevent contamination
>   1 dispenser of green soap
>   1 tube of hand lotion

Step 1 To get some idea of the number of transient organisms on the hands, the scrubber will scrub all surfaces of each hand with a sterile surgical scrub brush for 30 seconds into Basin A. No green soap will be used for this step. The successful performance of this step will depend on

- spending the same amount of time on each hand (30 seconds),
- maintaining the same amount of activity on each hand, and
- scrubbing under the fingernails, as well as working over their surfaces.

After completion of this 60-second scrub, notify Group A that their basin is ready for the inoculations.

Step 2 Using the *same* brush as above, begin scrubbing with green soap for 2 minutes, using cool tap water to moisten and rinse the hands. One minute is devoted to each hand.

The assistant will make one application of green soap to each hand as it is being scrubbed.

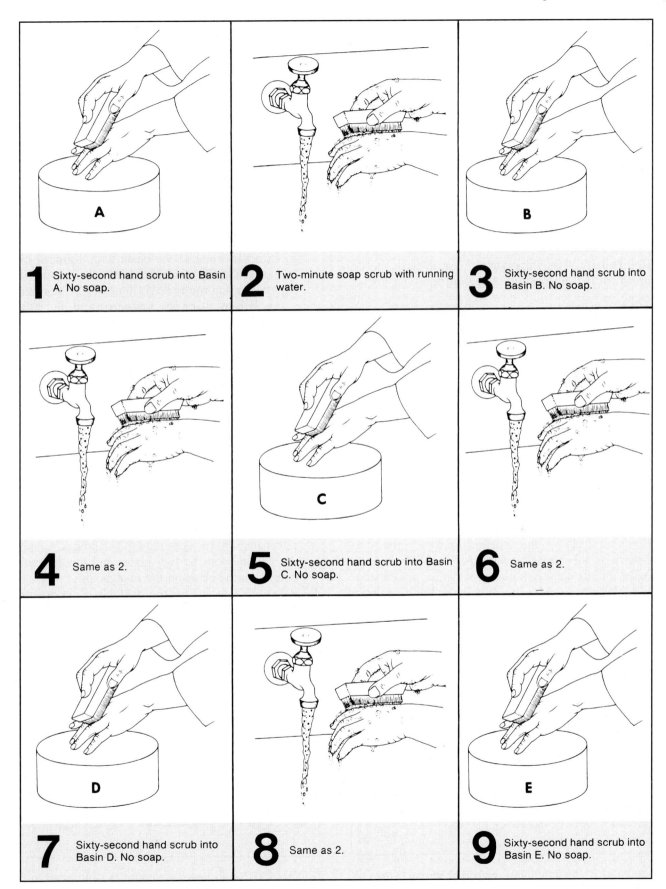

1 Sixty-second hand scrub into Basin A. No soap.

2 Two-minute soap scrub with running water.

3 Sixty-second hand scrub into Basin B. No soap.

4 Same as 2.

5 Sixty-second hand scrub into Basin C. No soap.

6 Same as 2.

7 Sixty-second hand scrub into Basin D. No soap.

8 Same as 2.

9 Sixty-second hand scrub into Basin E. No soap.

Figure 44.1 Hand scrubbing routine

Rinse both hands for 5 seconds under tap water at the completion of the scrub.

Discard the brush.

Note: This same procedure will be followed exactly in steps 4, 6, and 8 of figure 44.1.

Step 3   With a *fresh* sterile brush, scrub the hands into Basin B in a manner that is identical to step 1. Don't use soap. Notify Group B when this basin is ready.

Note: Exactly the same procedure is used in steps 5, 7, and 9 of figure 44.1, using Basins C, D, and E.

Remember: It is important to use a fresh sterile brush for the preparation of each of these basins.

After Scrubbing   After all scrubbing has been completed, the scrubber should dry his or her hands and apply hand lotion.

Making the Pour Plates

While the scrub is being performed, the rest of the class will be divided into five groups (A, B, C, D, and E) by the instructor. Each group will make six plate inoculations from one of the five basins (A, B, C, D, or E). It is the function of these groups to determine the bacterial count per milliliter in each basin. In this way we hope to determine, in a relative way, the effectiveness of scrubbing in bringing down the total bacterial count of the skin.

Materials:

30 veal infusion agar pours—6 per group

1 ml pipettes

30 sterile Petri plates—6 per group

70% alcohol

L-shaped glass stirring rod (optional)

1. Liquefy six pours of veal infusion agar and cool to 50° C. While the medium is being liquefied, label two plates each: 0.1 ml, 0.2 ml, and 0.4 ml. Also, indicate your group designation on the plate.
2. As soon as the scrubber has prepared your basin, take it to your table and make your inoculations as follows:
   a. Stir the water in the basin with a pipette or an L-shaped stirring rod for 15 seconds. If the stirring rod is used (figure 44.2), sterilize it before using by immersing it in 70% alcohol and flaming. *For consistency of results all groups should use the same method of stirring.*
   b. Deliver the proper amounts of water from the basin to the six Petri plates with a sterile serological pipette. Refer to figure 44.3. If a pipette was used for stirring, it may be used for the deliveries.
   c. Pour a tube of veal infusion agar, cooled to 50° C, into each plate, rotate to get good distribution of organisms, and allow to cool.
   d. Incubate the plates at 37° C for 24 hours.
3. After the plates have been incubated, select the pair that has the best colony distribution with no fewer than 30 or more than 300 colonies. Count the colonies on the two plates and record your counts on the chart on the chalkboard.
4. After all data are on the chalkboard, complete the table and graph on the Laboratory Report.

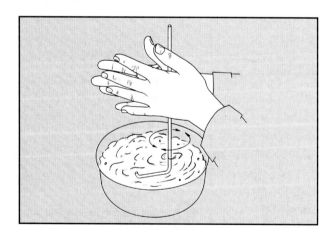

Figure 44.2   An alternative method of stirring utilizes an L-shaped glass stirring rod.

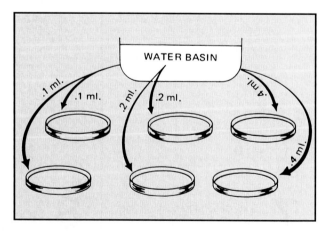

Figure 44.3   Scrub water for count is distributed to 6 Petri plates in amounts as shown.

PART 8 Identification of Unknown Bacteria

One of the most interesting experiences in introductory microbiology is to attempt to identify an unknown microorganism that has been assigned to you as a laboratory problem. The next seven exercises pertain to this phase of microbiological work. You will be given one or more cultures of bacteria to identify. The only information that might be given to you about your unknowns will pertain to their sources and habitats. All the information needed for identification will have to be acquired by you through independent study.

The first step in the identification procedure is to accumulate information that pertains to the organism's morphological, cultural, and physiological (biochemical) characteristics. This involves making different kinds of slides for cellular studies and the inoculation of various types of media to note the growth characteristics and types of enzymes produced. As this information is accumulated, it is recorded in an orderly manner on **Descriptive Charts,** which are located toward the back of the manual with the Laboratory Reports.

After sufficient information has been recorded, the next step is to consult a taxonomic key, which enables one to identify the organism. For this final step *Bergey's Manual of Systematic Bacteriology* will be used. Copies of volumes 1 and 2 of this book will be available in the laboratory, library, or both. In addition, a CD-ROM computer simulation program called *Identibacter interactus* may be available, which can be used for identifying and reporting your unknown. Exercise 51 pertains to the use of *Bergey's Manual* and *Identibacter interactus.*

Success in this endeavor will require meticulous techniques, intelligent interpretation, and careful recordkeeping. Your mastery of aseptic methods in the handling of cultures and the performance of inoculations will show up clearly in your results. Contamination of your cultures with unwanted organisms will yield false results, making identification hazardous speculation. If you have reason to doubt the validity of the results of a specific test, repeat it; *don't rely on chance!* As soon as you have made an observation or completed a test, record the information on the Descriptive Chart. Do not trust your memory—record data immediately!

45

Preparation and Care of Stock Cultures

Your unknown cultures will be used for making many different kinds of slides and inoculations. Despite meticulous aseptic practice on your part, the chance of contamination of these cultures increases with frequency of use. If you were to attempt to make all your inoculations from the single tube given to you, it is very likely that somewhere along the way contamination would result.

Another problem that will arise is aging of the culture. Two or three weeks may be necessary for the performance of all tests. In this period of time, the organisms in the broth culture may die, particularly if the culture is kept very long at room temperature. To ensure against the hazards of contamination or death of your organisms, it is essential that you prepare stock cultures before any slides or routine inoculations are made.

Different types of organisms require different kinds of stock media, but for those used in this unit, nutrient agar slants will suffice. For each unknown, you will inoculate two slants. One of these will be your reserve stock and the other one will be your working stock.

The **reserve stock culture** will *not* be used for making slides or routine inoculations; instead, it will be stored in the refrigerator after incubation until some time later when a transfer may be made from it to another reserve stock or working stock culture.

The **working stock culture** will be used for making slides and routine inoculations. When it becomes too old to use or has been damaged in some way, replace it with a fresh culture that is made from the reserve stock.

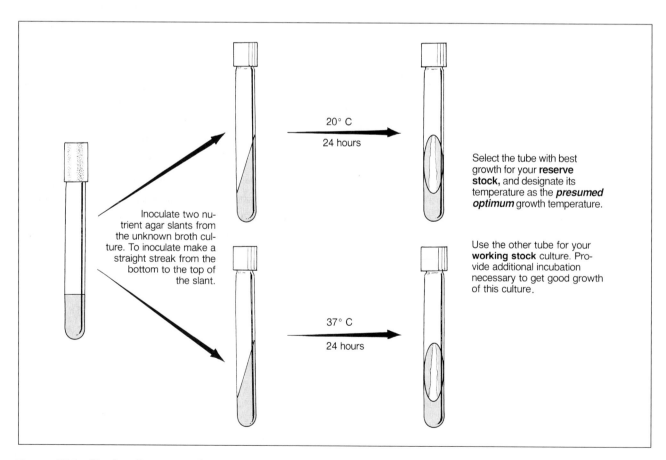

Figure 45.1 Stock culture procedure

Note in figure 45.1 that one slant will be incubated at 20° C and the other at 37° C. This will enable you to learn something about the optimum growth temperature of your unknown, which will be pertinent in Exercise 47. Proceed as follows:

First Period

Inoculate two nutrient agar slants from each of your unknowns as follows:

Materials:
for each unknown:
2 nutrient agar slants (screw-cap type)
gummed labels

1. Label two slants with the code number of the unknown and your initials. Use gummed labels. Also, mark one tube 20° C and the other 37° C.
2. With a loop, inoculate each slant with a straight streak *from the bottom to the top.* Since these slants will be used for your cultural study in Exercise 47, a straight streak is more useful than one that is spread over the entire surface.
3. Place the two slants in separate baskets on the demonstration table that are designated with labels for the two temperatures (20° C and 37° C).

    Although the 20° C temperature is thought of as "room temperature," it should be incubated in a biological incubator instead of leaving it out at laboratory room temperature. Laboratory temperatures are often quite variable in a 24-hour period.

Second Period

After 24 hours incubation, evaluate the slants made from each of your unknowns, as follows:

1. Examine the slants to note the extent of growth. Some organisms require close examination to see the growth, especially if the growth is thin and translucent.

2. Determine which temperature seems to promote the best growth.
3. Record on the Descriptive Chart the *presumed* optimum temperature. (Obviously, this may not be the actual optimum growth temperature, but for all practical purposes, it will suffice for this exercise.)
4. If there is no growth visible on either slant, there are several possible explanations:
    • It may be that the culture you were issued was not viable.
    • Another possibility might be that the organism grows too slowly to be visible at this time.
    • Or, possibly, neither temperature was suitable!

    Think through these possibilities and decide what you should do to circumvent the problem.
5. Label the tube with the best growth **reserve stock.** Label the other tube **working stock.**
6. If both tubes have good growth, place them in the refrigerator until needed.

    If one tube has very scanty growth, refrigerate the good one (reserve stock) and incubate the other one at the more desirable temperature for another 24 hours, then refrigerate.
7. Remember these points concerning your stock cultures:
    • Most stock cultures will keep for 4 weeks in the refrigerator. Some fastidious pathogens will survive for only a few days. Although none of the organisms issued in this unit are of the extremely delicate type, don't wait 4 weeks to make a new reserve stock culture; instead, make fresh transfers every 10 days.
    • Don't use your reserve stock culture for making slides or routine inoculations.
    • Don't store either of your stock cultures in your desk drawer or a cupboard. After the initial incubation period cultures must be refrigerated. After 2 or 3 days at room temperature, cultures begin to deteriorate. Some die out completely.
8. Answer the questions on the Laboratory Report.

The first step in the identification of an unknown bacterial organism is to learn as much as possible about its morphological characteristics. One needs to know whether the organism is rod-, coccus-, or spiral-shaped; whether or not it is pleomorphic; its reaction to gram staining; and the presence or absence of endospores, capsules, or granules. All this morphological information provides a starting point in the categorization of an unknown.

Figure 46.1 illustrates the steps that will be followed in determining morphological characteristics of your unknown. Note that fresh broth and slant cultures will be needed to make the various slides and perform motility tests. Since most of the slide techniques were covered in Part 3, you will find it necessary to refer to that section from time to time. Note that gram staining, motility testing, and measurements will be made from the broth culture; gram staining and other stained slides will also be made from the agar slant. The rationale as to the choice of broth or agar slants will be explained as each technique is performed.

As soon as morphological information is acquired be sure to record your observations on the Descriptive Chart at the back of the manual. Proceed as follows:

Materials:
> gram-staining kit
> spore-staining kit
> acid-fast staining kit
> Loeffler's methylene blue stain
> nigrosine or india ink
> tubes of nutrient broth and nutrient agar
> gummed labels for test tubes

New Inoculations

For all of these staining techniques you will need 24–48 hour cultures of your unknown. If your working stock slant is a fresh culture, use it. If you don't have a fresh broth culture of your unknown inoculate a tube of nutrient broth and incubate it at its estimated optimum temperature for 24 hours.

Gram's Stain

Since a good gram-stained slide will provide you with more valuable information than any other slide, this is the place to start. Make gram-stained slides from both the broth and agar slants, and compare them under oil immersion.

Two questions must be answered at this time: (1) Is the organism gram-positive, or is it gram-negative? and (2) Is the organism rod- or coccus-shaped? If your staining technique is correct, you should have no problem with the Gram reaction. If the organism is a long rod, the morphology question is easily settled; however, if your organism is a very short rod, you may incorrectly decide it is coccus-shaped.

Keep in mind that short rods with round ends (coccobacilli) look like cocci. If you have what seems to be a coccobacillus, examine many cells before you make a final decision. Also, keep in mind that *while rod-shaped organisms frequently appear as cocci under certain growth conditions, cocci rarely appear as rods.* (*Streptococcus mutans* is unique in forming rods under certain conditions.) Thus, it is generally safe to assume that if you have a slide on which you see both coccuslike cells and short rods, the organism is probably rod-shaped. This assumption is valid, however, only if you are not working with a contaminated culture!

Record the shape of the organism and its reaction to the stain on the Descriptive Chart.

Cell Size

Once you have a good gram-stained slide, determine the size of the organism with an ocular micrometer. Refer to Exercise 5. If the size is variable, determine the size range. Record this information on the Descriptive Chart.

Motility and Cellular Arrangement

If your organism is a nonpathogen make a wet mount or hanging drop slide from the broth culture. Refer to Exercise 19. This will enable you to determine whether the organism is motile, and it will allow you to confirm the cellular arrangement. By making this

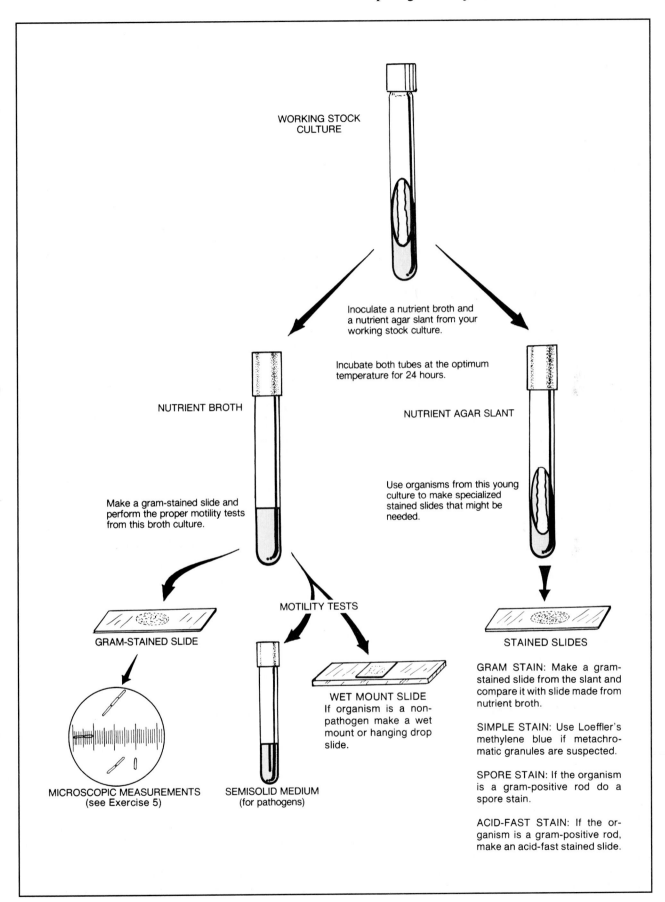

WORKING STOCK CULTURE

Inoculate a nutrient broth and a nutrient agar slant from your working stock culture.

Incubate both tubes at the optimum temperature for 24 hours.

NUTRIENT BROTH

NUTRIENT AGAR SLANT

Make a gram-stained slide and perform the proper motility tests from this broth culture.

Use organisms from this young culture to make specialized stained slides that might be needed.

GRAM-STAINED SLIDE

MOTILITY TESTS

STAINED SLIDES

WET MOUNT SLIDE
If organism is a non-pathogen make a wet mount or hanging drop slide.

GRAM STAIN: Make a gram-stained slide from the slant and compare it with slide made from nutrient broth.

SIMPLE STAIN: Use Loeffler's methylene blue if metachromatic granules are suspected.

MICROSCOPIC MEASUREMENTS
(see Exercise 5)

SEMISOLID MEDIUM
(for pathogens)

SPORE STAIN: If the organism is a gram-positive rod do a spore stain.

ACID-FAST STAIN: If the organism is a gram-positive rod, make an acid-fast stained slide.

Figure 46.1 **Procedure for morphological study**

slide from broth instead of the agar slant, the cells will be well dispersed in natural clumps. Note whether the cells occur singly, in pairs, masses, or chains. *Remember to place the slide preparation in a beaker of disinfectant when finished with it.*

If your organism happens to be a pathogen do not make a slide preparation of the organisms; instead, stab the organism into a tube of semisolid or SIM medium to determine motility (Exercise 19). Incubate for 48 hours.

Be sure to record your observations on the Descriptive Chart.

Endospores

If your unknown is a gram-positive rod, check for endospores. *Only rarely is a coccus or gram-negative rod a spore-former.* Examination of your gram-stained slide made from the agar slant should provide a clue, since endospores show up as transparent holes in gram-stained spore-formers. Endospores can also be seen on unstained organisms if studied with phase-contrast optics.

If there seems to be evidence that the organism is a spore-former, make a slide using one of the spore-staining techniques you used in Exercise 15. *Since some spore-formers require at least a week's time of incubation before forming spores, it is prudent to double-check for spores in older cultures.*

Record on the Descriptive Chart whether the spore is terminal, subterminal, or in the middle of the rod.

Acid-Fast Staining

If your organism is a gram-positive, non-spore-forming rod, you should determine whether or not it is acid-fast. Although some bacteria require 4 or 5 days growth to exhibit acid-fastness, most species become acid-fast within 2 days. For best results, therefore, do not use cultures that are too old.

Another point to keep in mind is that most acid-fast bacteria do not produce cells that are 100% acid-fast. An organism is considered acid-fast if only portions of the cells exhibit this characteristic. Refer to Exercise 16 for this staining technique.

A final bit of advice: If you feel insecure about your adeptness at Gram staining and think that you might *possibly* have a gram-positive organism, even though your organism seems to be gram-negative, make an acid-fast stained slide. Many students find (much to their chagrin later) that they didn't do acid-fast staining because their organism seemed to be gram-negative. An improperly gram-stained slide can be very misleading when it comes to unknown identification.

Other Structures

If the protoplast in gram-stained slides stains unevenly, you might wish to do a simple stain with Loeffler's methylene blue (Exercise 12) for evidence of metachromatic granules.

Although a capsule stain (Exercise 13) may be performed at this time, it might be better to wait until a later date when you have the organism growing on blood agar. Capsules usually are more apparent when the organisms are grown on this medium.

Laboratory Report

There is no Laboratory Report to fill out for this exercise. All information is recorded on the Descriptive Chart.

Cultural Characteristics

47

The cultural characteristics of an organism pertain to its macroscopic appearance on different kinds of media. Descriptive terms, which are familiar to all bacteriologists, and are used in *Bergey's Manual,* must be used in recording cultural characteristics. The most frequently used media for a cultural study are nutrient agar, nutrient broth, and nutrient gelatin. For certain types of unknowns it is also desirable to inoculate a blood agar plate; if necessary, this plate can be inoculated later. In addition to these media, you will be inoculating a fluid thioglycollate medium to determine the oxygen requirements of your unknown.

First Period

(Inoculations)

During this period one nutrient agar plate, one nutrient gelatin deep, two nutrient broths, and one tube of fluid thioglycollate medium will be inoculated. Inoculations will be made with the original broth culture of your unknown. The reason for inoculating two tubes of nutrient broth here is to recheck the optimum growth temperature of your unknown. In Exercise 45 you incubated your nutri-

ent agar slants at 20° C and 37° C. It may well be that the optimum growth temperature is closer to 30° C. It is to check out this intermediate temperature that an extra nutrient broth is being inoculated. Proceed as follows:

Materials:
for each unknown:
> 1 nutrient agar pour
> 1 nutrient gelatin deep
> 2 nutrient broths
> 1 fluid thioglycollate medium (FTM)
> 1 Petri plate

1. Pour a Petri plate of nutrient agar for each unknown and streak it with a method that will give good isolation of colonies. Use the original broth culture for streaking.
2. Inoculate the tubes of nutrient broth with a loop.
3. Make a stab inoculation into the gelatin deep by stabbing the inoculating needle (straight wire) directly down into the medium to the bottom of the tube and pulling it straight out. The medium must not be disturbed laterally.
4. Inoculate the tube of FTM with a loopful of your unknown. Mix the organisms throughout the tube by rolling the tube between your palms.

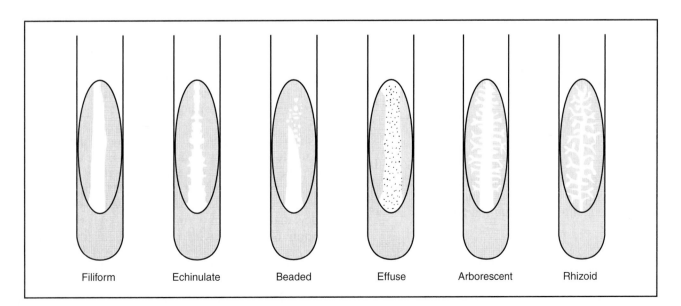

Figure 47.1 Types of bacterial growth on nutrient agar slants

151

5. Place all tubes except one nutrient broth into a basket and incubate for 24 hours at the temperature that seemed best in Exercise 45. Incubate the remaining tube of nutrient broth separately at 30° C. Incubate the agar plate, inverted, at the presumed best temperature.

Second Period
(Evaluation)

After the cultures have been properly incubated, *carry them to your desk in a careful manner* to avoid disturbing the growth pattern in the nutrient broths and FTM. Before studying any of the tubes or plates, place the tube of nutrient gelatin in an ice water bath. It will be studied later. Proceed as follows to study each type of medium and record the proper descriptive terminology on the Descriptive Chart.

Materials:
> reserve stock agar slant of unknown
> spectrophotometer and cuvettes
> hand lens
> ice water bath near sink

Nutrient Agar Slant (Reserve Stock)

Examine your reserve stock agar slant of your unknown that has been stored in the refrigerator since the last laboratory period. Evaluate it in terms of the following criteria:

Amount of Growth The abundance of growth may be described as *none, slight, moderate,* and *abundant.*

Color Pigmentation should be looked for on the organisms and within the medium. Most organisms will lack chromogenesis, exhibiting a white growth; others are various shades of different colors. Some bacteria produce soluble pigments that diffuse into the medium. Hold the slant up to a strong light to examine it for diffused pigmentation.

Opacity Organisms that grow prolifically on the surface of a medium will appear more opaque than those that exhibit a small amount of growth. Degrees of opacity may be expressed in terms of *opaque, transparent,* and *translucent* (partially transparent).

Form The gross appearance of different types of growth are illustrated in figure 47.1. The following descriptions of each type will help in differentiation:

> *Filiform:* characterized by uniform growth along the line of inoculation

> *Echinulate:* margins of growth exhibit toothed appearance
> *Beaded:* separate or semiconfluent colonies along the line of inoculation
> *Effuse:* growth is thin, veil-like, unusually spreading
> *Arborescent:* branched, treelike growth
> *Rhizoid:* rootlike appearance

Nutrient Broth

The nature of growth on the surface, subsurface, and bottom of the tube is significant in nutrient broth cultures. Describe your cultures as thoroughly as possible on the Descriptive Chart with respect to these characteristics:

Surface Figure 47.2 illustrates different types of surface growth. A *pellicle* type of surface differs from the *membranous* type in that the latter is much thinner. A *flocculent* surface is made up of floating adherent masses of bacteria.

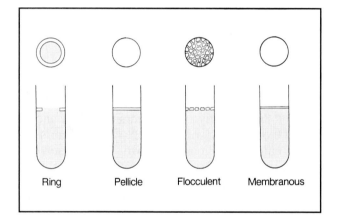

Ring   Pellicle   Flocculent   Membranous

Figure 47.2 **Types of surface growth in nutrient broth**

Subsurface Below the surface, the broth may be described as *turbid* if it is cloudy; *granular* if specific small particles can be seen; *flocculent* if small masses are floating around; and *flaky* if large particles are in suspension.

Sediment The amount of sediment in the bottom of the tube may vary from none to a great deal. To describe the type of sediment, agitate the tube, putting the material in suspension. The type of sediment can be described as *granular, flocculent, flaky,* and *viscid.* Test for viscosity by probing the bottom of the tube with a sterile inoculating loop.

Amount of Growth To determine the amount of growth, it is necessary to shake the tube to disperse the organisms. Terms such as *slight* (scanty), *moderate,* and *abundant* adequately describe the amount.

Optimum Temperature To determine which temperature produced the best growth, pour the contents from each tube of nutrient broth into separate cuvettes and measure their percent transmittances on the spectrophotometer. If the percent transmittance is less at 30° C than at the other presumed optimum temperature, revise the optimum temperature on your Descriptive Chart.

Fluid Thioglycollate Medium

Since the primary purpose of inoculating a tube of fluid thioglycollate medium is to determine oxygen requirements of your unknown, examine the tube to note the position of growth in the tube. Compare your tube with figure 22.5 on page 88 to make your analysis. Designate your organism as being *aerobic, microaerophilic, facultative,* or *anaerobic* on the Descriptive Chart.

Gelatin Stab Culture

Remove your tube of nutrient gelatin from the ice water bath and examine it. Check first to see if liquefaction has occurred. Organisms that are able to liquefy gelatin produce the enzyme *gelatinase.*

Liquefaction Tilt the tube from side to side to see if a portion of the medium is still liquid. If liquefaction has occurred, check the configuration with figure 47.3 to see if any of the illustrations match your tube. A description of each type follows:

> *Crateriform:* saucer-shaped liquefaction
> *Napiform:* turniplike
> *Infundibuliform:* funnel-like or inverted cone
> *Saccate:* elongate sac, tubular, cylindrical
> *Stratiform:* liquefied to the walls of the tube in the upper region

Note: The configuration of liquefaction is not as significant as the mere fact that liquefaction takes place. If your organism liquefies gelatin, but you are unable to determine the exact configuration, don't worry about it. However, be sure to record on the Descriptive Chart the *presence* or *absence* of gelatinase production.

Another important point: Some organisms produce gelatinase at a very slow rate. Tubes that are negative should be incubated for another 4 or 5 days to see if gelatinase is produced slowly.

Type of Growth (No Liquefaction) If no liquefaction has occurred, check the tube to see if the organism grows in nutrient gelatin (some do, some don't). If growth has occurred compare the growth with the left-hand illustration in figure 47.3. It should be pointed out, however, that, from a categorization standpoint, the nature of growth in gelatin is not very important.

Nutrient Agar Plate Culture

Colonies grown on plates of nutrient agar should be studied with respect to size, color, opacity, form, elevation, and margin. With a dissecting microscope or hand lens study individual colonies carefully. Refer to figure 47.4 for descriptive terminology. Record your observations on the Descriptive Chart.

Laboratory Report

There is no Laboratory Report for this exercise. Record all information on the Descriptive Chart.

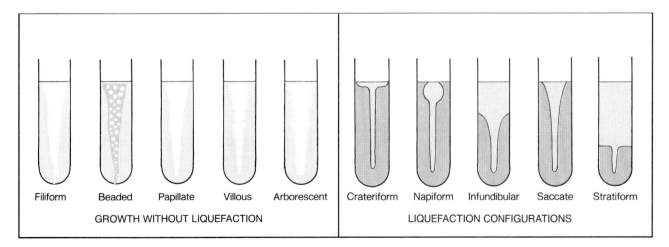

Filiform	Beaded	Papillate	Villous	Arborescent	Crateriform	Napiform	Infundibular	Saccate	Stratiform

GROWTH WITHOUT LIQUEFACTION | LIQUEFACTION CONFIGURATIONS

Figure 47.3 Growth in gelatin stabs

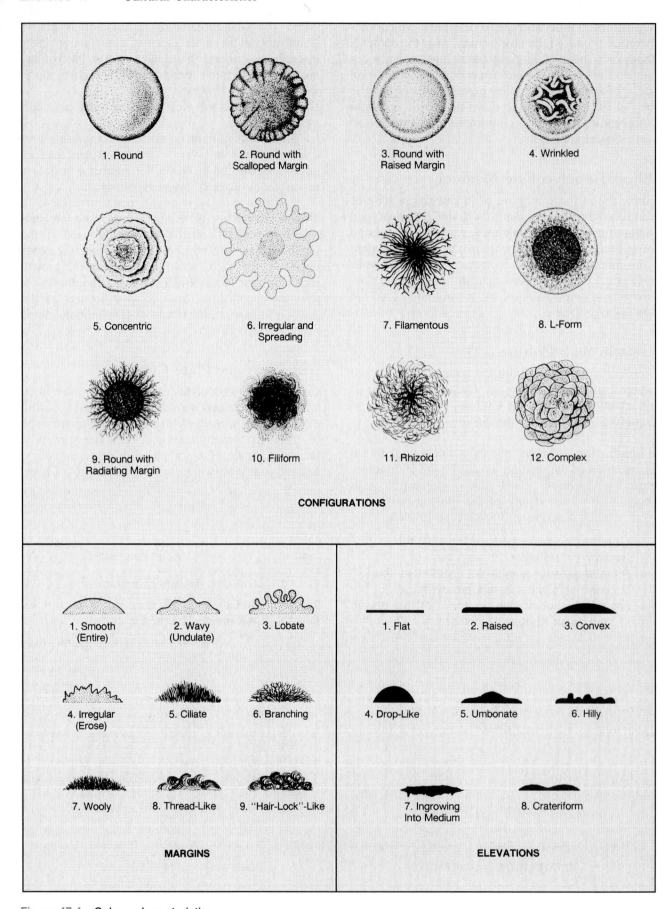

Figure 47.4 Colony characteristics

Physiological Characteristics:
Biooxidations

The assemblage of morphological and cultural characteristics on your Descriptive Chart during the past few laboratory periods may be leading you to believe that you already know the name of your unknown. Students at this stage often begin to draw premature conclusions. To provide you with a clearer perspective of where you are in the categorization process, refer to the separation outlines in figures 51.1 and 51.2, pages 172 and 173. Note that morphological and cultural characteristics can lead you to 11 separate groups of genera. It is very likely that one of these groups contains the genus that includes your unknown.

Although morphological and cultural characteristics are essential in getting to the genus, species determination requires a good deal more information. The physiological information that will be accumulated here and in the next two exercises will make species identification possible.

Since all physiological (biochemical) reactions in organisms are catalyzed by enzymes, and since each enzyme is produced by individual genes, we are, essentially, formulating a genetic profile of an organism as we discover what enzymes are produced. The physiological characteristics of concern in this exercise pertain to the chemistry of biooxidations. The enzymatic reactions that fall in this category pertain to respiration and fermentation.

Biooxidation reactions in bacteria pertain to the manner in which they get their energy: some are oxidative and others are fermentative. Strict aerobes that oxidize organic substances to produce the end products carbon dioxide and water are said to be *oxidative.* By utilizing organic compounds as electron donors, with oxygen as the ultimate electron (and hydrogen) acceptor, they produce CO_2 and water as end products to release energy. The ability to utilize free oxygen is accomplished by a cytochrome enzyme system. This process is also called *respiration.*

Fermentative bacteria are organisms that also utilize organic compounds for energy but lack a cytochrome system. Instead of producing only CO_2 and H_2O, they produce complex end products, such as acids, aldehydes, and alcohols, that are oxidizable and reducible. In these organisms oxygen is not the ultimate electron acceptor, and the reactions occur under anaerobic conditions. Various gases, such as carbon dioxide, hydrogen, and methane, are also produced. In fermentative bacteria the organic compounds act both as electron donors and electron acceptors.

Sugars, particularly glucose, are the compounds most widely used by fermenting organisms. Other substances such as organic acids, amino acids, purines, and pyrimidines also can be fermented by some bacteria. The end products of a particular fermentation are determined by the nature of the organism, the characteristics of the substrate, and environmental conditions such as temperature and pH.

Although fermentation and respiration represent two different types of energy-yielding biooxidations, they can both be present in the same organism, as is true of facultative anaerobes. It was pointed out in Exercise 22 that in the presence of molecular oxygen these organisms shift from fermentation to respiration. An exception, however, is seen in the lactic acid bacteria where fermentation occurs in the presence of air (O_2).

Six types of biooxidation reactions will be studied in this exercise: (1) Durham tube sugar fermentations, (2) mixed acid fermentation, (3) butanediol fermentation, (4) catalase production, (5) oxidase production, and (6) nitrate reduction.

The performance of all these tests on your unknown will involve a considerable number of inoculations because a set of positive test controls will also be needed. Although photographs of positive test results are provided in this exercise, seeing the actual test results in test tubes will make it more meaningful.

As you perform these various tests, attempt to keep in mind what groups of bacteria relate to each test. Although some tests are not very specific in pointing the way to unknown identification, others are very narrow in application.

One last comment of importance: *It is not routine practice to perform all these tests in identifying every unknown.* Although it might appear that our prime concern here is to identify an organism, our most important goal is to learn about the various types of tests for biooxidation enzymes that are available. The use of unknown bacteria to learn about them simply makes it more of a challenge. In actual practice physiological tests are used very selectively. The "shotgun approach" employed here is used to expose you to the multitude of tests that are available.

First Period

(Inoculations)

The following two sets of inoculations (unknown and test controls) may be done separately or combined into one operation. The media for each set of inoculations are listed separately under each heading.

Unknown Inoculations

Figure 48.1 illustrates the procedure for inoculating seven test tubes and one Petri plate with your unknown. Since your instructor may want you to inoculate some different sugar broths, blanks have been provided in the materials list for write-ins. *If different media are distinguished from each other with differently colored tube caps, write down the colors after each medium below.*

Materials:
for each unknown:
 Durham tubes with phenol red indicator
  1 glucose broth
  1 lactose broth
  1 mannitol broth

  _____ _____

  _____ _____

 2 MR-VP medium
 1 nitrate broth
 1 nutrient agar slant
 1 Petri plate of trypticase soy agar (TSA)

1. Label each tube with the number of your unknown and an identifying letter as designated in figure 48.1.
2. Label one half of the Petri plate UNKNOWN and the other half *P. AERUGINOSA.*
3. Inoculate all broths and the slant with a loop. Inoculate one half of the TSA plate with your unknown, using an isolation technique.

Test Control Inoculations

Figure 48.3 on page 159 illustrates the procedure that will be used for inoculating five test tubes to be used for positive test controls. The Petri plate shown on the right side is the same one that is shown in figure 48.1; thus, it will not be listed in the materials list.

Materials:
 1 glucose broth (Durham tube)
 2 MR-VP medium
 1 nitrate broth
 1 nutrient agar slant
 nutrient broth cultures of *Escherichia coli,*
  Enterobacter aerogenes, Staphylococcus
  aureus, and *Pseudomonas aeruginosa*

1. Label each tube with the code letter assigned to it as listed:

  glucose broth A^1
  MR-VP medium D^1
  MR-VP medium E^1
  nitrate broth F^1
  nutrient agar slant G^1

2. Inoculate each of these tubes with a loopful of the appropriate test organism according to figure 48.3.
3. Inoculate the other half of the TSA plate with *P. aeruginosa.*

Incubation

Except for tube E (MR-VP), all the unknown inoculations should be incubated for 24–48 hours at the unknown's optimum temperature. Tube E should be incubated for 3–5 days at the optimum temperature.

 Except for Tube E^1 of the test controls, incubate all the test control tubes and the TSA plate at 37° C for 24–48 hours. Tube E^1 should be incubated at 37° C for 3–5 days.

Second Period

(Test Evaluations)

After 24 to 48 hours incubation, arrange all your tubes (except tubes E and E^1) in a test-tube rack in alphabetical order, with the unknown tubes in one row and the test controls in another row. As you interpret the results, record the information on the Descriptive Chart immediately. Don't trust your memory. Any result that is not properly recorded will have to be repeated.

Durham Tube Sugar Fermentations

When we use a bank of Durham tubes containing various sugars, we are able to determine what sugars an organism is able to ferment. If an organism is able to ferment a particular sugar, acid will be produced and gas *may* be produced. The presence of acid is detectable with the color change of a pH indicator in the medium. Gas production is revealed by the formation of a void in the inverted vial of the

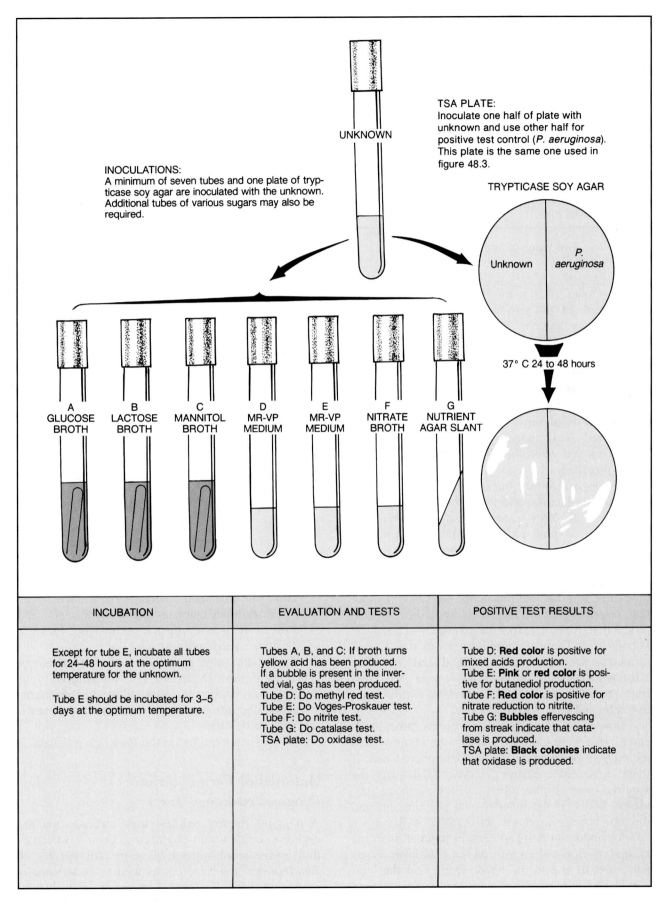

INOCULATIONS:
A minimum of seven tubes and one plate of trypticase soy agar are inoculated with the unknown. Additional tubes of various sugars may also be required.

UNKNOWN

TSA PLATE:
Inoculate one half of plate with unknown and use other half for positive test control (*P. aeruginosa*). This plate is the same one used in figure 48.3.

TRYPTICASE SOY AGAR

Unknown

P. aeruginosa

37° C 24 to 48 hours

| A GLUCOSE BROTH | B LACTOSE BROTH | C MANNITOL BROTH | D MR-VP MEDIUM | E MR-VP MEDIUM | F NITRATE BROTH | G NUTRIENT AGAR SLANT |

INCUBATION	EVALUATION AND TESTS	POSITIVE TEST RESULTS
Except for tube E, incubate all tubes for 24–48 hours at the optimum temperature for the unknown. Tube E should be incubated for 3–5 days at the optimum temperature.	Tubes A, B, and C: If broth turns yellow acid has been produced. If a bubble is present in the inverted vial, gas has been produced. Tube D: Do methyl red test. Tube E: Do Voges-Proskauer test. Tube F: Do nitrite test. Tube G: Do catalase test. TSA plate: Do oxidase test.	Tube D: **Red color** is positive for mixed acids production. Tube E: **Pink** or **red color** is positive for butanediol production. Tube F: **Red color** is positive for nitrate reduction to nitrite. Tube G: **Bubbles** effervescing from streak indicate that catalase is produced. TSA plate: **Black colonies** indicate that oxidase is produced.

Figure 48.1 **Procedure for unknown biooxidation tests**

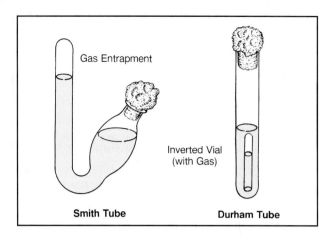

Gas Entrapment

Inverted Vial
(with Gas)

Smith Tube   **Durham Tube**

Figure 48.2   Two types of fermentation tubes

Durham tube. If it were important to know the composition of the gas, we would have to use a Smith tube as shown in figure 48.2. For our purposes here the Durham tube is preferable.

Media   The sugar broths used here contain 0.5% of the specific carbohydrate plus sufficient amounts of beef extract and peptone to satisfy the nitrogen and mineral needs of most bacteria. The pH indicator phenol red is included for acid detection. This indicator is red when the pH is above 7 and yellow below this point.

Although there are many sugars that one might use, glucose, lactose, and mannitol are logical ones to begin with. Your instructor may have had you include one or more additional kinds, and it is very likely that you may wish to use some others later.

Interpretation   Examine the glucose test control tube (tube A^1) that you inoculated with *E. coli*. Note that the phenol red has turned yellow, indicating acid production. Also, note that the inverted vial has a gas bubble in it. These observations tell us that *E. coli* ferments glucose to produce acid and gas. The left-hand illustration in figure 48.4, page 161, illustrates how this positive tube compares with a negative tube and an uninoculated one.

Now examine the three sugar broths (tubes A, B, and C) that were inoculated with your unknown and record your observations on the Descriptive Chart. If there is no color change, record NONE after the specific sugar. If the tube is yellow with no gas, record ACID. If the inverted vial contains gas and the tube is yellow, record ACID AND GAS.

An important point to keep in mind at this time is that *a negative result on an unknown is as important as a positive result.* Don't feel that you have failed in your technique if many of your tubes are negative!

Mixed Acid Fermentation (Methyl Red Test)

A considerable number of gram-negative intestinal bacteria can be differentiated on the basis of the end products produced when they ferment the glucose in MR-VP medium. Genera of bacteria such as *Escherichia, Salmonella, Proteus,* and *Aeromonas* ferment glucose to produce large amounts of lactic, acetic, succinic, and formic acids, plus CO_2, H_2, and ethanol. The accumulation of these acids lowers the pH of the medium to 5.0 and less.

If methyl red is added to such a culture, the indicator turns red, an indication that the organism is a *mixed acid fermenter.* These organisms are generally great gas producers, too, because they produce the enzyme *formic hydrogenylase,* which splits formic acid into equal parts of CO_2 and H_2.

$$HCOOH \xrightarrow{\text{formic hydrogenylase}} CO_2 + H_2$$

Medium   MR-VP medium is essentially a glucose broth with some buffered peptone and dipotassium phosphate.

Test Procedure   Perform the methyl red test first on your test control tube (D^1) and then on your unknown (tube D). Proceed as follows:

Materials:
dropping bottle of methyl red indicator

1. Add three or four drops of methyl red to test control tube D^1, which was inoculated with *E. coli.* The tube should become red immediately.

    A **reddish color,** as shown in the left-hand tube of the middle illustration of figure 48.4 is a positive methyl red test.

2. Repeat the same procedure with your unknown culture (tube D) of MR-VP medium. If your unknown culture becomes yellow like the right-hand tube in figure 48.4, your unknown is negative for this test.

3. Record your results on the Descriptive Chart.

Butanediol Fermentation (Voges-Proskauer Test)

A negative methyl red test may indicate that the organism being tested produced a lot of 2,3 butanediol and ethanol instead of acids. All species of *Enterobacter* and *Serratia,* as well as some species of *Erwinia, Bacillus,* and *Aeromonas,* do just that. The production of these non-acid end products re-

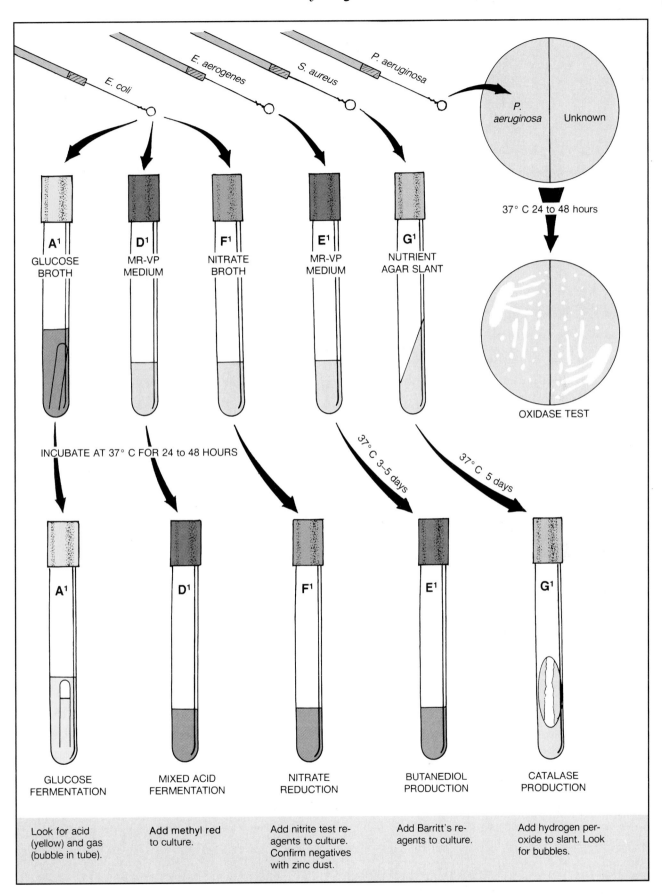

Figure 48.3   Test control procedure for biooxidations

sults in less lowering of the pH in MR-VP medium, causing the methyl red test to be negative.

Unfortunately, there is no satisfactory test for 2,3 butanediol; however, acetoin (acetylmethylcarbinol), a precursor of 2,3 butanediol, is easily detected with Barritt's reagent.

Barritt's reagent consists of alpha-naphthol and KOH. When added to a 3- to 5-day culture of MR-VP medium, and allowed to stand for some time, the medium changes to pink or red in the presence of acetoin. Since acetoin and 2,3 butanediol are always simultaneously present, the test is valid. This indirect method of testing for 2,3 butanediol is called the *Voges-Proskauer* test.

Test Procedure   Perform the Voges-Proskauer test on your unknown and test control tubes of MR-VP medium (tubes E and E^1). Note that the test control tube was inoculated with *E. aerogenes*. Follow this procedure:

Materials:
   Barritt's reagents
   2 pipettes (1 ml size)
   2 empty test tubes

1. Label one empty test tube **E** (for unknown) and the other **E^1** (for control).
2. Pipette 1 ml from culture tube **E** to the empty tube **E** and 1 ml from culture tube **E^1** to the empty tube **E^1**. Use separate pipettes for each tube.
3. Add 18 drops (about 0.5 ml) of Barritt's solution A (alpha-naphthol) to each of these tubes that contain 1 ml of culture.
4. Add an equal amount of Barritt's solution B (KOH) to the same tubes.

5. Shake the tubes vigorously every 20 seconds until the control tube (E^1) turns pink or red. Let the tubes stand for one or two hours to see if the unknown turns red. *Vigorous shaking is very important* to achieve complete aeration.

A positive Voges-Proskauer reaction is **pink** or **red.** The left-hand tube in the right-hand illustration of figure 48.4 shows what a positive result looks like.

6. Record your results on the Descriptive Chart.

Catalase Production

Most aerobes and facultatives that utilize oxygen produce hydrogen peroxide, which is toxic to their own enzyme systems. Their survival in the presence of this antimetabolite is possible because they produce an enzyme called *catalase,* which converts the hydrogen peroxide to water and oxygen:

$$2H_2O_2 \xrightarrow{\text{catalase}} 2H_2O + O_2$$

It has been postulated that the death of strict anaerobes in the presence of oxygen may be due to the suicidal act of H_2O_2 production in the absence of catalase production. The presence or absence of catalase production is an important means of differentiation between certain groups of bacteria.

Test Procedure   To determine whether or not catalase is produced, all that is necessary is to place a few drops of 3% hydrogen peroxide on the organisms of a slant culture. If the hydrogen peroxide effervesces, the organism is catalase-positive.

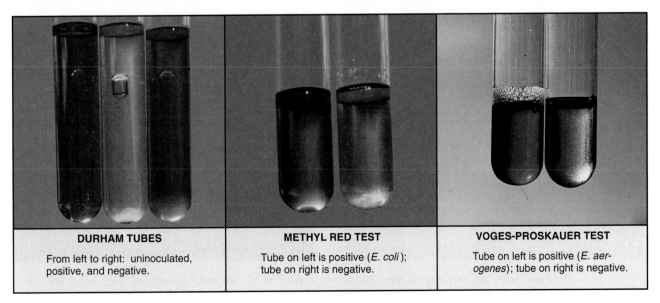

DURHAM TUBES	METHYL RED TEST	VOGES-PROSKAUER TEST
From left to right: uninoculated, positive, and negative.	Tube on left is positive (*E. coli*); tube on right is negative.	Tube on left is positive (*E. aerogenes*); tube on right is negative.

Figure 48.4   Durham tubes, mixed acid, and butanediol fermentation tests

Materials:

    3% hydrogen peroxide

    test control tube G^1 with *S. aureus* growth and
        unknown tube G

1. While holding test control tube G^1 at an angle, allow a few drops of H_2O_2 to flow slowly down over the *S. aureus* growth on the slant. Note how bubbles emerge from the organisms.
2. Repeat the test on your unknown (tube G) and record your results on the Descriptive Chart.

Oxidase Production

The production of oxidase is one of the most significant tests we have for differentiating certain groups of bacteria. For example, all the Enterobacteriaceae are oxidase-negative and most species of *Pseudomonas* are oxidase-positive. Another important group, the *Neisseria,* are oxidase producers.

Two methods are described here for performing this test. The first method utilizes the entire TSA plate; the second method is less demanding in that only a loopful of organisms from the plate is used. The two methods are equally reliable.

Materials:

    TSA plate streaked with unknown and *P.*
        aeruginosa

    oxidase test reagents (1% solution of
        dimethyl-*p*-phenylenediamine
        hydrochloride)

    Whatman No. 2 filter paper

    Petri dish

Entire Plate Method   Onto the TSA plate that you streaked your unknown and *P. aeruginosa,* pour some of the oxidase test reagent, covering the colonies of both organisms.

Observe that the *Pseudomonas* colonies first become **pink,** then change to **maroon, dark red,** and finally **black.** Refer to figure 48.5. If your unknown follows the same color sequence, it, too, is oxidase-positive. Record your results on the Descriptive Chart.

Filter Paper Method   On a piece of Whatman No. 2 filter paper in a Petri dish, place several drops of oxidase test reagent. Remove a loopful of the organisms from one of the colonies and smear the organisms over a small area of the paper. The positive color reaction described above will show up within 10–15 seconds. Record your results on the Descriptive Chart.

Nitrate Reduction

Many facultative bacteria are able to use the oxygen in nitrate as a hydrogen acceptor in anaerobic respiration, thus converting nitrate to nitrite. This enzymatic reaction is controlled by an inducible enzyme called *nitratase.*

$$NO_3^- + 2e^- + 2H^+ \xrightarrow{\text{nitratase}} NO_2^- + H_2O$$

Since the presence of free oxygen prevents nitrate reduction, actively multiplying organisms will use up the oxygen first and then utilize the nitrate. In culturing some organisms, it is desirable to use anaerobic methods to ensure nitrate reduction.

Test Procedure   The nitrate broth used in this test consists of beef extract, peptone, and potassium nitrate. To test for nitrite after incubation, we use two reagents designated as A and B.

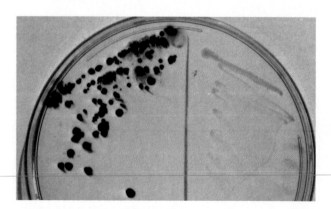

Figure 48.5 **Oxidase Test:** The colonies on the left are positive; the ones on the right are negative.

Figure 48.6 **Nitrate Reduction Test:** Tube on left is positive (*E. coli*); tube on right is negative.

Reagent A contains sulfanilic acid and reagent B contains dimethyl-alpha-naphthylamine. In the presence of nitrite these reagents cause the culture to turn red. Negative results must be confirmed as negative with zinc dust.

Materials:

    nitrate broth cultures of unknown (tube F) and
        test control *E. coli* (tube F^1)
    nitrite test reagents (solutions A and B)
    zinc dust

1. Add two or three drops of nitrite test solution A (sulfanilic acid) and an equal amount of solution B (dimethyl-alpha-naphthylamine) to the nitrate broth culture of *E. coli* (tube F^1).

    A **red color** should appear almost immediately (see figure 48.6), indicating that nitrate reduction has occurred.

    Caution: Since the agent in solution B is carcinogenic, avoid skin contact.

2. Repeat this procedure with your unknown (tube F). If the red color does not develop, your unknown is negative for nitrate reduction. All negative results should be confirmed as being negative as follows:

    Negative Confirmation: Add a pinch of zinc dust to the tube and shake it vigorously. If the tube becomes red, the test is confirmed as being negative. Zinc causes this reaction by reducing nitrate to nitrite; the newly formed nitrite reacts with the reagents to produce the red color.

3. Record your results on the Descriptive Chart.

Laboratory Report

Answer the questions on Laboratory Report 48–50 that pertain to biooxidations.

Physiological Characteristics: Hydrolysis

Many bacteria produce exoenzymes called *hydrolases,* which split complex organic compounds into smaller units. All hydrolytic enzymes accomplish this molecular splitting in the presence of water. We have already observed one example of protein hydrolysis in Exercise 47: gelatin hydrolysis by gelatinase. In this exercise we shall observe the hydrolysis of starch, casein, fat, tryptophan, and urea. Each test plays an important role in the identification of certain types of bacteria. This exercise will be performed in the same manner as the previous one, with test controls being made for comparisons.

Figure 49.1 illustrates the general procedure to be used. Three agar plates and four test tubes will be inoculated. After incubation, some of the plates and tubes will have test reagents added to them; others will reveal the presence of hydrolysis by changes that have occurred during incubation. Proceed as follows:

First Period

(Inoculations)

If each student is working with only one unknown, students can work in pairs to share Petri plates. Note in figure 49.1 how each plate can serve for two unknowns with the test control organism streaked down the middle. If each student is working with two unknowns, the plates will not be shared. Whether the two tubes for test controls will be shared depends on the availability of materials.

Materials:
per pair *of students with one unknown each, or for* one *student with two unknowns:*
    1 starch agar plate
    1 skim milk agar plate
    1 spirit blue agar plate
    3 urea broths
    3 tryptone broths
nutrient broth cultures of *B. subtilis, E. coli, S. aureus,* and *P. vulgaris*

1. Label and streak the three different agar plates in the manner shown in figure 49.1. Note that straight-line streaks are made on each plate. Indicate, also, the type of medium in each plate.

2. Label a tube of urea broth *P. VULGARIS* and a tube of tryptone broth *E. COLI.* These will be your test controls for urea and tryptophan hydrolysis. Inoculate each tube accordingly.
3. For each unknown, label one tube of urea broth and one tube of tryptone broth with the code number of your unknown. Inoculate each tube with the appropriate unknown.
4. Incubate the plates and two test control tubes at 37° C. Incubate the unknown tubes of urea broth and tryptone broth at the optimum temperatures for the unknowns.

Second Period

(Evaluation of Tests)

After 24 to 48 hours incubation of unknowns and test controls, compare your unknowns with the test controls, recording all data on the Descriptive Chart.

Starch Hydrolysis

Since many bacteria are capable of hydrolyzing starch, this test has fairly wide application. The starch molecule is a large one consisting of two constituents: amylose, a straight chain polymer of 200 to 300 glucose units, and amylopectin, a larger branched polymer with phosphate groups. Bacteria that hydrolyze starch produce *amylases* that yield molecules of maltose, glucose, and dextrins.

Materials:
    Gram's iodine
    starch agar culture plate

Iodine solution (Gram's) is an indicator of starch. When iodine comes in contact with a medium containing starch, it turns blue. If starch is hydrolyzed and starch is no longer present, the medium will have a **clear zone** next to the growth.

By pouring Gram's iodine over the growth on the medium, one can see clearly where starch has been hydrolyzed. If the area immediately adjacent to the growth is clear, amylase is produced.

Pour enough iodine over each streak to completely wet the entire surface of the plate. Rotate and tilt the plate gently to spread the iodine. Compare

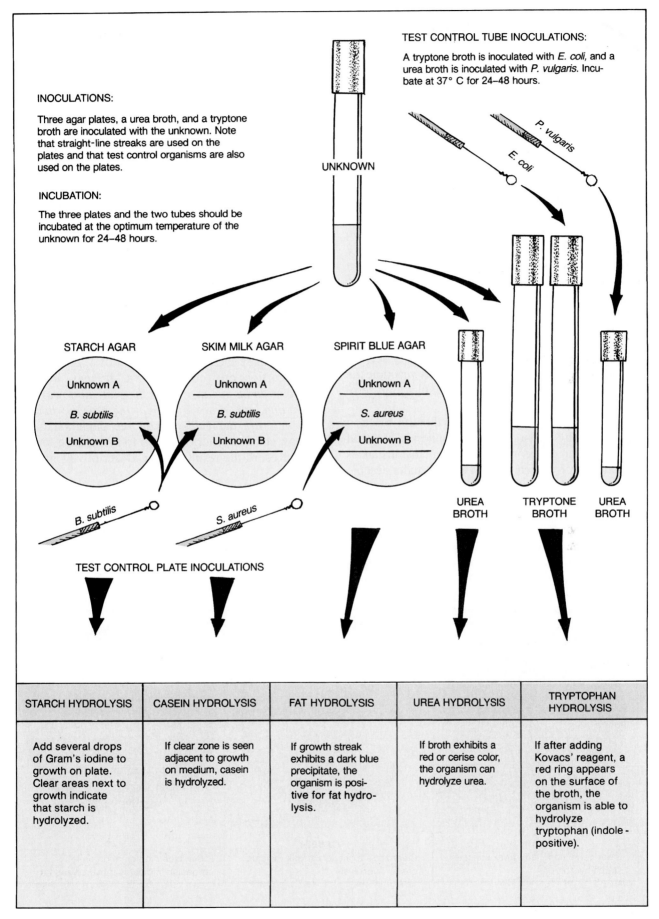

INOCULATIONS:

Three agar plates, a urea broth, and a tryptone broth are inoculated with the unknown. Note that straight-line streaks are used on the plates and that test control organisms are also used on the plates.

INCUBATION:

The three plates and the two tubes should be incubated at the optimum temperature of the unknown for 24–48 hours.

TEST CONTROL TUBE INOCULATIONS:

A tryptone broth is inoculated with *E. coli*, and a urea broth is inoculated with *P. vulgaris*. Incubate at 37° C for 24–48 hours.

UNKNOWN

P. vulgaris

E. coli

STARCH AGAR

Unknown A

B. subtilis

Unknown B

SKIM MILK AGAR

Unknown A

B. subtilis

Unknown B

SPIRIT BLUE AGAR

Unknown A

S. aureus

Unknown B

UREA BROTH

TRYPTONE BROTH

UREA BROTH

B. subtilis

S. aureus

TEST CONTROL PLATE INOCULATIONS

STARCH HYDROLYSIS	CASEIN HYDROLYSIS	FAT HYDROLYSIS	UREA HYDROLYSIS	TRYPTOPHAN HYDROLYSIS
Add several drops of Gram's iodine to growth on plate. Clear areas next to growth indicate that starch is hydrolyzed.	If clear zone is seen adjacent to growth on medium, casein is hydrolyzed.	If growth streak exhibits a dark blue precipitate, the organism is positive for fat hydrolysis.	If broth exhibits a red or cerise color, the organism can hydrolyze urea.	If after adding Kovacs' reagent, a red ring appears on the surface of the broth, the organism is able to hydrolyze tryptophan (indole-positive).

Figure 49.1   Procedure for doing hydrolysis tests on unknowns

your unknowns with the positive result seen along the growth of *B. subtilis*. The left-hand illustration of figure 49.2 illustrates what it looks like.

Casein Hydrolysis

Casein is the predominant protein in milk. Its presence causes milk to have its characteristic white appearance. Many bacteria produce the exoenzyme *caseinase,* which hydrolyzes casein to produce more soluble, transparent derivatives. Protein hydrolysis also is referred to as *proteolysis,* or *peptonization.*

Examine the streaks on the skim milk agar plates. Note that a **clear zone** exists adjacent to the growth of *B. subtilis*. This is evidence of casein hydrolysis. The middle illustration in figure 49.2 shows what it looks like. Compare your unknown with this positive result and record the results on the Descriptive Chart.

Fat Hydrolysis

The ability of organisms to hydrolyze fat is accomplished with the enzyme *lipase.* In this reaction the fat molecule is split to form one molecule of glycerol and three fatty acid molecules:

The glycerol and fatty acids produced in this reaction can be used by the organism to synthesize bacterial fats and other cell components. In many instances they are even oxidized to yield energy under aerobic conditions. This ability of bacteria to decompose fats plays a role in the rancidity of certain foods, such as margarine.

Spirit blue agar contains a vegetable oil that, when hydrolyzed by most organisms, results in the lowering of the pH sufficiently to produce a **dark**

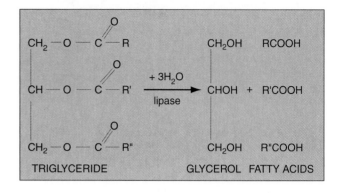

blue precipitate. Unfortunately, the hydrolytic action of some organisms on this medium does not produce a blue precipitate because the pH is not lowered sufficiently.

Examine the *S. aureus* growth carefully. You should be able to see this dark blue reaction. The right-hand illustration in figure 49.2 exhibits what it should look like.

Compare the positive reaction of *S. aureus* with the reaction on your unknown. *If your unknown appears to be negative, hold the plate up toward the light and look for a region near the growth where oil droplets are depleted.* If you see depletion of oil drops, consider your organism to be positive for this test. Record the results on the Descriptive Chart.

Tryptophan Hydrolysis

Certain bacteria such as *E. coli* have the ability to split the amino acid tryptophan into indole and pyruvic acid. The enzyme that causes this hydrolysis is *tryptophanase.* Indole can be easily detected with

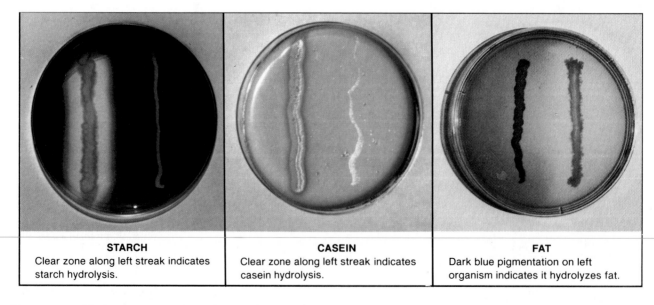

STARCH	**CASEIN**	**FAT**
Clear zone along left streak indicates starch hydrolysis.	Clear zone along left streak indicates casein hydrolysis.	Dark blue pigmentation on left organism indicates it hydrolyzes fat.

Figure 49.2 **Hydrolysis Test Plates: Starch, casein, and fat**

Kovacs' reagent. This test is particularly useful in differentiating *E. coli* from some closely related enteric bacteria.

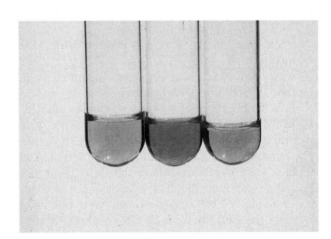

Tryptone broth (1%) is used for this test because it contains a great deal of tryptophan. Tryptone is a peptone derived from casein by pancreatic digestion.

Materials:

Kovacs' reagent
tryptone broth cultures of unknown and *E. coli*

To test for indole add 10 to 12 drops of Kovacs' reagent to the *E. coli* culture in tryptone broth. A **red layer** should form at the top of the culture, as shown in figure 49.3. Repeat the test on your unknown and record the results on the Descriptive Chart.

Urea Hydrolysis

The differentiation of gram-negative enteric bacteria is greatly helped if one can demonstrate that the unknown can produce *urease*. This enzyme splits off ammonia from the urea molecule, as shown below. Note in the separation outline in figure 51.3 that three genera (*Proteus, Providencia,* and *Morganella*) are positive for the production of this hydrolytic enzyme.

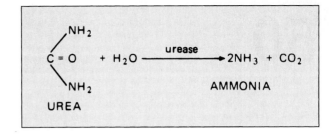

Urea broth is a buffered solution of yeast extract and urea. It also contains phenol red as a pH indicator. Since urea is unstable and breaks down in the autoclave at 15 psi steam pressure, it is often sterilized by filtration. It is tubed in small amounts to hasten the visibility of the reaction.

When urease is produced by an organism in this medium, the ammonia that is released raises the pH. As the pH becomes higher, the phenol red changes from a yellow color (pH 6.8) to a **red** or **cerise color** (pH 8.1 or more).

Examine your tube of urea broth that was inoculated with *Proteus vulgaris*. Compare your unknown with this standard. Figure 49.4 reveals how positive and negative results of the test should appear. *If your unknown is negative incubate the tube for a total of 7 days to check for a slow urease producer.* Record your result on the Descriptive Chart.

Figure 49.3 **Indole Test:** Tube on left is positive (*E. coli*); tube on right is negative.

Figure 49.4 **Urease Test:** From left to right—uninoculated, positive (*Proteus*), and negative.

Physiological Characteristics:
Miscellaneous Tests

There are several additional physiological tests used in unknown identification that are best grouped separately as "miscellaneous tests." They include tests for hydrogen sulfide production, citrate utilization, phenylalanine deaminization, and litmus milk reactions. During the first period, inoculations of four kinds of media will be made for these tests. An explanation of the value of the IMViC tests will also be included.

First Period
(Inoculations)

Since test controls are included in this exercise, two sets of inoculations will be made. For economy of materials, one set of test controls will be made by students working in pairs.

Materials:

for test controls, per pair of students:
    1 Kligler's iron agar deep or SIM medium
    1 Simmons citrate agar slant
    1 phenylalanine agar slant
    nutrient broth cultures of *Proteus vulgaris,*
        Staphylococcus aureus, and *Enterobacter*
        aerogenes

per unknown, per student:
    1 Kligler's iron agar deep or SIM medium
    1 Simmons citrate agar slant
    1 phenylalanine agar slant
    1 litmus milk

1. Label one tube of Kligler's iron agar (or SIM medium) *P. VULGARIS* and additional tubes with your unknown numbers. Inoculate each tube by stabbing with a straight wire.
2. Label one tube of Simmons citrate agar *E. AEROGENES* and additional tubes with your unknown numbers. Use a straight wire to streak-stab each slant; i.e., streak the slant first, and then stab into the middle of the slant.
3. Label one tube of phenylalanine agar slant *P. VULGARIS* and the other with your unknown code number. Streak each slant with the appropriate organisms.
4. With a loop, inoculate one tube of litmus milk with your unknown. (**Note:** A test control for this

medium will not be made. Figure 50.2 will take its place.)
5. Incubate the unknowns at their optimum temperatures. Incubate the test controls at 37° C for 24–48 hours.

Second Period
(Evaluation of Tests)

After 24 to 48 hours incubation, examine the tubes to evaluate according to the following discussion. Record all results on the Descriptive Chart.

Hydrogen Sulfide Production

Certain bacteria, such as *Proteus vulgaris,* produce hydrogen sulfide from the amino acid cysteine. These organisms produce the enzyme *cysteine desulfurase,* which works in conjunction with the coenzyme pyridoxyl phosphate. The production of H_2S is the initial step in the deamination of cysteine as indicated below:

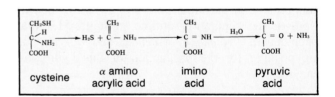

Kligler's iron agar or SIM medium is used here to detect hydrogen sulfide production. Both of these media contain iron salts that react with H_2S to form a **dark precipitate** of iron sulfide.

Kligler's iron agar also contains glucose, lactose, and phenol red. When this medium is used in slants it is an excellent medium for detecting glucose and lactose fermentation. SIM medium, on the other hand, can also be used for determining motility and testing for indole production.

Examine the tube of one of these media that was inoculated with *P. vulgaris.* If it is Kligler's iron agar it will look like the left-hand tube in figure 50.1. A positive reaction in SIM medium will look like the small tube on the right.

Compare your unknown with this control tube and record your results on the Descriptive Chart.

Citrate Utilization

The ability of some organisms, such as *E. aerogenes* and *Salmonella typhimurium,* to utilize citrate as a sole source of carbon can be a very useful differentiation characteristic in working with intestinal bacteria. Koser's citrate medium and Simmons citrate agar are two media that are used to detect this ability in bacteria. In both of these synthetic media sodium citrate is the sole carbon source; nitrogen is supplied by ammonium salts instead of amino acids.

Examine the test control slant of this medium that was inoculated with *E. aerogenes.* Note the distinct **Prussian blue color change** that has occurred. Refer to the right hand illustration in figure 50.1. Record your results on the Descriptive Chart.

Phenylalanine Deamination

A few bacteria, such as *Proteus, Morganella,* and *Providencia,* produce the deaminase *phenylalanase,* that deaminizes the amino acid phenylalanine to produce phenylpyruvic acid (PPA). This characteristic is used to help differentiate these three genera from other genera of the Enterobacteriaceae. The reaction is as follows:

PHENYLALANINE $\xrightarrow{\text{phenylalanase}}$ PHENYLPYRUVIC ACID $+ NH_3$

Proceed as follows to test for the production of phenylpyruvic acid which is evidence that the enzyme phenylalanase has been produced:

Materials:

dropping bottle of 10% ferric chloride

Allow 5–10 drops of 10% ferric chloride to flow down over the slants of the test control (*P. vulgaris*) and your unknowns. To hasten the reaction, use a loop to emulsify the organisms into solution. A deep **green color** should appear on the test control slant in 1–5 minutes. Refer to the middle illustration in Figure 50.1. Compare your unknown with the control and record your results on the Laboratory Report.

The IMViC Tests

In the differentiation of *E. aerogenes* and *E. coli,* as well as some other related species, four physiological tests have been grouped together into what are called the IMViC tests. The *I* stands for indole; the *M* and *V* stand for methyl red and Voges-Proskauer tests; *i* simply facilitates pronunciation; and the *C* signifies citrate utilization. In the differentiation of the two coliforms *E. coli* and *E. aerogenes,* the test results appear as charted below, revealing completely opposite reactions for the two organisms on all tests.

	I	M	V	C
E. coli	+	+	−	−
E. aerogenes	−	−	+	+

The significance of these tests is that when testing drinking water for the presence of the sewage indicator *E. coli,* one must be able to rule out *E. aerogenes,* which has many of the morphological and physiological characteristics of *E. coli.* Since *E. aerogenes* is not always associated with sewage, its

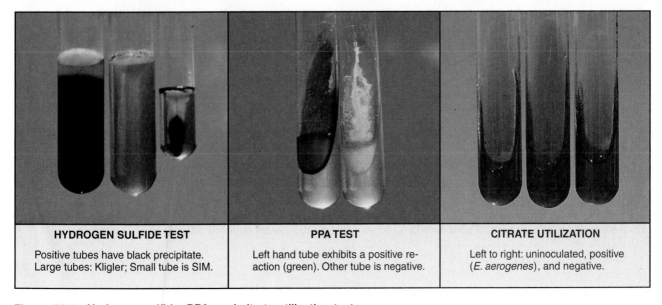

HYDROGEN SULFIDE TEST	PPA TEST	CITRATE UTILIZATION
Positive tubes have black precipitate. Large tubes: Kligler; Small tube is SIM.	Left hand tube exhibits a positive reaction (green). Other tube is negative.	Left to right: uninoculated, positive (*E. aerogenes*), and negative.

Figure 50.1 Hydrogen sulfide, PPA, and citrate utilization tests

presence in water would not necessarily indicate sewage contamination.

If you are attempting to identify a gram-negative, facultative, rod-shaped bacterial organism, group these series of tests together in this manner to see how your unknown fits this combination of tests.

Litmus Milk Reactions

Litmus milk contains 10% powdered skim milk and a small amount of litmus as a pH indicator. When the medium is made up, its pH is adjusted to 6.8. It is an excellent growth medium for many organisms and can be very helpful in unknown characterization. In addition to revealing the presence or absence of fermentation, it can detect certain proteolytic characteristics in bacteria. A number of facultative bacteria with strong reducing powers are able to utilize litmus as an alternative electron acceptor to render it colorless. Figure 50.2 reveals the color changes that cover the spectrum of litmus milk changes. Since some of the reactions take 4 to 5 days to occur, the cultures should be incubated for at least this period of time; they should be examined every 24 hours, however. Look for the following reactions:

Acid Reaction Litmus becomes pink. Typical of fermentative bacteria.

Alkaline Reaction Litmus turns blue or purple. Many proteolytic bacteria cause this reaction in the first 24 hours.

Litmus Reduction Culture becomes white; actively reproducing bacteria reduce the O/R potential of medium.

Coagulation Curd formation. Solidification is due to protein coagulation. Tilting tube at 45° will indicate whether or not this has occurred.

Peptonization Medium becomes translucent. It often turns brown at this stage. Caused by proteolytic bacteria.

Ropiness Thick, slimy residue in bottom of tube. Ropiness can be demonstrated with sterile loop.

Record the litmus milk reactions of your unknown on the Descriptive Chart.

Laboratory Report

Complete Laboratory Report 48–50, which reviews all physiological tests performed in the last three exercises.

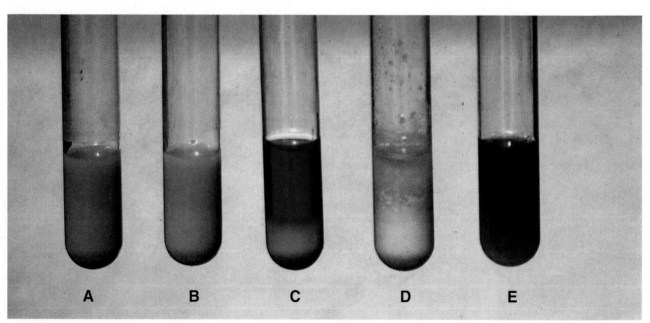

Figure 50.2 Litmus milk reactions: (A) Alkaline. (B) Acid. (C) Upper transparent portion is peptonization; solid white portion in bottom is coagulation and litmus reduction; overall redness is interpreted as acid. (D) Coagulation and litmus reduction in lower half; some peptonization (transparency) and acid in top portion. (E) Litmus indicator is masked by production of soluble pigment (*Pseudomonas*); some peptonization is present but difficult to see in photo.

Once you have recorded all the data on your Descriptive Chart pertaining to morphological, cultural, and physiological characteristics of your unknown, you are ready to determine its genus and species. Determination of the genus should be relatively easy; species differentiation, however, is considerably more difficult.

The most important single source of information we have for the identification of bacteria is *Bergey's Manual of Systematic Bacteriology.* This monumental achievement, which consists of four volumes, replaced a single-volume eighth edition of *Bergey's Manual of Determinative Bacteriology.* Although the more recent publication consists of four volumes, only volumes 1 and 2 will be used for the identification of the unknowns in this course.

In addition to using *Bergey's Manual* you may have an opportunity to use a computer simulation program called *Identibacter interactus,* which is available on a CD-ROM disc. Details of the application of this computer program are discussed on page 175 and Appendix F.

Bergey's Manual is a worldwide collaborative effort that has an editorial board of 13 trustees. Over 200 specialists from 19 countries are listed as contributors to the first two volumes. One of the purposes of this exercise is to help you glean the information from these two volumes that is needed to identify your unknown. Before we get into the mechanics of using *Bergey's Manual,* a few comments are in order pertaining to the problems of bacterial classification.

Classification Problems

Compared with the classification of bacteria, the classification of plants and animals has been relatively easy. In these higher forms, a hierarchy of orders, families, and genera is based, primarily, on evolutionary evidence revealed by fossils laid down in sedimentary layers of Earth's crust. Some of the earlier editions of *Bergey's Manual* attempted to use the same hierarchial system, but the attempt had to be abandoned when the eighth edition was published; without paleontological information to support the system it literally fell apart.

The present system of classification in *Bergey's Manual* uses a list of "Sections" that separate the various groups. Each section is described in common terms so that it is easily understood (even for beginners). For example, Section 1 is entitled **The Spirochaetes.** Section 4 pertains to **Gram-Negative Aerobic Rods and Cocci.** If one scans the Table of Contents in each volume after having completed all tests, it is possible, usually, to find a section that contains the unknown being studied.

A perusal of these sections will reveal that some sections have a semblance of hierarchy in the form of orders, families, and genera. Other sections list only genera.

Thus, we see that the classification system of bacteria, as developed in *Bergey's Manual,* is not the tidy system we see in higher forms of life. The important thing is that it works.

Our dependency over the years on *Bergey's Manual* has led many to think of its classification system as the "Official Classification." Staley and Krieg in their Overview in Volume 1 emphasize that no official classification of bacteria exists; in other words, the system offered in *Bergey's Manual* is simply a workable system, but in no sense of the word should it be designated as the official classification system.

Presumptive Identification

The place to start in identifying your unknown is to determine what genus it fits into. If *Bergey's Manual* is available, scan the Tables of Contents in Volumes 1 and 2 to find the section that seems to describe your unknown. If these books are not immediately available you can determine the genus by referring to the separation outlines in figures 51.1 and 51.2. Note that seven groups of gram-positive bacteria are winnowed out in figure 51.1 and four groups of gram-negative bacteria in figure 51.2.

To determine which genus in the group best fits the description of your unknown, compare the genera descriptions provided below. Note that each group has a section designation to identify its position in *Bergey's Manual.*

Group I (Section 13, Vol. 2) Although there are only three genera listed in this group, Section 13 in *Bergey's Manual* lists three additional genera, one of which is *Sporosarcina,* a coccus-shaped organism (see Group V). Most members of Group I are motile and differentiation is based primarily on oxygen needs.

Bacillus   Although most of these organisms are aerobic, some are facultative anaerobes. Catalase is usually produced. For comparative characteristics of the 34 species in this genus refer to Table 13.4 on pages 1122 and 1123.

Clostridium   While most of members of this genus are strict anaerobes, some may grow in the presence of oxygen. Catalase is not usually produced. An excellent key for presumptive species identification is provided on pages 1143–1148. Species characterization tables are also provided on pages 1149–1154.

Sporolactobacillus   Microaerophilic and catalase-negative. Nitrates are not reduced and indole is not formed. Spore formation occurs very infrequently (1% of cells).

Since there is only one species in this genus, one needs only to be certain that the unknown is definitely of this genus. Table 13.11 on page 1140 can be used to compare other genera that are similar to this one.

Group II (Section 16, Vol. 2)   This group consists of Family Mycobacteriaceae, with only one genus: *Mycobacterium.* Fifty-four species are listed in Sec-

tion 16. Differentiation of species within this group depends to some extent on whether the organism is classified as a slow or a fast grower. Tables on pages 1439–1442 can be used for comparing the characteristics of the various species.

Group III (Section 14, Vol. 2)   Of the seven diverse genera listed in Section 14, only three have been included here in this group.

Lactobacillus   Non-spore-forming rods, varying from long and slender to coryneform (club-shaped) coccobacilli. Chain formation is common. Only rarely motile. Facultative anaerobic or microaerophilic. Catalase-negative. Nitrate usually not reduced. Gelatin not liquefied. Indole and H_2S not produced.

Listeria   Regular, short rods with rounded ends; occur singly and in short chains. Aerobic and facultative anaerobic. Motile when grown at 20–25° C. Catalase-positive and oxidase-negative. Methyl red positive. Voges-Proskauer positive. Negative for citrate utilization, indole production, urea hydrolysis, gelatinase production, and casein hydrolysis. Table 14.12 on page 1241 provides information pertaining to species differentiation in this genus.

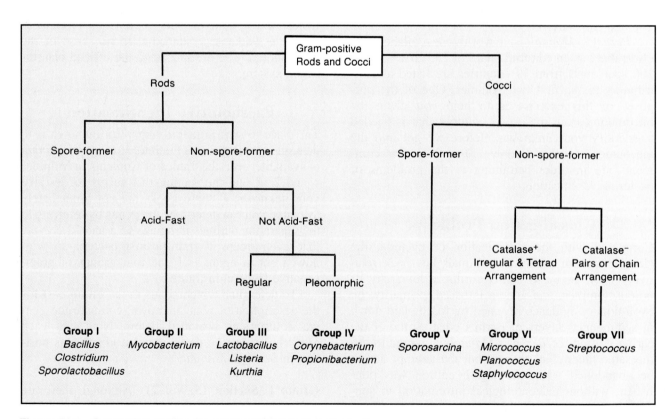

Figure 51.1   Separation outline for gram-positive rods and cocci

Kurthia Regular rods, 2–4 µm long with rounded ends; in chains in young cultures; coccoidal in older cultures. Strictly aerobic. Catalase-positive, oxidase-negative. Also negative for gelatinase production and nitrate reduction. Only two species in this genus.

Group IV (Section 15, Vol. 2) Although there are 21 genera listed in this section of *Bergey's Manual,* only two genera concern us here.

Corynebacterium Straight to slightly curved rods with tapered ends. Sometimes club-shaped. Palisade arrangements common due to snapping division of cells. Metachromatic granules formed. Facultative anaerobic. Catalase-positive. Most species produce acid from glucose and some other sugars. Often produce pellicle in broth. Table 15.3 on page 1269 provides information for species characterization.

Proprionibacterium Pleomorphic rods, often diphtheroid or club-shaped with one end rounded and the other tapered or pointed. Cells may be coccoid, bifid (forked, divided), or even branched. Nonmotile. Some produce clumps of cells with "Chinese character" arrangements. Anaerobic to aerotolerant. Generally catalase-positive. Produce large amounts of proprionic and acetic acids. All produce acid from glucose.

Group V (Section 13, Vol. 2) This group, which has only one genus in it, is closely related to genus *Bacillus.*

Sporosarcina Cells are spherical or oval when single. Cells may adhere to each other when dividing to produce tetrads or packets of eight or more. Endospores formed (see photomicrographs on page 1203). Strictly aerobic. Generally motile. Only two species: *S. ureae* and *S. halophila.*

Group VI (Section 12, Vol. 2) This section contains two families and fifteen genera. Our concern here is with only three genera in this group. Oxygen requirements and cellular arrangement are the principal factors in differentiating the genera. Most of these genera are not closely related.

Micrococcus Spheres, occurring as singles, pairs, irregular clusters, tetrads, or cubical packets. Usually nonmotile. Strict aerobes (one species is facultative anaerobic). Catalase- and oxidase-positive. Most species produce caretenoid pigments. All species will grow in media containing 5% NaCl. For species differentiation see Table 12.4 on page 1007.

Planococcus Spheres, occurring singly, in pairs, in groups of three cells, occasionally in tetrads. Although cells are generally gram-positive, they may be gram-variable. Motility is present. Catalase- and gelatinase-positive. Carbohydrates not attacked. Do not

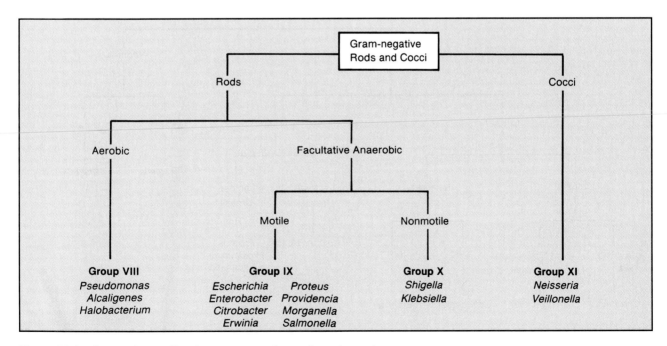

Figure 51.2 Separation outline for gram-negative rods and cocci

hydrolyze starch or reduce nitrate. Refer to Table 12.9 for species differentiation.

Staphylococcus Spheres, occurring as singles, pairs, and irregular clusters. Nonmotile. Facultative anaerobes. Usually catalase-positive. Most strains grow in media with 10% NaCl. Susceptible to lysis by lysostaphin. Glucose fermentation: acid, no gas. Coagulase production by some. Refer to Exercise 77 for species differentiation, or to Table 12.10 on pages 1016 and 1017.

Group VII (Section 12, Vol. 2) Note that the single genus of this group is included in the same section of *Bergey's Manual* as the three genera above. Members of the genus *Streptococcus* have spherical to ovoid cells that occur in pairs or chains when grown in liquid media. Some species, notably, *S. mutans,* will develop short rods when grown under certain circumstances. Facultative anaerobes. Catalase-negative. Carbohydrates are fermented to produce lactic acid without gas production. Many species are commensals or parasites of humans or animals. Refer to Exercise 78 for species differentiation of pathogens. Several tables in *Bergey's Manual* provide differentiation characteristics of all the streptococci.

Group VIII (Section 4, Vol. 1) Although there are many genera of gram-negative aerobic rod-shaped

bacteria, only three genera are likely to be encountered here.

Pseudomonas Generally motile. Strict aerobes. Catalase-positive. Some species produce soluble fluorescent pigments that diffuse into the agar of a slant. Many tables are available in *Bergey's Manual* for species differentiation.

Alcaligenes Rods, coccal rods, or cocci. Motile. Obligate aerobes with some strains capable of anaerobic respiration in presence of nitrate or nitrite.

Halobacterium Cells may be rod- or disk-shaped. Cells divide by constriction. Most are strict aerobes; a few are facultative anaerobes. Catalase- and oxidase-positive. Colonies are pink, red, or red to orange. Gelatinase not produced. Most species require high NaCl concentrations in media. Cell lysis occurs in hypotonic solutions.

Groups IX and X (Section 5, Vol. 1) Section 5 in *Bergey's Manual* lists three families and 34 genera; of these 34 only 10 genera of family Enterobacteriaceae have been included in these two groups. If your unknown appears to fall into one of these groups, use the separation outline in figure 51.3 to determine the genus. Another useful separation outline is provided in figure 79.1 on page 260. *Keep in*

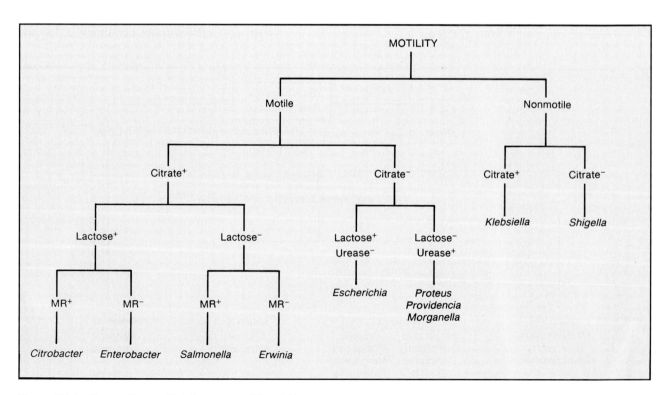

Figure 51.3 Separation outline for groups IX and X

mind, when using these separation outlines, that there are some minor exceptions in the applications of these tests. The diversity of species within a particular genus often presents some problematical exceptions to the rule. Your final decision can be made only after checking the species characteristics tables for each genus in *Bergey's Manual*.

Group XI These genera are morphologically quite similar, yet physiologically they are quite different.

Neisseria (Section 4, Vol. 1) Cocci, occurring singly, but more often in pairs (diplococci); adjacent sides are flattened. One species (*N. elongata*) consists of short rods. Nonmotile. Except for *N. elongata,* all species are oxidase- and catalase-positive. Aerobic.

Veillonella (Section 8, Vol. 1) Cocci, appearing as diplococci, masses, and short chains. Diplococci have flattening at adjacent surfaces. Nonmotile. All are oxidase- and catalase-negative. Nitrate is reduced to nitrite. Anaerobic.

Problem Analysis

If you have identified your unknown by following the above procedures . . . congratulations! Not everyone succeeds at first attempt. If you are having difficulty, consider the following possibilities:

• You may have been given the wrong unknown! Although this is a remote possibility, it does happen at times. Occasionally, clerical errors are made when unknowns are put together.

• Your organism may be giving you a "false-negative" result on a test. This may be due to an incorrectly prepared medium, faulty test reagents, or improper testing technique.

• Your unknown organisms may not match the description *exactly* as stated in *Bergey's Manual*. By now you are aware that the words *generally, usually,* and *sometimes* are frequently used in the book. It is entirely possible for one of these words to be inadvertently left out in *Bergey's*

assignment of certain test results to a species. *In other words, test results, as stated in the manual, may not always apply!*

• Your culture may be contaminated. If you are not working with a pure culture, all tests are unreliable.

• You may not have performed enough tests. Check the various tables in *Bergey's Manual* to see if there is some other test that will be helpful. In addition, double check the tables to make sure that you have read them correctly.

Confirmation of Results

There are several ways to confirm your presumptive identification. One method is to apply serological techniques, if your organism is one for which typing serum is available. Another alternative is to use one of the miniature multitest systems that are described in the next section of this manual. Your instructor will indicate which of these alternatives, if any, will be available.

Identibacter Interactus

Identibacter interactus is a computer simulation program on a CD-ROM disc that was developed by Allan Konopka, Paul Furbacher, and Clark Gedney at Purdue University. The program is copyrighted by the Purdue Research Foundation, and WCB/McGraw-Hill Co. has exclusive distribution rights.

If your laboratory is set up with a computer that has this program installed in it, use the program to confirm the identification of your unknown. A group of 50 tests can be pulled down from menus. The tests are shown in color exactly as you would see them when performed in the laboratory. It is your responsibility to interpret each test as applied to your unknown. Before you attempt to use this program, read the comments pertaining to it in Appendix F.

Laboratory Report

There is no Laboratory Report for this exercise.

PART 9 Miniaturized Multitest Systems

Having run a multitude of tests in Exercises 45 through 50 in an attempt to identify an unknown, you undoubtedly have become aware of the tremendous amount of media, glassware, and preparation time that is involved just to set up the tests. And then, after performing all of the tests and meticulously following all the instructions, you discover that finding the specific organism in "Encyclopedia Bergey" is not exactly the simplest task you have accomplished in this course. The question must arise occasionally: "There's got to be an easier way!" Fortunately, there is: *miniaturized multitest systems.*

Miniaturized systems have the following advantages over the macromethods you have used to study the physiological characteristics of your unknown: (1) minimum media preparation, (2) simplicity of performance, (3) reliability, (4) rapid results, and (5) uniform results. These advantages have resulted in widespread acceptance of these systems by microbiologists.

Since it is not possible to describe all of the systems that are available, only four have been selected here: two by Analytab Products and two by Becton-Dickinson. All four of these products are designed specifically to provide rapid identification of medically important organisms, often within 5 hours. Each method consists of a plastic tube or strip that contains many different media to be inoculated and incubated. To facilitate rapid identification, these systems utilize numerical coding systems that can be applied to charts or computer programs.

The four multitest systems described in this unit have been selected to provide several options. Exercises 52 and 53 pertain to the identification of gram-negative *oxidase-negative* bacteria (Enterobacteriaceae). Exercise 54 (Oxi/Ferm Tube) is used for identifying gram-negative *oxidase-positive* bacteria. Exercise 55 (Staph-Ident) is a rapid system for the differentiation of the staphylococci.

As convenient as these systems are, one must not assume that the conventional macromethods of Part 8 are becoming obsolete. Macromethods must still be used for culture studies and confirmatory tests; confirmatory tests by macromethods are often necessary when a particular test on a miniaturized system is in question. Another point to keep in mind is that all of the miniaturized multitest systems have been developed for the identification of *medically important* microorganisms. If one is trying to identify a saprophytic organism of the soil, water, or some other habitat, there is no substitute for the conventional methods.

If these systems are available to you in this laboratory, they may be used to confirm your conclusions that were drawn in Part 8 or they may be used in conjunction with some of the exercises in Part 14. Your instructor will indicate what applications will be made.

52

Enterobacteriaceae Identification:
The API 20E System

The **API 20E System** is a miniaturized version of conventional tests that is used for the identification of members of the family Enterobacteriaceae and other gram-negative bacteria. It was developed by Analytab Products, of Plainview, New York. This system utilizes a plastic strip (figure 52.1) with 20 separate compartments. Each compartment consists of a depression, or *cupule,* and a small *tube* that contains a specific dehydrated medium (see illustration 4, figure 52.2). The system has a capacity of 23 biochemical tests.

To inoculate each compartment it is necessary to first make up a saline suspension of the unknown organism; then, with the aid of a Pasteur pipette, each compartment is filled with the bacterial suspension. The cupule receives the suspension and allows it to flow into the tube of medium. The dehydrated medium is reconstituted by the saline. To provide anaerobic conditions for some of the compartments it is necessary to add sterile mineral oil to them.

After incubation for 18–24 hours, the reactions are recorded, test reagents are added to some compartments, and test results are tabulated. Once the test results are tabulated, a *profile number* (7 or 9 digits) is computed. By finding the profile number in a code book, the *Analytical Profile Index,* one is able to determine the name of the organism. If no *Analytical Profile Index* is available, characterization can be done by using Chart III in Appendix D.

Although this system is intended for the identification of nonenterics, as well as the Enterobacteriaceae, only the identification of the latter will be pursued in this experiment. Proceed as follows to use the API 20E System to identify your unknown enteric.

First Period

Two things will be accomplished during this period: (1) the oxidase test will be performed if it has not been previously performed, and (2) the API 20E test strip will be inoculated. All steps are illustrated in figure 52.2. Proceed as follows to use this system:

Materials:
   agar slant or plate culture of unknown
   test tube of 5 ml 0.85% sterile saline

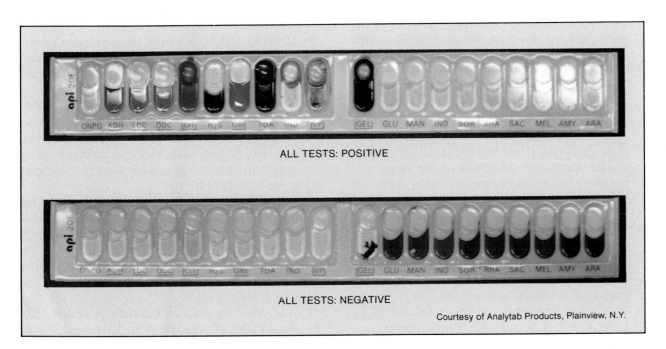

ALL TESTS: POSITIVE

ALL TESTS: NEGATIVE

Courtesy of Analytab Products, Plainview, N.Y.

Figure 52.1   Positive and negative test results on API 20E test strips

API 20E test strip
API incubation tray and cover
squeeze bottle of tap water
test tube of 5 ml sterile mineral oil
Pasteur pipettes (5 ml size)
oxidase test reagent
Whatman No. 2 filter paper
empty Petri dish
Vortex mixer

1. If you haven't already done the **oxidase test** on your unknown, do so at this time. It must be established that your unknown is definitely oxidase-negative before using this system. Use the filter paper method that is described on page 162.

2. Prepare a **saline suspension** of your unknown by transferring organisms from the center of a well-established colony on an agar plate (or from a slant culture) to a tube of 0.85% saline solution. Disperse the organisms well throughout the saline.

3. Label the end strip of the API 20E tray with your name and unknown number. See illustration 2, figure 52.2.

4. Dispense about 5 ml of tap water into the tray with a squeeze bottle. Note that the bottom of the tray has numerous depressions to accept the water.

5. Remove an API 20E test strip from the sealed pouch and place it into the tray (see illustration 3). Be sure to reseal the pouch to protect the remaining strips.

6. Vortex mix the saline suspension to get uniform dispersal, and fill a sterile Pasteur pipette with the suspension. *Take care not to spill any of the organisms on the table or yourself. You may have a pathogen!*

7. Inoculate all the tubes on the test strip with the pipette by depositing the suspension into the cupules as you tilt the API tray (see illustration 4, figure 52.2).
   Important: Slightly *underfill* ADH, LDC, ODC, H₂S, and URE. (Note that the labels for these compartments are underlined on the strip.) Underfilling these compartments leaves room for oil to be added and facilitates interpretation of the results.

8. Since the media in |CIT|, |VP|, and |GEL| compartments require oxygen, *completely fill both the cupule and tube* of these compartments. Note that the labels on these three compartments are bracketed as shown here.

9. To provide anaerobic conditions for the ADH, LDC, ODC, H₂S, and URE compartments, dispense sterile **mineral oil** to the cupules of these

compartments. Use another sterile Pasteur pipette for this step.

10. Place the lid on the incubation tray and incubate at 37° C for 18 to 24 hours. Refrigeration after incubation is not recommended.

Second Period
(Evaluation of Tests)

During this period all reactions will be recorded on the Laboratory Report, test reagents will be added to four compartments, and the seven-digit profile number will be determined so that the unknown can be looked up in the *API 20E Analytical Profile Index*. Proceed as follows:

Materials:
   incubation tray with API 20E test strip
   10% ferric chloride
   Barritt's reagents A and B
   Kovacs' reagent
   nitrite test reagents A and B
   zinc dust or 20-mesh granular zinc
   hydrogen peroxide (1.5%)
   API 20E Analytical Profile Index
   Pasteur pipettes

1. Before any test reagents are added to any of the compartments, consult Chart I, Appendix D, to determine the nature of positive reactions of each test, except TDA, VP, and IND.

2. Refer to Chart II, Appendix D, for an explanation of the 20 symbols that are used on the plastic test strip.

3. Record the results of these tests on the Laboratory Report.

4. **If GLU test is negative** (blue or blue-green), **and there are fewer than three positive reactions** before adding reagents, do not progress any further with this test as outlined here in this experiment. Organisms that are GLU-negative are nonenterics.

   For nonenterics, additional incubation time is required. If you wish to follow through on an organism of this type consult your instructor for more information.

5. **If GLU test is positive** (yellow), **or there are more than three positive reactions,** proceed to add reagents as indicated in the following steps.

6. Add one drop of **10% ferric chloride** to the TDA tube. A positive reaction (brown-red), if it occurs, will occur immediately. A negative reaction color is yellow.

7. Add one drop each of **Barritt's A** and **B solutions** to the VP tube. Read the VP tube within 10 minutes. The pale pink color that occurs imme-

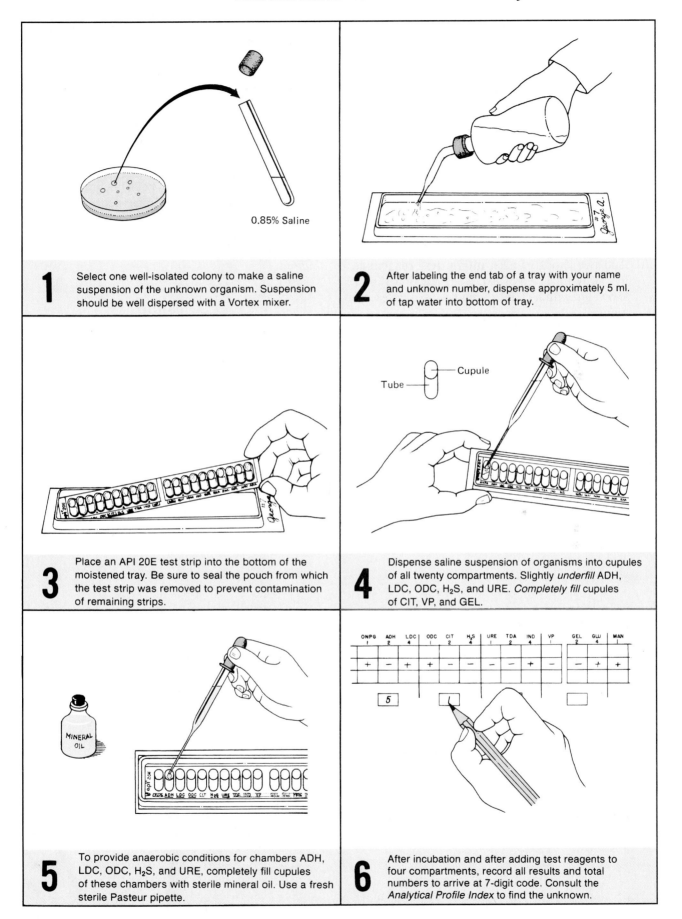

0.85% Saline

1 Select one well-isolated colony to make a saline suspension of the unknown organism. Suspension should be well dispersed with a Vortex mixer.

2 After labeling the end tab of a tray with your name and unknown number, dispense approximately 5 ml. of tap water into bottom of tray.

Cupule

Tube

3 Place an API 20E test strip into the bottom of the moistened tray. Be sure to seal the pouch from which the test strip was removed to prevent contamination of remaining strips.

4 Dispense saline suspension of organisms into cupules of all twenty compartments. Slightly *underfill* ADH, LDC, ODC, H₂S, and URE. *Completely fill* cupules of CIT, VP, and GEL.

MINERAL OIL

ONPG	ADH	LDC	ODC	CIT	H₂S	URE	TDA	IND	VP	GEL	GLU	MAN
1	2	4	1	2	4	1	2	4	1	2	4	1
+	−	+	+	−	−	−	−	+	−	−	+	+

5

5 To provide anaerobic conditions for chambers ADH, LDC, ODC, H₂S, and URE, completely fill cupules of these chambers with sterile mineral oil. Use a fresh sterile Pasteur pipette.

6 After incubation and after adding test reagents to four compartments, record all results and total numbers to arrive at 7-digit code. Consult the *Analytical Profile Index* to find the unknown.

Figure 52.2 **The API 20E procedure**

diately has no significance. A positive reaction is dark pink or red and may take 10 minutes before it appears.

8. Add one drop of **Kovacs' reagent** to the IND tube. Look for a positive (red ring) reaction within 2 minutes.

   After several minutes the acid in the reagent reacts with the plastic cupule to produce a color change from yellow to brownish-red, which is considered negative.

9. Examine the GLU tube closely for evidence of bubbles. Bubbles indicate the reduction of nitrate and the formation of N_2 gas. Note on the Laboratory Report that there is a place to record the presence of this gas.

10. Add two drops of each **nitrite test reagent** to the GLU tube. A positive (red) reaction should show up within 2 to 3 minutes if nitrates are reduced.

    If this test is negative, confirm negativity with **zinc dust** or 20-mesh granular zinc. A pink-orange color after 10 minutes confirms that nitrate reduction did not occur. A yellow color results if N_2 was produced.

11. Add one drop of **hydrogen peroxide** to each of the MAN, INO, and SOR cupules. If catalase is produced, gas bubbles will appear within 2 minutes. Best results will be obtained in tubes that have no gas from fermentation.

Final Confirmation

After all test results have been recorded and the seven-digit profile number has been determined, according to the procedures outlined on the Laboratory Report, identify your unknown by looking up the profile number in the *API 20E Analytical Profile Index*.

Cleanup

When finished with the test strip be sure to place it in a container of disinfectant that has been designated for test strip disposal.

Enterobacteriaceae Identification: The Enterotube II System

The **Enterotube II** miniaturized multitest system was developed by Becton-Dickinson of Cockeysville, Maryland, for rapid identification of Enterobacteriaceae. It incorporates 12 different conventional media and 15 biochemical tests into a single ready-to-use tube that can be simultaneously inoculated in a moment's time with a minimum of equipment.

If you have an unknown gram-negative rod or coccobacillus that appears to be one of the Enterobacteriaceae, you may wish to try this system on it. Before applying this test, however, *make certain that your unknown is oxidase-negative,* since with only a few exceptions, all Enterobacteriaceae are oxidase-negative. If you have a gram-negative rod that is oxidase-positive you might try the *Oxi/Ferm Tube II* instead, which is featured in the next exercise.

Figure 53.1 illustrates an uninoculated tube (upper) and a tube with all positive reactions (lower). Figure 53.2 outlines the entire procedure for utilizing this system.

Each of the 12 compartments of an Enterotube II contains a different agar-based medium. Compartments that require aerobic conditions have openings for access to air. Those compartments that require anaerobic conditions have layers of paraffin wax over the media. Extending through all compartments of the entire tube is an inoculating wire. To inoculate the media, one simply picks up some organisms on the end of the wire and pulls the wire through each of the chambers in a single, rotating action.

After incubation, the reactions in all the compartments are noted and the indole test is performed. The Voges-Proskauer test may also be performed as a confirmation test. Positive reactions are given numerical values, which are totaled to arrive at a five-digit code. Identification of the unknown is achieved by consulting a coding manual, the *Enterotube II Interpretation Guide,* which lists these numerical codes for the Enterobacteriaceae. Proceed as follows to use an Enterotube II in the identification of your unknown.

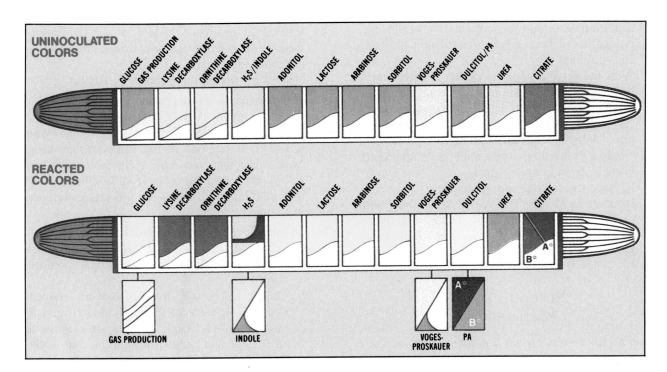

Figure 53.1 Enterotube II color differences between uninoculated and positive tests
Courtesy of Becton-Dickinson, Cockeysville, Maryland.

First Period
Inoculation and Incubation

The Enterotube II can be used to identify Enterobacteriaceae from colonies on agar that have been inoculated from urine, blood, sputum, etc. The culture may be taken from media such as MacConkey, EMB, SS, Hektoen enteric, or trypticase soy agar.

Materials:
> culture plate of unknown
> 1 Enterotube II

1. Write your initials or unknown number on the white paper label on the side of the tube.
2. Unscrew both caps from the Enterotube II. The tip of the inoculating end is under the white cap.
3. *Without heat-sterilizing* the exposed inoculating wire, insert it into a well-isolated colony.
4. Inoculate each chamber by first twisting the wire and then withdrawing it through all 12 compartments. Rotate the wire as you pull it through. See illustration 2, figure 53.2.
5. Again, *without sterilizing,* reinsert the wire, and with a turning motion, force it through all 12 compartments until the notch on the wire is aligned with the opening of the tube. (The notch is about 1⅝″ from handle end of wire.) The tip of the wire should be visible in the citrate compartment. See illustration 3, figure 53.2.
6. Break the wire at the notch by bending, as shown in step 4, figure 53.2. The portion of the wire remaining in the tube maintains anaerobic conditions essential for fermentation of glucose, production of gas, and decarboxylation of lysine and ornithine.
7. With the retained portion of the needle, punch holes through the thin plastic coverings over the small depressions on the sides of the last eight compartments (adonitol, lactose, arabinose, sorbitol, Voges-Proskauer, dulcitol/PA, urea, and citrate). These holes will enable aerobic growth in these eight compartments.
8. Replace the caps at both ends.
9. Incubate at 35° to 37° C for 18 to 24 hours with the Enterotube II lying on its flat surface. *When incubating several tubes together, allow space between them to allow for air circulation.*

Second Period
Reading Results

Reading the results on the Enterotube may be done in one of two ways: (1) by simply comparing the results with information on Chart IV, Appendix D, or (2) by finding the five-digit code number you compute for your unknown in the *Enterotube II Interpre-* tation Guide. Of the two methods, the latter is much preferred. The chart in the appendix should be used *only* if the *Interpretation Guide* is not available.

Whether or not the *Interpretation Guide* is available, these three steps will be performed during this period to complete this experiment: (1) positive test results must *first* be recorded on the Laboratory Report, (2) the indole test, a presumptive test, is performed on compartment 4, and (3) confirmatory tests, if needed, are performed. The Voges-Proskauer test falls in the latter category. Proceed as follows:

Materials:
> Enterotube II, inoculated and incubated
> Kovacs' reagent
> 10% KOH with 0.3% creatine solution
> 5% alpha-naphthol in absolute ethyl alcohol
> syringes with needles, or disposable Pasteur pipettes
> test-tube rack
> Enterotube II Results Pad (optional)
> coding manual: *Enterotube II Interpretation Guide*

1. Compare the colors of each compartment of your Enterotube II with the lower tube illustrated in figure 53.1. With a pencil, mark a small plus (+) or minus (−) near each compartment symbol on the white label on the side of the tube.
2. With a pencil, mark a small plus (+) or minus (−) near each compartment symbol on the white label on the side of the tube.
3. Consult table 53.1 for information as to the significance of each compartment label.
4. Record the results of the tests on the Laboratory Report. *All results must be recorded before doing the indole test.*
5. Record results on the Laboratory Report.

   Important: If at this point you discover that your unknown is GLU-negative, proceed no further with the Enterotube II because your unknown is not one of the Enterobacteriaceae. Your unknown may be *Acinobacter* sp. or *Pseudomonas maltophilia.* If an Oxi/Ferm Tube is available, try it, using the procedure outlined in the next exercise.

6. **Indole Test:** Perform the indole test as follows:
   a. Place the Enterotube II into a test-tube rack with the GLU-GAS compartment pointing upward.
   b. Inject one or two drops of Kovacs' reagent onto the surface of the medium in the H_2S/indole compartment. This may be done with a syringe and needle through the thin Mylar plastic film that covers the flat surface, or with a disposable Pasteur pipette through a small hole made in the Mylar film with a hot inoculating needle.

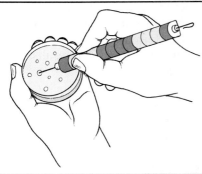

1 Remove organisms from a well-isolated colony. Avoid touching the agar with the wire. To prevent damaging Enterotube II media, do not heat-sterilize the inoculating wire.

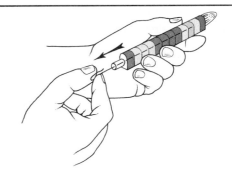

2 Inoculate each compartment by first twisting the wire and then withdrawing it all the way out through the 12 compartments, using a turning movement.

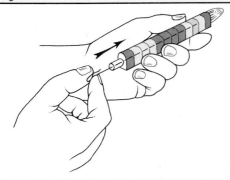

3 Reinsert the wire (without sterilizing), using a turning motion through all 12 compartments until the notch on the wire is aligned with the opening of the tube.

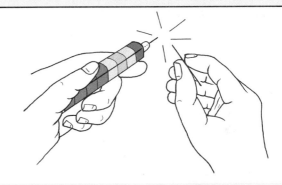

4 Break the wire at the notch by bending. The portion of the wire remaining in the tube maintains anaerobic conditions essential for true fermentation.

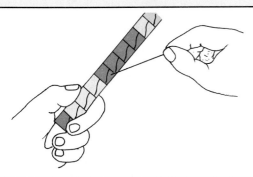

5 Punch holes with broken off part of wire through the thin plastic covering over depressions on sides of the last eight compartments (adonitol through citrate). Replace caps and incubate at 35° C for 18-24 hours.

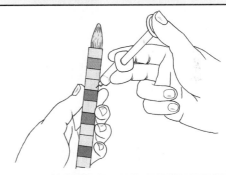

6 After interpreting and recording positive results on the sides of the tube, perform the indole test by injecting 1 or 2 drops of Kovacs' reagent into the H₂S/Indole compartment.

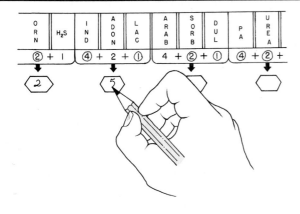

7

Perform the Voges-Proskauer test, if needed for confirmation, by injecting the reagents into the H₂S/indole compartment.

After encircling the numbers of the positive tests on the Laboratory Report, total up the numbers of each bracketed series to determine the 5-digit code number. Refer to the *Enterotube II Interpretation Guide* for identification of the unknown by using the code number.

Figure 53.2 The Enterotube II procedure

c. A positive test is indicated by the development of a **red color** on the surface of the medium or Mylar film within 10 seconds.

7. **Voges-Proskauer Test:** Since this test is used as a confirmatory test, it should be performed *only* when called for in the *Enterotube II Interpretation Guide*. If it is called for, perform the test in the following manner:

a. Use a syringe or Pasteur pipette to inject two drops of potassium hydroxide containing creatine into the V-P section.

b. Inject three drops of 5% alpha-naphthol.

c. A positive test is indicated by a **red color** within 10 minutes.

8. Record the indole and V-P results on the Laboratory Report.

Laboratory Report

Determine the name of your unknown by following the instructions on the Laboratory Report. Note that two methods of making the final determination are given.

Table 53.1 Biochemical Reactions of Enterotube II

SYMBOL	UNINOCULATED COLOR	REACTED COLOR	TYPE OF REACTION
GLU-GAS			**Glucose (GLU)** The end products of bacterial fermentation of glucose are either acid or acid and gas. The shift in pH due to the production of acid is indicated by a color change from red (alkaline) to yellow (acidic). Any degree of yellow should be interpreted as a positive reaction; orange should be considered negative.
			Gas Production (GAS) Complete separation of the wax overlay from the surface of the glucose medium occurs when gas is produced. The amount of separation between the medium and overlay will vary with the strain of bacteria.
LYS			**Lysine Decarboxylase** Bacterial decarboxylation of lysine, which results in the formation of the alkaline end product cadaverine, is indicated by a change in the color of the indicator from pale yellow (acidic) to purple (alkaline). Any degree of purple should be interpreted as a positive reaction. The medium remains yellow if decarboxylation of lysine does not occur.
ORN			**Ornithine Decarboxylase** Bacterial decarboxylation of ornithine causes the alkaline end product putrescine to be produced. The acidic (yellow) nature of the medium is converted to purple as alkalinity occurs. Any degree of purple should be interpreted as a positive reaction. The medium remains yellow if decarboxylation of ornithine does not occur.
H2S/IND			**H$_2$S Production** Hydrogen sulfide, liberated by bacteria that reduce sulfur-containing compounds such as peptones and sodium thiosulfate, reacts with the iron salts in the medium to form a black precipitate of ferric sulfide usually along the line of inoculation. Some **Proteus** and **Providencia** strains may produce a diffuse brown coloration in this medium, which should not be confused with true H$_2$S production.
			Indole Formation The production of indole from the metabolism of tryptophan by the bacterial enzyme tryptophanase is detected by the development of a pink to red color after the addition of Kovac's reagent.

Table 53.1 Biochemical Reactions of Enterotube II

SYMBOL	UNINOCULATED COLOR	REACTED COLOR	TYPE OF REACTION
ADON			**Adonitol** Bacterial fermentation of adonitol, which results in the formation of acidic end products, is indicated by a change in color of the indicator present in the medium from red (alkaline) to yellow (acidic). Any sign of yellow should be interpreted as a positive reaction; orange should be considered negative.
LAC			**Lactose** Bacterial fermentation of lactose, which results in the formation of acidic end products, is indicated by a change in color of the indicator present in the medium from red (alkaline) to yellow (acidic). Any sign of yellow should be interpreted as a positive reaction; orange should be considered negative.
ARAB			**Arabinose** Bacterial fermentation of arabinose, which results in the formation of acidic end products, is indicated by a change in color from red (alkaline) to yellow (acidic). Any sign of yellow should be interpreted as a positive reaction; orange should be considered negative.
SORB			**Sorbitol** Bacterial fermentation of sorbitol, which results in the formation of acidic end products, is indicated by a change in color from red (alkaline) to yellow (acidic). Any sign of yellow should be interpreted as a positive reaction; orange should be considered negative.
V.P.			**Voges-Proskauer** Acetylmethylcarbinol (acetoin) is an intermediate in the production of butylene glycol from glucose fermentation. The presence of acetoin is indicated by the development of a red color within 20 minutes. Most positive reactions are evident within 10 minutes.
DUL-PA			**Dulcitol** Bacterial fermentation of dulcitol, which results in the formation of acidic end products, is indicated by a change in color of the indicator present in the medium from green (alkaline) to yellow or pale yellow (acidic). **Phenylalanine Deaminase** This test detects the formation of pyruvic acid from the deamination of phenylalanine. The pyruvic acid formed reacts with a ferric salt in the medium to produce a characteristic black to smoky gray color.
UREA			**Urea** The production of urease by some bacteria hydrolyzes urea in this medium to produce ammonia, which causes a shift in pH from yellow (acidic) to reddish-purple (alkaline). This test is strongly positive for **Proteus** in 6 hours and weakly positive for **Klebsiella** and some **Enterobacter** species in 24 hours.
CIT			**Citrate** Organisms that are able to utilize the citrate in this medium as their sole source of carbon produce alkaline metabolites that change the color of the indicator from green (acidic) to deep blue (alkaline). Any degree of blue should be considered positive.

Courtesy of Becton-Dickinson, Cockeysville, Maryland.

54

O/F Gram-Negative Rods Identification:
The Oxi/Ferm Tube II System

The Oxi/Ferm Tube II, produced by Becton-Dickinson, takes care of the identification of the oxidase-positive, gram-negative bacteria that cannot be identified by using the Enterotube II system. The two multitest systems were developed to work together. If an unknown gram-negative rod is oxidase-negative, the Enterotube II is used. If the organism is oxidase-positive, the Oxi/Ferm Tube II must be used. Whenever an oxidase-negative gram-negative rod turns out to be glucose-negative on the Enterotube II test, one must move on to use the Oxi/Ferm Tube II.

The Oxi/Ferm Tube II system is intended for the identification of non-fastidious species of oxidative-fermentative gram-negative rods from clinical specimens. This includes the following genera: *Aeromonas, Plesiomonas, Vibrio, Achromobacter, Alcaligenes, Bordetella, Moraxella,* and *Pasteurella.* Some other gram-negative bacteria can also be identified with additional biochemical tests. The system incorporates

12 different conventional media that can be inoculated simultaneously in a moment's time with a minimum of equipment. A total of 14 physiological tests are performed.

Like the Enterotube II system, the Oxi/Ferm Tube II has an inoculating wire that extends through all 12 compartments of the entire tube. To inoculate the media, one simply picks up some organisms on the end of the wire and pulls the wire through each of the chambers in a rotating action.

After incubation, the results are recorded and Kovacs' reagent is injected into one of the compartments to perform the indole test. Positive reactions are given numerical values that are totaled to arrive at a five-digit code. By looking up the code in an Oxi/Ferm *Biocode Manual,* one can quickly determine the name of the unknown and any tests that might be needed to confirm the identification.

Figure 54.1 illustrates an uninoculated tube and a tube with all positive reactions. Figure 54.2 illus-

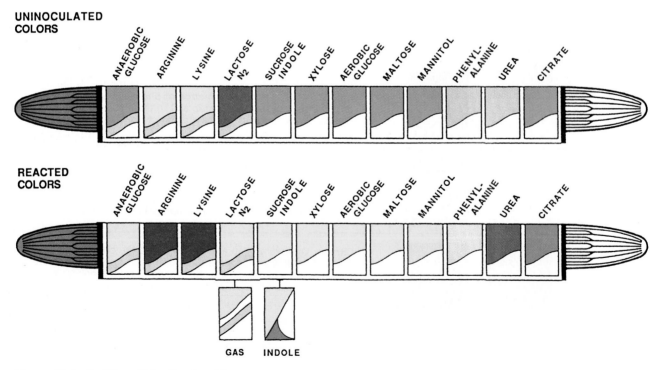

Figure 54.1 Oxi/Ferm Tube II color differences between uninoculated and positive tests
Courtesy of Becton-Dickinson, Cockeysville, Maryland.

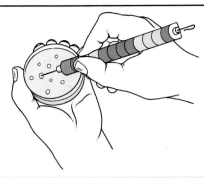

1 Remove organisms from a well-isolated colony. Avoid touching the agar with the wire. To prevent damaging Enterotube II media, do not heat-sterilize the inoculating wire.

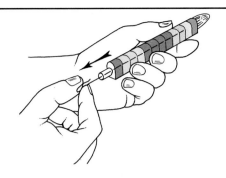

2 Inoculate each compartment by first twisting the wire and then withdrawing it all the way out through the 12 compartments, using a turning movement.

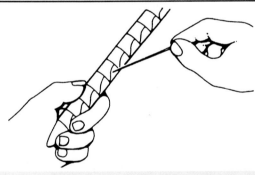

3 Reinsert the wire (without sterilizing), using a turning motion through all 12 compartments until the notch on the wire is aligned with the opening of the tube.

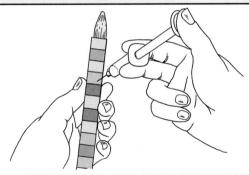

4 Break the wire at the notch by bending. The portion of the wire remaining in the tube maintains anaerobic conditions essential for true fermentation.

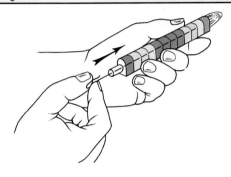

5 Punch holes with broken off part of wire through the thin plastic covering over depressions on sides of the last eight compartments (sucrose/indole through citrate). Replace caps and incubate at 35° C for 18-24 hours.

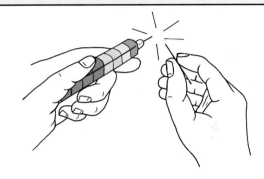

6 After interpreting and recording positive results on the sides of the tube, perform the indole test by injecting 1 or 2 drops of Kovacs' reagent into the sucrose/Indole compartment.

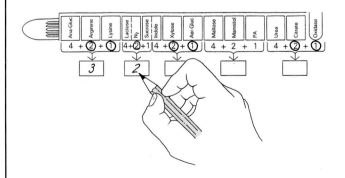

7

After encircling the numbers of the positive tests on the Laboratory Report, total up the numbers of each bracketed series to determine the 5-digit code number. Refer to the *Biocode Manual* for identification of the unknown by using the code number.

Figure 54.2 The Oxi/Ferm Tube II procedure

trates the entire procedure for utilizing this system. A minimum of two periods are required to use this system. Proceed as follows:

First Period
Inoculation and Incubation

The Oxi/Ferm Tube II must be inoculated with a large inoculum from a well-isolated colony. Culture purity, of course, is of paramount importance. If there is any doubt of purity, a TSA plate should be inoculated and incubated at 35° C for 24 hours, followed by 24 hours incubation at room temperature. If no growth occurs on TSA, but growth does occur on blood agar, the organism has special growth requirements. *Such organisms are too fastidious and cannot be identified with the Oxi/Ferm Tube II.*

Materials:
>   culture plate of unknown
>   1 Oxi/Ferm Tube II
>   1 plate of trypticase soy agar (TSA)
>   (for purity check, if needed)

1. Write your initials or unknown number on the side of the tube.
2. Unscrew both caps from the Oxi/Ferm Tube II. The tip of the inoculating end is under the white cap.
3. *Without heat-sterilizing* the exposed inoculating wire, insert it into a well-isolated colony. Do not puncture the agar.
4. Inoculate each chamber by first twisting the wire and then withdrawing it through all 12 compartments. Rotate the wire as you pull it through. See illustration 2, figure 54.2.
5. If a purity check of the culture is necessary, streak a Petri plate of TSA with the inoculating wire that has just been pulled through the tube. **Do not flame.**
6. Again, *without sterilizing,* reinsert the wire, and with a turning motion, force it through all 12 compartments until the notch on the wire is aligned with the opening of the tube. (The notch is about 1⅝″ from the handle end of the wire.) The tip of the wire should be visible in the citrate compartment. See illustration 3, figure 54.2.
7. Break the wire at the notch by bending, as noted in step 4, figure 54.2. The portion of the wire remaining in the tube maintains anaerobic conditions essential for true fermentation.
8. With the retained portion of the needle, punch holes through the thin plastic coverings over the small depressions on the sides of the last eight compartments (sucrose/indole, xylose, aerobic glucose, maltose, mannitol, phenylalanine, urea, and citrate). These holes will enable aerobic growth in these eight compartments.

9. Replace both caps on the tube.
10. Incubate at 35° to 37° C for 24 hours, with the tube lying on its flat surface or upright. At the end of 24 hours inspect the tube to check results and continue incubation for another 24 hours. The 24-hour check may be needed for doing confirmatory tests as required in the *Biocode Manual.* Occasionally, an Oxi/Ferm Tube II should be incubated longer than 48 hours.

Second Period
Evaluation of Tests

During this period you will record the results of the various tests on your Oxi/Ferm Tube II, do an indole test, tabulate your results, use the *Biocode Manual,* and perform any confirmatory tests called for. Proceed as follows:

Materials:
>   Oxi/Ferm Tube II, inoculated and incubated
>   Kovacs' reagent
>   syringes with needles, or disposable Pasteur pipettes
>   Becton-Dickinson *Biocode Manual* (a booklet)

1. Compare the colors of each compartment of your Oxi/Ferm Tube II with the lower tube illustrated in figure 54.1.
2. With a pencil, mark a small plus (+) or minus (−) near each compartment symbol on the white label on the side of the tube.
3. Consult Table 54.1 for information as to the significance of each compartment label.
4. Record the results of all the tests on the Laboratory Report. *All results must be recorded before doing the indole test.*
5. **Indole Test** (illustration 6, figure 54.2): Do an indole test by injecting two or three drops of Kovacs' reagent through the flat, plastic surface into the sucrose/indole compartment. Release the reagent onto the inside flat surface and allow it to drop down onto the agar.

   If a Pasteur pipette is used instead of a syringe needle, it will be necessary to form a small hole in the Mylar film with a hot inoculating needle to admit the tip of the Pasteur pipette.

   A positive test is indicated by the development of a **red color** on the surface of the medium or Mylar film within 10 seconds.
6. Record the results of the indole test on the Laboratory Report.

Laboratory Report

Follow the instructions on the Laboratory Report for determining the five-digit code. Use the *Biocode Manual* booklet for identifying your unknown.

Table 54.1   Biochemical Reactions of the Oxi/Ferm Tube II

Anaerobic glucose			Positive fermentation is shown by change in color from green (neutral) to yellow (acid). Most oxidative-fermentative, gram-negative rods are negative.
Arginine dihydrolase			Decarboxylation of arginine results in the formation of alkaline end products that changes bromcresol purple from yellow (acid) to purple (alkaline). Gray is negative.
Lysine			Decarboxylation of lysine results in the formation of alkaline end products that changes bromcresol purple from yellow (acid) to purple (alkaline). Gray is negative.
Lactose			Fermentation of lactose changes the color of the medium from red (neutral) to yellow (acid). Most O/F gram-negative rods are negative.
N_2 gas production			Gas production causes separation of wax overlay from medium. Occasionally, the gas will also cause separation of the agar from the compartment wall.
Sucrose			Bacterial oxidation of sucrose causes a change in color from green (neutral) to yellow (acid).
Indole			The bacterial enzyme tryptophanase metabolizes tryptophan to produce indole. Detection is by adding Kovacs' reagent to the compartment 48 hours after incubation.
Xylose			Bacterial oxidation of xylose causes a color change of green (neutral) to yellow (acid).
Aerobic glucose			Bacterial oxidation of glucose causes a color change of green (neutral) to yellow (acid).
Maltose			Bacterial oxidation of maltose causes a color change of green (neutral) to yellow (acid).
Mannitol			Bacterial oxidation of this carbohydrate is evidenced by a change in color from green (neutral) to yellow (acid).
Phenylalanine			Pyruvic acid is formed by deamination of phenylalanine. The pyruvic acid reacts with a ferric salt to produce a brownish tinge.
Urea			The production of ammonia by the action of urease on urea increases the alkalinity of the medium. The phenol red in this medium changes from beige (acid) to pink or purple. Pale pink should be considered negative.
Citrate			Organisms that grow on this medium are able to utilize citrate as their sole source of carbon. Utilization of citrate raises the alkalinity of the medium. The color changes from green (neutral) to blue (alkaline).

55

Staphylococcus Identification: The API Staph-Ident System

The **API Staph-Ident System,** produced by Analytab Products of Plainview, New York, was developed to provide a rapid (5-hour) method for identifying 13 of the most clinically important species of staphylococci. This system consists of 10 microcupules that contain dehydrated substrates and/or nutrient media. Except for the coagulase test, all the tests that are needed for the identification of staphylococci are included on the strip.

Figure 55.1 illustrates two inoculated strips: the lower one just after inoculation and the upper one with all positive reactions. Note that the appearance of each microcupule undergoes a pronounced color change when a positive reaction occurs.

Figure 55.2 illustrates the overall procedure. The first step is to make a saline suspension of the organism from an isolated colony. A Staph-Ident strip is then placed in a tray that has a small amount of water added to it to provide humidity during incubation. Next, a sterile Pasteur pipette is used to dispense two to three drops of the bacterial suspension to each microcupule. The inoculated tray is then covered and incubated aerobically at 35° to 37° C for 5 hours. After incubation, a few drops of Staph-Ident reagent are added to the tenth microcupule and the results are read immediately. Finally, a four-digit profile is computed that is used to determine the species from a chart in Appendix D.

As simple as this system might seem, there are a few limitations that one must keep in mind. Final species determination by a competent microbiologist must take into consideration other factors such as the source of the specimen, the catalase reaction, colony characteristics, and antimicrobial susceptibility pattern. Very often there are confirmatory tests that must also be made.

If you have been working with an unknown that appears to be one of the staphylococci, use this system to confirm your conclusions. If you have already done the coagulase test and have learned that your organism is coagulase-negative, this system will enable you to identify one of the numerous coagulase-negative species that are not identifiable by the procedures in Exercise 77.

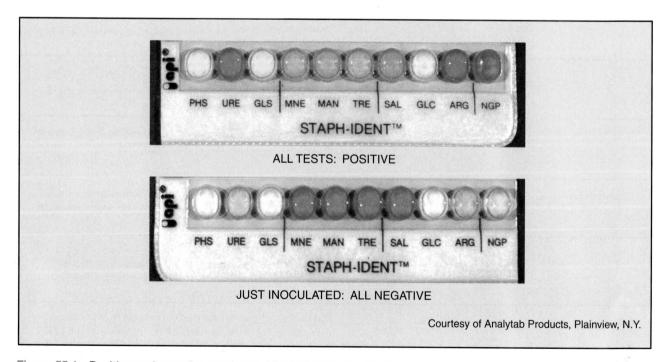

ALL TESTS: POSITIVE

JUST INOCULATED: ALL NEGATIVE

Figure 55.1 Positive and negative results on API Staph-Ident test strips

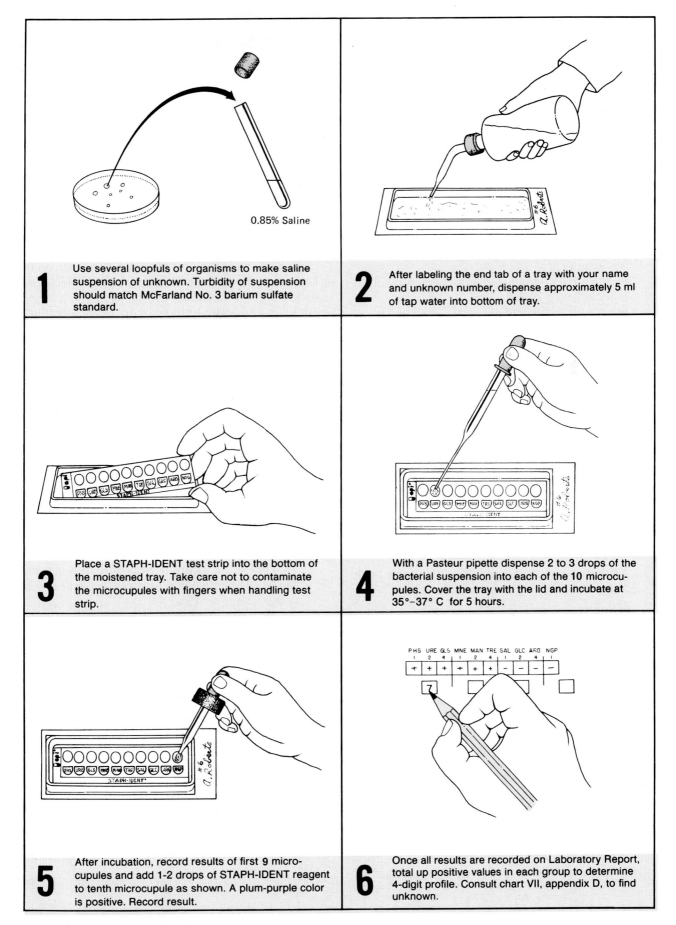

1 Use several loopfuls of organisms to make saline suspension of unknown. Turbidity of suspension should match McFarland No. 3 barium sulfate standard.

0.85% Saline

2 After labeling the end tab of a tray with your name and unknown number, dispense approximately 5 ml of tap water into bottom of tray.

3 Place a STAPH-IDENT test strip into the bottom of the moistened tray. Take care not to contaminate the microcupules with fingers when handling test strip.

4 With a Pasteur pipette dispense 2 to 3 drops of the bacterial suspension into each of the 10 microcupules. Cover the tray with the lid and incubate at 35°–37° C for 5 hours.

5 After incubation, record results of first 9 microcupules and add 1-2 drops of STAPH-IDENT reagent to tenth microcupule as shown. A plum-purple color is positive. Record result.

6 Once all results are recorded on Laboratory Report, total up positive values in each group to determine 4-digit profile. Consult chart VII, appendix D, to find unknown.

Figure 55.2 The API Staph-Ident procedure

First Period

(Inoculations and Coagulase Test)

Before setting up this experiment, take into consideration that it must be completed at the end of 5 hours. Holding the test strips overnight is not recommended.

Materials:

> API Staph-Ident test strip
> API incubation tray and cover
> blood agar plate culture of unknown (must not have been incubated over 30 hours)
> blood agar plate (if needed for purity check)
> serological tube of 2 ml sterile saline
> test-tube rack
> sterile swabs (optional in step 2 below)
> squeeze bottle of tap water
> tubes containing McFarland No. 3 ($BaSO_4$) standard (see Appendix B)
> sterile Pasteur pipette (5 ml size)

1. If the **coagulase test** has not been performed, refer to Exercise 77, page 250, for the procedure and perform it on your unknown.

2. Prepare a saline suspension of your unknown by transferring organisms to a tube of sterile saline from one or more colonies with a loop or sterile swab. Turbidity of the suspension should match a tube of No. 3 McFarland barium sulfate standard.

   Important: Do not allow the bacterial suspension to go unused for any great length of time. Suspensions older than 15 minutes become less effective.

3. Label the end strip of the tray with your name and unknown number. See illustration 2, figure 55.2.

4. Dispense about 5 ml of tap water into the bottom of the tray with a squeeze bottle. Note that the bottom of the tray has numerous depressions to accept the water.

5. Remove the API test strip from its sealed envelope and place the strip in the bottom of the tray.

6. After shaking the saline suspension to disperse the organisms, fill a sterile Pasteur pipette with the bacterial suspension.

7. Inoculate each of the microcupules with two or three drops of the suspension. If a purity check is necessary, use the excess suspension to inoculate another blood agar plate.

8. Place the plastic lid on the tray and incubate the strip aerobically for 5 hours at 35° to 37° C.

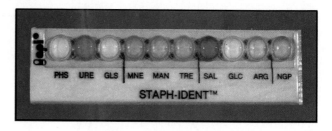

Figure 55.3 **Test results of a strip inoculated with S. aureus.**
Courtesy of Analytab Products.

Second Period

(Five Hours Later)

During this period the results will be recorded on the Laboratory Report, the profile number will be determined, and the unknown will be identified by looking up the number on the *Staph-Ident Profile Register* (or Chart VII, Appendix D).

Materials:

> API Staph-Ident test strip (incubated 5 hours)
> 1 bottle of Staph-Ident reagent (room temperature)
> *Staph-Ident Profile Register*

1. After 5 hours incubation, refer to Chart V, Appendix D, to interpret and record the results of the first nine microcupules (PHS through ARG).

2. Record the results on the Profile Determination Table on the Laboratory Report. Chart VI, Appendix D, reveals the biochemistry involved in these tests.

3. Add one or two drops of **Staph-Ident reagent** to the NGP microcupule. Allow 30 seconds for the color change to occur.

   A positive test results in a change of color to plum-purple. Record the results of this test.

4. Construct the profile number according to the instructions on the Laboratory Report and determine the name of your unknown.

   If no recent *Profile Register* is available, use Chart VII, Appendix D. Since the *API Register* is constantly being updated, the one in the appendix may be out of date.

Disposal

Once all the information has been recorded be sure to place the entire incubation unit in a receptacle that is to be autoclaved.

PART 10 Microbiology of Soil

With ideal temperature and moisture conditions, soils provide excellent culture media for many kinds of microorganisms. This is especially true of cultivated and improved soils. In many different ways these organisms contribute to the fertility of the very medium they inhabit. The action of certain autotrophic protists on minerals produces substances, organic and inorganic, that are available to plants. Maintaining a proper balance of available nitrogen to photosynthetic plants is one of the most important activities of some forms of bacteria. The decomposition of lifeless plant and animal tissues returns materials to soils in a form that is reusable by plants. Since higher plants are indispensable to the welfare of animal life, it is evident that their complete destruction would result in the obliteration of animal life as we know it on this planet.

Soils contain enormous numbers and kinds of micro-organisms. In addition to the multitudes of bacteria, there are protozoans, yeasts, molds, algae, and microscopic worms in unbelievable numbers. Types that predominate will depend on the composition of the soil, moisture, pH, and other related environmental factors. No one technique can be used for counting all organisms since such great variability in types exists.

In this exercise we will use the plate count procedure that was used in Exercise 23 to determine the numbers of bacteria, actinomycetes, and molds. It will be necessary to use different kinds of media for each group of organisms. For economy of time and materials, the class will be divided into three groups.

Materials:
> 1 bottle (50 ml) of nutrient agar (⅓ of class)
> 1 bottle (50 ml) of glucose peptone acid agar (⅓ of class)
> 1 bottle (50 ml) of glycerol yeast extract agar (⅓ of class)
> 3 sterile water blanks (99 ml) (per pair of students)
> 4 sterile Petri plates per student
> 1.1 ml dilution pipettes
> soil sample

1. Liquefy and cool to 50° C a bottle of medium to be used for the organisms that you will attempt to count. The chart below indicates your assignment:

2. Label four Petri plates according to type of organisms and dilutions. Since the numbers of each type will vary, different dilutions are necessary.

Bacteria	Actinomycetes	Molds
1:10,000	1:1000	1:100
1:100,000	1:10,000	1:1000
1:1,000,000	1:100,000	1:10,000
1:10,000,000	1:1,000,000	1:100,000

3. Label three 99 ml sterile water blanks as you did in Exercise 23.

4. Add 1 gram of soil to blank A, shake vigorously for 5 minutes, and carry out the dilution of blanks B and C.

5. With 1.1 ml pipettes, distribute the proper amounts of water from the blanks to the plates for your final dilutions. See figure 23.1. For the 1:1000 and 1:100 plates, you will need 0.1 ml and 1.0 ml, respectively, from blank A.

6. Pour the appropriate medium into each plate and allow them to cool.

7. Incubate the plates in your locker for 3 to 7 days.

8. Count the colonies, using the procedures outlined in Exercise 23. Record the results on the first portion of Laboratory Report 56, 57.

Student Number	Organisms	Medium
1, 4, 7, 10, 13, 16, 19, 22, 25, 28	Bacteria	Nutrient agar
2, 5, 8, 11, 14, 17, 20, 23, 26, 29	Actinomycetes	Glycerol yeast extract agar
3, 6, 9, 12, 15, 18, 21, 24, 27, 30	Molds	Glucose peptone acid agar

The nitrogen in most plants and animals exists in the form of protein. When they die, the protein is broken down to amino acids, which, in turn, are deaminated to liberate ammonia. This process of the production of ammonia from organic compounds is called **ammonification.** Since most bacteria and plants can assimilate ammonia, this is a very important step in the nitrogen cycle. The majority of bacteria in soil are able to take part in this process.

In this exercise we will inoculate peptone broth with a sample of soil, incubate it for a few days, and test for ammonia production. After a total of 7 days incubation it will be tested again to see if the amount of ammonia has increased.

First Period
(Inoculation)

Materials:
   2 tubes of peptone broth
   rich garden soil

1. Inoculate one tube of peptone broth with a loopful of soil. Save the other tube for a control.
2. Incubate the tube at room temperature for 3–4 days and 7 days.

Second and Third Periods
(Ammonia Detection)

After 3 or 4 days, test the medium for ammonia with the following procedure. Repeat these tests again after a total of 7 days of incubation.

Materials:
   Nessler's reagent
   bromthymol blue and indicator chart
   spot plate

1. Deposit a drop of **Nessler's reagent** into two separate depressions of a spot plate.
2. Add a loopful of the inoculated peptone broth to one depression and a loopful from the sterile uninoculated tube in the other. Interpretation of ammonia presence is as follows:

   Faint yellow color—small amount of ammonia
   Deep yellow—more ammonia
   Brown precipitate—large amount of ammonia

3. Check the pH of the two tubes by placing several loopfuls of each in separate depressions on the spot plate and adding one drop of **bromthymol blue** to each one. Compare the color with a color chart or set of indicator tubes to determine the pH.
4. Record results on the last portion of Laboratory Report 56, 57.

58

Nitrification in Soil

The conversion of ammonia to nitrate is a process called **nitrification.** This is an oxidation reaction that occurs in two steps and is accomplished by two genera of bacteria: *Nitrosomonas* and *Nitrobacter.*

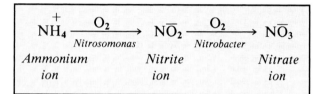

Nitrosomonas and *Nitrobacter* are autotrophic and appear to be the principal organisms in soil that can perform these reactions. Although there are some heterotrophs that can produce nitrites and nitrates in similar reactions, the amounts that they produce are insignificant. These organisms are very small gram-negative rods that do best in a completely inorganic medium, using carbon dioxide as their carbon source. Many kinds of organic matter are toxic to them, especially those substances that have free amino groups.

First Period
(Inoculations)

In this exercise we will inoculate two separate media with soil. A medium containing only the ammonium

ion as a nitrogen source will be used to detect the presence of nitrite-producing bacteria. For the nitrate-producing forms, a medium containing the nitrite ion will be used. The compositions of these two media are shown in the box at the bottom of this page.

Materials:
  6 oz prescription bottle with 20 ml of
      ammonium medium
  6 oz prescription bottle with 20 ml of nitrite
      medium

1. Inoculate one of each bottle of media with several loopfuls of soil.
2. Incubate these bottles on their flat sides at room temperature for one week.

Second and Third Periods
(Tests for Nitrites and Nitrates)

After 7 days incubation, and again 7 days beyond that, perform tests for nitrites and nitrates as follows:

Materials:
  Nessler's reagent
  Trommsdorf's reagent
  sulfuric acid (one part concentrated acid, three
      parts water)
  sulfuric acid, concentrated
  diphenylamine reagent
  glass rod
  spot plate

Inorganic Compound	Ammonium Medium	Nitrite Medium
Ammonium sulfate [$(NH_4)_2SO_4$]	2.0 gm	
Sodium nitrite [$NaNO_3$]		1.0 gm
Magnesium sulfate [$MgSO_4 \cdot H_2O$]	0.5 gm	0.5 gm
Ferrous sulfate [$FeSO_4 \cdot 7H_2O$]	0.03 gm	0.03 gm
Sodium chloride [$NaCl$]	0.3 gm	0.3 gm
Magnesium carbonate [$MgCO_3$]	10.0 gm	
Sodium carbonate [Na_2CO_3]		1.0 gm
Dipotassium phosphate [K_2HPO_4]	1.0 gm	1.0 gm
Water	1000.0 ml	1000.0 ml

Nitrite Production Test the ammonium medium for the nitrite ion as follows:

1. Mix 3 drops of Trommsdorf's reagent with 1 drop of dilute sulfuric acid (1:3 conc) on a spot plate.
2. With a glass rod, transfer a drop of the culture from the ammonium medium to this reagent mixture and stir. To avoid false results, don't use a wire loop. If nitrites are present, an intense **blue-black color** will appear.
3. Do a test for ammonia with Nessler's reagent. When the medium becomes negative, all ammonium ions have been oxidized to nitrite.
4. Make a gram-stained slide each time tests are performed, and record your results on the Laboratory Report.

Nitrate Production The test for nitrates is performed with diphenylamine. This reagent produces a blue-black color in the presence of either nitrates or nitrites; hence, it is necessary to make sure that no nitrites are present when it is used as a test for nitrates.

1. First, test your nitrite medium culture with Trommsdorf's reagent to establish the absence of nitrites.
2. If nitrites are lacking, test for nitrates with diphenylamine by mixing 1 drop of diphenylamine, 2 drops of concentrated sulfuric acid, and 1 drop of the culture on the plate.

   The **blue-black color** will be evidence of nitrate production. Record your observations on the Laboratory Report.
3. Make a gram-stained slide each time a test is performed, and record your observations.

Laboratory Report

Complete the first portion of Laboratory Report 58, 59.

Among the most beneficial microorganisms of the soil are those that are able to convert gaseous nitrogen of the air to "fixed forms" of nitrogen that can be utilized by other bacteria and plants. Without these nitrogen-fixers, life on this planet would probably disappear within a relatively short period of time.

The utilization of free nitrogen gas by fixation can be accomplished by organisms that are able to produce the essential enzyme *nitrogenase*. This enzyme, in the presence of traces of molybdenum, enables the organisms to combine atmospheric nitrogen with other elements to form organic compounds in living cells. In organic combinations nitrogen is more reduced than when it is free. From these organic compounds, upon their decomposition, the nitrogen is liberated in a fixed form, available to plants either directly or through further microbial action.

The most important nitrogen-fixers belong to two families: **Azotobacteraceae** and **Rhizobiaceae.** Other organisms of less importance that have this ability are a few strains of *Klebsiella,* some species of *Clostridium,* the cyanobacteria, and photosynthetic bacteria.

In this exercise we will concern ourselves with two activities: the isolation of *Azotobacter* from garden soil and the demonstration of *Rhizobium* in root nodules of legumes.

Azotobacteraceae

Bergey's Manual of Systematic Bacteriology, volume 1, section 4, lists two genera of bacteria in family Azotobacteraceae that fix nitrogen as free-living organisms under aerobic conditions: *Azotobacter* and *Azomonas.* The basic difference between these two genera is that *Azotobacter* produces drought-resistant cysts and *Azomonas* does not. Aside from the presence or absence of cysts, these two genera are very similar. Both are large gram-negative motile rods that may be ovoid or coccoidal in shape (pleomorphic). Catalase is produced by both genera. There are six species of *Azotobacter* and three species of *Azomonas.*

Figure 59.1 illustrates the overall procedure that we will use for isolating Azotobacteraceae from garden soil. Note that a small amount of rich garden soil is added to a bottle of nitrogen-free medium that contains glucose as a carbon source. The bottle of medium is incubated in a horizontal position for 4 to 7 days at 30° C.

After incubation, a wet mount slide is made from surface growth to see if typical azotobacter-like organisms are present. If organisms are present, an agar plate of the same medium, less iron, is used to streak out for isolated colonies. After another 4 to 7 days incubation, colonies on the plate are studied and more slides are made in an attempt to identify the isolates.

The N_2-free medium used here contains glucose for a carbon source and is completely lacking in nitrogen. It is selective in that only organisms that can use nitrogen from the air and use the carbon in glucose will grow on it. All species of *Azotobacter* and *Azomonas* are able to grow on it. The metallic ion molybdenum is included to activate the enzyme nitrogenase, which is involved in this process.

First Period (Enrichment)

Proceed as follows to inoculate a bottle of the nitrogen-free glucose medium with a sample of garden soil.

Materials:
   1 bottle (50 ml) N_2-free glucose medium
   rich garden soil (neutral or alkaline)
   spatula

1. With a small spatula, put about 1 gm of soil into the bottle of medium. Cap the bottle and shake it sufficiently to mix the soil and medium.
2. Loosen the cap slightly and incubate the bottle at 30° C for 4 to 7 days. Since the organisms are strict aerobes, it is best to incubate the bottle horizontally to provide maximum surface exposure to air.

Second Period (Plating Out)

During this period a slide will be made to make certain that organisms have grown on the medium. If the culture has been successful, a streak plate will be made on nitrogen-free, iron-free agar. Proceed as follows:

Materials:
microscope slides and cover glasses
microscope with phase-contrast optics
1 agar plate of nitrogen-free, iron-free glucose medium

1. After 4 to 7 days incubation, carefully move the bottle of medium to your desktop *without agitating the culture.*
2. Make a wet mount slide with a few loopfuls from the surface of the medium and examine under oil immersion, preferably with phase-contrast optics. Look for large ovoid to rod-shaped organisms, singly and in pairs.

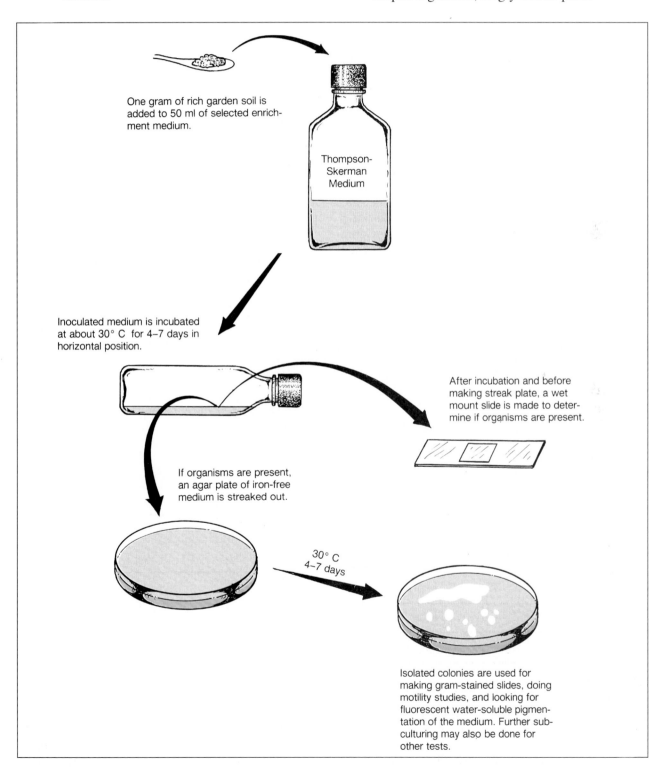

Figure 59.1 Enrichment and isolation procedure for *Azotobacter* and *Azomonas*

3. If azotobacter-like organisms are seen, note whether or not they are motile and if cysts are present. Cysts look much like endospores in that they are refractile. Since cysts often take 2 weeks to form, they may not be seen.

4. If the presence of azotobacter-like organisms is confirmed, streak an agar plate of nitrogen-free, iron-free medium, using a good isolation streak pattern. Ferrous sulfate has been left out of this medium to facilitate the detection of water-soluble pigments.

5. Incubate the plate at 30° C for 4 or 5 days. A longer period of incubation is desirable for cyst formation.

Third Period (Identification)

Azotobacter chroococcum is the type species of genus *Azotobacter*. The cells are 1.5–2.0 micrometers in diameter and pleomorphic, ranging from rods to coccoidal in shape. They occur singly, in pairs, and in irregular clumps. Motility exists with peritrichous flagella. Drought-resistant cysts are produced. They are strict aerobes. Catalase is produced and starch is hydrolyzed. Morphologically, the other five species of this genus look very much like this organism.

Azomonas agilis is the type species of genus *Azomonas*. Except for the absence of cysts, this species and the other two species in this genus are morphologically very similar to *Azotobacter chroococcum*. Practically all of them produce water-soluble fluorescent pigments.

Differentiation of the six species of the genus *Azotobacter* and three species of *Azomonas* is based primarily on the presence or absence of motility, the type of water-soluble pigment produced, and carbon source utilization. Table 59.1 reveals how the organisms can be differentiated. For presumptive identification, use the following character information to identify your isolate.

Materials:
  agar plate from previous period
  ultraviolet lamp

Motility Note in table 59.1 that four species of *Azotobacter* and all three species of *Azomonas* are motile.

Pigmentation Although these organisms produce both water-soluble and water-insoluble pigments, only the water-soluble ones (those capable of diffus-

Table 59.1 Differential characteristics of the Azotobacteraceae

		Cysts	Motility	Long Filaments	Brown-Black	Brown-Black to Red-Violet	Red-Violet	Green	Yellow-Green Fluorescent	Blue-White Fluorescent
						Water-Soluble Pigments				
Azotobacter	A. chroococcum	+	+	−	−	−	−	−	−	−
	A. vinelandii	+	+	−	−	−	d	d	+	−
	A. beijerinckii	+	−	−	−	−	−	−	−	
	A. nigricans	+	−	−	d	+	d	−		
	A. armeniacus	+	+	−	−	+	+	−		
	A. paspali	+	+	+	−	−	+	−	+	
Azomonas	A. agilis	−	+	−	−	−	−	−	+	+
	A. insignis	−	+	−	d¹	−	d	−	d	−
	A. macrocytogenes	−	+	−	−	−	−	−	d	d

d = 11%–89% positive
d¹ = 11%–89% positive on benzoate

From *Bergey's Manual of Systematic Bacteriology*, volume 1, section 4

ing into an agar medium) are important from the standpoint of species differentiation.

Note in table 59.1 that two of the water-soluble pigments are fluorescent: one is yellow-green and the other is blue-white. To observe fluorescence the cultures must be exposed to ultraviolet light (wavelength 364 nm) in a darkened room. The characteristics of pigment production in each species may be limited by certain factors, as indicated below:

Brown-black: If the colonies produce this hue of diffusible pigment without becoming red-violet, the organism is *A. nigricans.* Although the table indicates that *A. insignis* can produce the brown-black pigment, it can do so only if the medium contains benzoate.

Brown-black to red-violet: As indicated in the table, *A. nigricans* and *A. armeniacus* are the only genera that produce this type of pigment. Motility is a good way to differentiate these two species.

Red-violet: Although table 59.1 reveals that five species can produce this color of diffusible pigment, one (*A. insignis*) cannot produce it on the medium we used. A red-violet isolate is unlikely to be *A. paspali* because this organism has been isolated from the rhizosphere of only one species of grass (*Paspalum notatum*). Thus, isolates that produce this pigment are probably one of the other three in the table.

Green: Note that only *A. vinelandii* can produce this water-soluble pigment; however, only 11%–89% of them produce it.

Yellow-green fluorescent: *A. vinlandii, A. paspali,* and all species of *Azomonas* are able to produce this pigment on the medium we used. Check for fluorescence with an ultraviolet lamp in a darkened room.

Blue-white fluorescent: Note in table 59.1 that two species of *Azomonas* can produce this type of diffusible pigment; no *Azotobacter* are able to produce it. Check for fluorescence with an ultraviolet lamp in a darkened room.

Carbon Source The medium we used in this experiment contains 1% glucose, which can be utilized by all *Azotobacter* and *Azomonas*. Selectivity can be achieved by replacing the glucose with rhamnose, caproate, caprylate, *meso*-inositol, mannitol, malonate, or several other carbon sources. If more precise differentiation is desirable, the student is referred to Tables 4.48 and 4.49 on pages 231 and 232 in *Bergey's Manual,* volume 1.

Laboratory Report

Record your observations and conclusions for the Azotobacteraceae on the Laboratory Report.

Rhizobiaceae

Although the free-living Azotobacteraceae are beneficial nitrogen-fixers, their contribution to nitrogen enrichment of the soil is limited due to the fact that they would rather utilize NH_3 in soil than fix nitrogen. In other words, if ammonia is present in the soil, nitrogen fixation by these organisms is suppressed. By contrast, the symbiotic nitrogen-fixers of genus *Rhizobium,* family Rhizobiaceae, are the principal nitrogen enrichers of soil.

Bergey's Manual lists three genera in family Rhizobiaceae: *Rhizobium, Bradyrhizobium,* and *Agrobacterium.* Although the three genera are related, only genus *Rhizobium* fixes nitrogen. This genus of symbiotic nitrogen-fixers contains only three species. Differentiation of these species relies primarily on plant inoculation tests. A partial list of the host plants for each species is as follows:

R. leguminosarum: peas, vetch, lentils, beans, scarlet runner, and clover
R. meliloti: sweet clover, alfalfa, and fenugreek
R. loti: trefoil, lupines, kidney vetch, chickpea, mimosa, and a few others

All three of these species are gram-negative pleomorphic rods (bacteroids), often X-, Y-, star-, and club-shaped; some exhibit branching. Refractile granules are usually observed with phase-contrast optics. All are aerobic and motile. Our study of *Rhizobium* will be of crushed root nodules from whatever legume is available.

Materials:
washed nodules from the root of a legume
methylene blue stain
microscope slides

1. Place a nodule on a clean microscope slide and crush it by pressing another slide over it. Produce a thin smear by sliding the top slide over the lower one.
2. After air-drying and fixing with heat, stain the smear with **methylene blue** for **30 seconds.**
3. Examine under oil immersion and draw some of the organisms on the Laboratory Report. Look for typical bacteroids of various configurations.

Laboratory Report

Complete the last portion of Laboratory Report 58, 59.

60 Isolation of an Antibiotic Producer from Soil

The constant search of soils throughout the world has yielded an abundance of antibiotics of great value for the treatment of many infectious diseases. Pharmaceutical companies are in constant search for new strains of bacteria, molds, and *Actinomyces* that can be used for antibiotic production. Although many organisms in soil produce antibiotics, only a small portion of new antibiotics are suitable for medical use. In this experiment an attempt will be made to isolate an antibiotic-producing *Actinomyces* from soil. Students will work in pairs.

First Period
(Primary Isolation)

Unless the organisms in a soil sample are thinned out sufficiently, the isolation of potential antibiotic producers is nearly impossible. As indicated in figure 60.1, it will be necessary to use a series of six

dilution tubes to produce a final soil dilution of 10^{-6}. Proceed as follows:

Materials:
per pair of students:
    6 large test tubes
    1 bottle of physiological saline solution
    3 Petri plates of glycerol yeast extract agar
    L-shaped glass rod
    beaker of alcohol
    6 1 ml pipettes
    1 10 ml pipette

1. Label six test tubes 1 through 6, and with a 10 ml pipette, dispense 9 ml of saline into each tube.
2. Weigh out 1 gm of soil and deposit it into tube 1.
3. Vortex mix tube 1 until all soil is well dispersed throughout the tube.

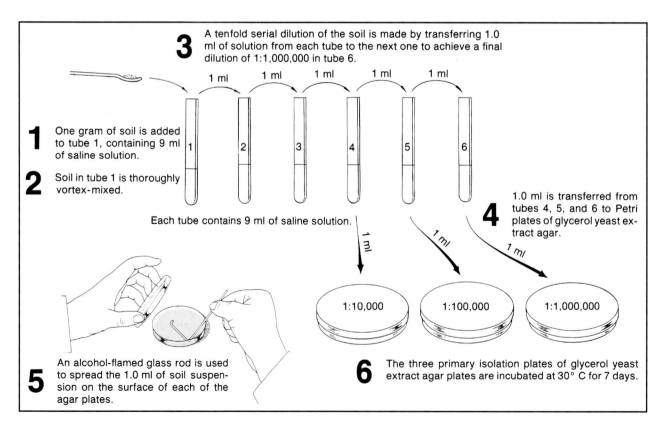

Figure 60.1 Primary isolation of antibiotic-producing *Actinomyces*

4. Make a tenfold dilution from tube 1 through tube 6 by transferring 1 ml from tube to tube. Use a fresh pipette for each transfer and be sure to pipette-mix thoroughly before each transfer.
5. Label three Petri plates with your initials and the dilutions to be deposited into them.
6. From each of the last three tubes transfer 1 ml to a plate of glycerol yeast extract agar.
7. Spread the organisms over the agar surfaces on each plate with an L-shaped glass rod that has been sterilized each time in alcohol and open flame. Be sure to cool rod before using.
8. Incubate the plates at 30° C for 7 days.

Second Period

(Colony Selection and Inoculation)

The objective in this laboratory period will be to select *Actinomyces*-like colonies that may be antibiotic producers. The organisms will be streaked on nutrient agar plates that have been seeded with *Staphylococcus epidermidis*. After incubation we will look for evidence of antibiosis. Students will continue to work in pairs. Figure 60.2 illustrates the procedure.

Materials:
per pair of students:
    4 trypticase soy agar pours (liquefied)
    4 sterile Petri plates
    TSB culture of *Staphylococcus epidermidis*
    1 ml pipette
    3 primary isolate plates from previous period
    water bath at student station (50° C)

1. Place four liquefied agar pours in water bath (50° C) to prevent solidification, and then inoculate each one with 1 ml of *S. epidermidis*.
2. Label the Petri plates with your initials and date.
3. Pour the contents of each inoculated tube into Petri plates. Allow agar to cool and solidify.
4. Examine the three primary isolation plates for the presence of *Actinomyces*-like colonies. They

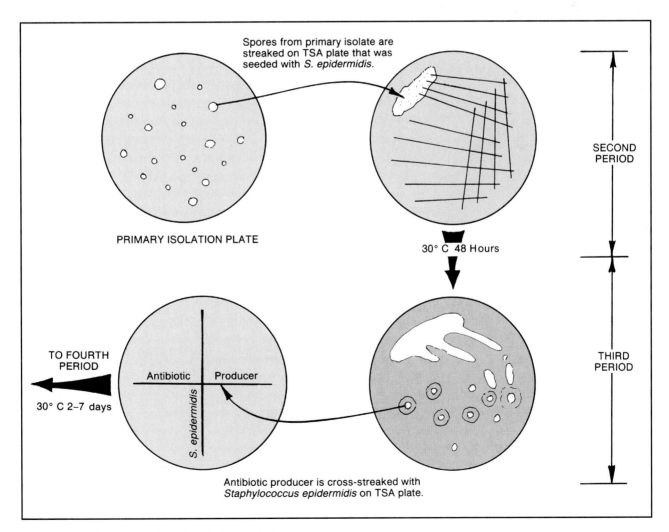

PRIMARY ISOLATION PLATE

Spores from primary isolate are streaked on TSA plate that was seeded with *S. epidermidis*.

30° C 48 Hours

SECOND PERIOD

THIRD PERIOD

TO FOURTH PERIOD

30° C 2–7 days

Antibiotic | Producer

S. epidermidis

Antibiotic producer is cross-streaked with *Staphylococcus epidermidis* on TSA plate.

Figure 60.2 Second and third period inoculations

have a dusty appearance due to the presence of spores. They may be white or colored. Your instructor will assist in the selection of colonies.

5. Using a sterile inoculating needle, scrape spores from *Actinomyces*-like colonies on the primary isolation plates to inoculate the seeded TSA plates. Use inoculum from a different colony for each of the four plates.

6. Incubate the plates at 30° C until the next laboratory period.

Third and Fourth Periods

(Evidence of Antibiosis and Confirmation)

Examine the four plates you streaked during the last laboratory period. If you see evidence of antibiosis (inhibition of *S. epidermidis* growth), proceed as follows to confirm results.

Materials:
   1 Petri plate of trypticase soy agar
   TSB culture of *S. epidermidis*

If antibiosis is present, make two streaks on the TSA plate as shown in figure 60.2. Make a straight line streak first with spores from the *Actinomyces* colony, using a sterile inoculating needle. Cross-streak with organisms from a culture of *S. epidermidis*. Incubate at 30° C until the next period.

Laboratory Report

After examining the cross-streaked plate during the fourth period, record your results on the Laboratory Report and answer all the questions.

PART 11 Microbiology of Water

The microorganisms of natural waters are extremely diverse. The numbers and types of bacteria present will depend on the amounts of organic matter present, the presence of toxic substances, the water's saline content, and environmental factors such as pH, temperature, and aeration. The largest numbers of heterotrophic forms will exist on the bottoms and banks of rivers and lakes where organic matter predominates. Open water in the center of large bodies of water, free of floating debris, will have small numbers of bacteria. Many species of autotrophic types are present, however, that require only the dissolved inorganic salts and minerals that are present.

The threat to human welfare by contamination of water supplies with sewage is a prime concern of everyone. The enteric diseases such as cholera, typhoid fever, and bacillary dysentery often result in epidemics when water supplies are not properly protected or treated. Thus, our prime concern in this unit is the sanitary phase of water microbiology. The American Public Health Association in its *Standard Methods for the Examination of Water and Wastewater* has outlined acceptable procedures for testing water for sewage contamination. The exercises of this unit are based on the procedures in that book.

Water that contains large numbers of bacteria may be perfectly safe to drink. The important consideration, from a microbiological standpoint, is the kinds of microorganisms that are present. Water from streams and lakes that contain multitudes of autotrophs and saprophytic heterotrophs is potable as long as pathogens for humans are lacking. The intestinal pathogens such as those that cause typhoid fever, cholera, and bacillary dysentery are of prime concern. The fact that human fecal material is carried away by water in sewage systems that often empty into rivers and lakes presents a colossal sanitary problem; thus, constant testing of municipal water supplies for the presence of fecal microorganisms is essential for the maintenance of water purity.

Routine examination of water for the presence of intestinal pathogens would be a tedious and difficult, if not impossible, task. It is much easier to demonstrate the presence of some nonpathogenic intestinal types such as *Escherichia coli* or *Streptococcus faecalis*. Since these organisms are always found in the intestines, and normally are not present in soil or water, it can be assumed that their presence in water indicates that fecal material has contaminated the water supply.

E. coli and *S. faecalis* are classified as good **sewage indicators.** The characteristics that make them good indicators of fecal contamination are (1) they are normally not present in water or soil, (2) they are relatively easy to identify, and (3) they survive a *little* longer in water than enteric pathogens. If they were hardy organisms, surviving a long time in water, they would make any water purity test too sensitive. Since both organisms are non-spore-formers, their survival in water is not extensive.

E. coli and *S. faecalis* are completely different organisms. *E. coli* is a gram-negative non-spore-forming rod; *S. faecalis* is a gram-positive coccus. The former is classified as a coliform; the latter is an enterococcus. Physiologically, they are also completely different.

The series of tests depicted in figure 61.1 is based on tests that will demonstrate the presence of a coliform in water. By definition, a **coliform** is a facultative anaerobe that ferments lactose to produce gas and is a gram-negative, non-spore-forming rod. *Escherichia coli* and *Enterobacter aerogenes* fit this

description. Since *S. faecalis* is not a coliform, a completely different set of tests must be used for it.

Note that three different tests are shown in figure 61.1: presumptive, confirmed, and completed. Each test exploits one or more of the characteristics of a coliform. A description of each test follows.

Presumptive Test   In the presumptive test a series of 9 or 12 tubes of lactose broth are inoculated with measured amounts of water to see if the water contains any lactose-fermenting bacteria that produce gas. If, after incubation, gas is seen in any of the lactose broths, it is *presumed* that coliforms are present in the water sample. This test is also used to determine the most probable number (MPN) of coliforms present per 100 ml of water.

Confirmed Test   In this test, plates of Levine EMB agar or Endo agar are inoculated from positive (gas-producing) tubes to see if the organisms that are producing the gas are gram-negative (another coliform characteristic). Both of these media inhibit the growth of gram-positive bacteria and cause colonies of coliforms to be distinguishable from noncoliforms. On EMB agar coliforms produce small colonies with dark centers (nucleated colonies). On Endo agar coliforms produce reddish colonies. The presence of coliform-like colonies confirms the presence of a lactose-fermenting gram-negative bacterium.

Completed Test   In the completed test our concern is to determine if the isolate from the agar plates truly matches our definition of a coliform. Our media for this test include a nutrient agar slant and a Durham tube of lactose broth. If gas is produced in the lactose tube and a slide from the agar slant reveals that we have a gram-negative non-spore-forming rod, we can be certain that we have a coliform.

The completion of these three tests with positive results establishes that coliforms are present; however, there is no certainty that *E. coli* is the coliform present. The organism might be *E. aerogenes*. Of the two, *E. coli* is the better sewage indicator since *E. aerogenes* can be of nonsewage origin. To differentiate these two species, one must perform the **IMViC**

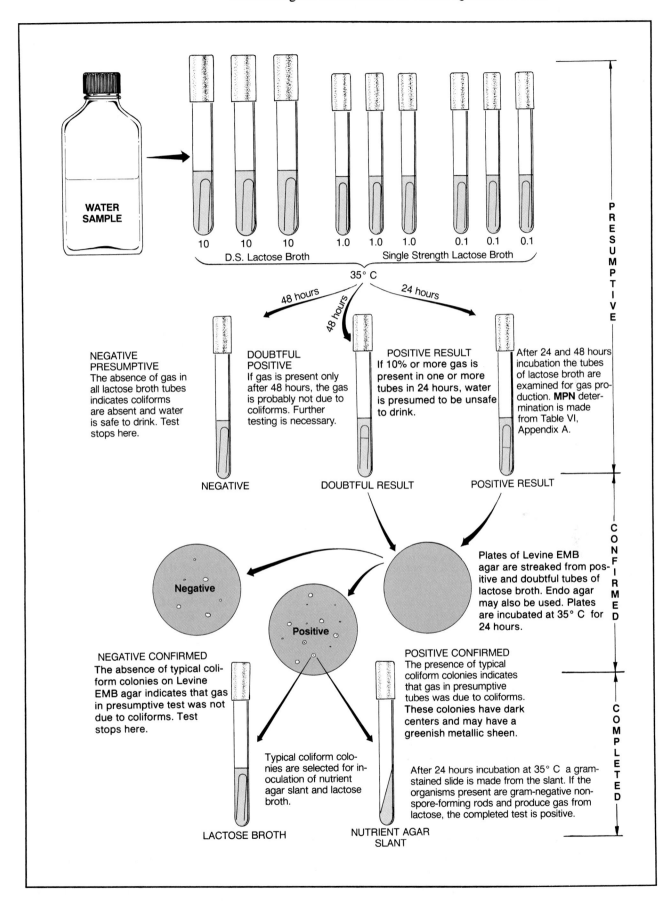

WATER SAMPLE

10 10 10   1.0 1.0 1.0   0.1 0.1 0.1

D.S. Lactose Broth   Single Strength Lactose Broth

35° C

48 hours   48 hours   24 hours

NEGATIVE PRESUMPTIVE
The absence of gas in all lactose broth tubes indicates coliforms are absent and water is safe to drink. Test stops here.

DOUBTFUL POSITIVE
If gas is present only after 48 hours, the gas is probably not due to coliforms. Further testing is necessary.

POSITIVE RESULT
If 10% or more gas is present in one or more tubes in 24 hours, water is presumed to be unsafe to drink.

After 24 and 48 hours incubation the tubes of lactose broth are examined for gas production. **MPN** determination is made from Table VI, Appendix A.

NEGATIVE   DOUBTFUL RESULT   POSITIVE RESULT

P R E S U M P T I V E

Negative

Positive

Plates of Levine EMB agar are streaked from positive and doubtful tubes of lactose broth. Endo agar may also be used. Plates are incubated at 35° C for 24 hours.

C O N F I R M E D

NEGATIVE CONFIRMED
The absence of typical coliform colonies on Levine EMB agar indicates that gas in presumptive test was not due to coliforms. Test stops here.

Typical coliform colonies are selected for inoculation of nutrient agar slant and lactose broth.

POSITIVE CONFIRMED
The presence of typical coliform colonies indicates that gas in presumptive tubes was due to coliforms. These colonies have dark centers and may have a greenish metallic sheen.

After 24 hours incubation at 35° C a gram-stained slide is made from the slant. If the organisms present are gram-negative non-spore-forming rods and produce gas from lactose, the completed test is positive.

C O M P L E T E D

LACTOSE BROTH   NUTRIENT AGAR SLANT

Figure 61.1 Bacteriological analysis of water

209

tests, which are described on page 169 in Exercise 50.

In this exercise water will be tested from local ponds, streams, swimming pools, and other sources supplied by students and instructor. Enough known positive samples will be evenly distributed throughout the laboratory so that all students will be able to see positive test results. All three tests in figure 61.1 will be performed. If time permits, the IMViC tests may also be performed.

The Presumptive Test

As stated earlier, the presumptive test is used to determine if gas-producing lactose fermenters are present in a water sample. If clear surface water is being tested, nine tubes of lactose broth will be used as shown in figure 61.1. For turbid surface water an additional three tubes of single strength lactose broth will be inoculated.

In addition to determining the presence or absence of coliforms, we can also use this series of lactose broth tubes to determine the **most probable number** (MPN) of coliforms present in 100 ml of water. A table for determining this value from the number of positive lactose tubes is provided in Appendix A.

Before setting up your test, determine whether your water sample is clear or turbid. Note that a separate set of instructions is provided for each type of water.

Clear Surface Water

If the water sample is relatively clear, proceed as follows:

Materials:
    3 Durham tubes of DSLB
    6 Durham tubes of SSLB
    1 10 ml pipette
    1 1 ml pipette
    Note: DSLB designates double strength lactose broth. It contains twice as much lactose as SSLB (single strength lactose broth).

1. Set up 3 DSLB and 6 SSLB tubes as illustrated in figure 61.1. Label each tube according to the amount of water that is to be dispensed to it: *10 ml, 1.0 ml,* and *0.1 ml,* respectively.
2. Mix the bottle of water to be tested by shaking 25 times.
3. With a 10 ml pipette, transfer 10 ml of water to each of the DSLB tubes.
4. With a 1.0 ml pipette, transfer 1 ml of water to each of the middle set of tubes, and 0.1 ml to each of the last three SSLB tubes.

5. Incubate the tubes at 35° C for 24 hours.
6. Examine the tubes and record the number of tubes in each set that have 10% gas or more.
7. Determine the MPN by referring to Table VI, Appendix A. Consider the following:

   Example: If you had gas in the first three tubes and gas only in one tube of the second series, but none in the last three tubes, your test would be read as 3–1–0. Table VI indicates that the MPN for this reading would be 43. This means that this particular sample of water would have approximately 43 organisms per 100 ml with 95% probability of there being between 7 and 210 organisms. *Keep in mind that the MPN figure of 43 is only a statistical probability figure.*

8. Record the data on the Laboratory Report.

Turbid Surface Water

If your water sample appears to have considerable pollution, do as follows:

Materials:
    3 Durham tubes of DSLB
    9 Durham tubes of SSLB
    1 10 ml pipette
    2 1 ml pipettes
    1 water blank (99 ml of sterile water)
    Note: See comment in previous materials list concerning DSLB and SSLB.

1. Set up three DSLB and nine SSLB tubes in a test-tube rack, with the DSLB tubes on the left.
2. Label the three DSLB tubes *10 ml;* the next three SSLB tubes *1.0 ml;* the next three SSLB tubes *0.1 ml;* and the last three tubes *0.01 ml.*
3. Mix the bottle of water to be tested by shaking 25 times.
4. With a 10 ml pipette, transfer 10 ml of water to each of the DSLB tubes.
5. With a 1.0 ml pipette, transfer 1 ml to each of the next three tubes, and 0.1 ml to each of the third set of tubes.
6. With the same 1 ml pipette, transfer 1 ml of water to the 99 ml blank of sterile water and shake 25 times.
7. *With a fresh 1 ml pipette,* transfer 1.0 ml of water from the blank to the remaining tubes of SSLB. This is equivalent to adding 0.01 ml of full strength water sample.
8. Incubate the tubes at 35° C for 24 hours.
9. Examine the tubes and record the number of tubes in each set that have 10% gas or more.
10. Determine the MPN by referring to Table VI, Appendix A. This table is set up for only 9 tubes. To apply a 12-tube reading to it, do as follows:

a. Select the three consecutive sets of tubes that have at least one tube with no gas.

b. If the first set of tubes (10 ml tubes) are not used, multiply the MPN by 10.

Example: Your tube reading was 3−3−3−1. What is the MPN?

The first set of tubes (10 ml) is ignored and the figures 3−3−1 are applied to the table. The MPN for this series is 460. Multiplying this by 10, the MPN becomes 4600.

Example: Your tube reading was 3−2−2−0. What is the MPN?

The first three numbers are applied to the table. The MPN is 210. Since the last set of tubes is ignored, 210 is the MPN.

The Confirmed Test

Once it has been established that gas-producing lactose fermenters are present in the water, it is *presumed* to be unsafe. However, gas formation may be due to noncoliform bacteria. Some of these organisms, such as *Clostridium perfringens,* are gram-positive. To confirm the presence of gram-negative lactose fermenters, the next step is to inoculate media such as Levine eosin–methylene blue agar or Endo agar from positive presumptive tubes.

Levine EMB agar contains methylene blue, which inhibits gram-positive bacteria. Gram-negative lactose fermenters (coliforms) that grow on this medium will produce "nucleated colonies" (dark centers). Colonies of *E. coli* and *E. aerogenes* can be differentiated on the basis of size and the presence of a greenish metallic sheen. *E. coli* colonies on this medium are small and have this metallic sheen, whereas *E. aerogenes* colonies usually lack the sheen and are larger. Differentiation in this manner is not completely reliable, however. It should be remembered that *E. coli* is the more reliable sewage indicator since it is not normally present in soil, while *E. aerogenes* has been isolated from soil and grains.

Endo agar contains a fuchsin sulfite indicator that makes identification of lactose fermenters relatively easy. Coliform colonies and the surrounding medium appear red on Endo agar. Nonfermenters of lactose, on the other hand, are colorless and do not affect the color of the medium.

In addition to these two media, there are several other media that can be used for the confirmed test. Brilliant green bile lactose broth, Eijkman's medium,

and EC medium are just a few examples that can be used.

To demonstrate the confirmation of a positive presumptive in this exercise, the class will use Levine EMB agar and Endo agar. One half of the class will use one medium; the other half will use the other medium. Plates will be exchanged for comparisons.

Materials:
  1 Petri plate of Levine EMB agar (odd-numbered students)
  1 Petri plate of Endo agar (even-numbered students)

1. Select one positive lactose broth tube from the presumptive test and streak a plate of medium according to your assignment. Use a streak method that will produce good isolation of colonies. If all your tubes were negative, borrow a positive tube from another student.

2. Incubate the plate for 24 hours at 35° C.

3. Look for typical coliform colonies on both kinds of media. Record your results on the Laboratory Report. If no coliform colonies are present, the water is considered bacteriologically safe to drink.

Note: In actual practice, confirmation of all presumptive tubes would be necessary to ensure accuracy of results.

The Completed Test

A final check of the colonies that appear on the confirmatory media is made by inoculating a nutrient agar slant and a Durham tube of lactose broth. After incubation for 24 hours at 35° C, the lactose broth is examined for gas production. A gram-stained slide is made from the slant, and the slide is examined under oil immersion optics.

If the organism proves to be a gram-negative, non-spore-forming rod that ferments lactose, we know that coliforms were present in the tested water sample. If time permits, complete these last tests and record the results on the Laboratory Report.

The IMViC Tests

Review the discussion of the IMViC tests on page 169. The significance of these tests should be much more apparent at this time. Your instructor will indicate whether these tests should also be performed if you have a positive completed test.

In addition to the multiple tube test, a method utilizing the membrane filter has been recognized by the United States Public Health Service as a reliable method for the detection of coliforms in water. These filter disks are 150 micrometers thick, have pores of 0.45 micrometer diameter, and have 80% area perforation. The precision of manufacture is such that bacteria larger than 0.47 micrometer cannot pass through. Eighty percent area perforation facilitates rapid filtration.

To test a sample of water, the water is passed through one of these filters. All bacteria present in the sample will be retained directly on the filter's surface. The membrane filter is then placed on an absorbent pad saturated with liquid nutrient medium and incubated for 22 to 24 hours. The organisms on the filter disk will form colonies that can be counted under the microscope. If a differential medium such as *m* Endo MF broth is used, coliforms will exhibit a characteristic golden metallic sheen.

The advantages of this method over the multiple tube test are (1) higher degree of reproducibility of results; (2) greater sensitivity since larger volumes of water can be used; and (3) shorter time (one-fourth) for getting results.

Figure 62.1 illustrates the procedure we will use in this experiment.

Materials:

vacuum pump or water faucet aspirators
membrane filter assemblies (sterile)
side-arm flask, 1000 ml size, and rubber hose
sterile graduates (100 ml or 250 ml size)
sterile, plastic Petri dishes, 50 mm dia
   (Millipore #PD10 047 00)
sterile membrane filter disks (Millipore
   #HAWG 047 AO)
sterile absorbent disks (packed with filters)
sterile water
5 ml pipettes
bottles of *m* Endo MF broth (50 ml)*
water samples

1. Prepare a small plastic Petri dish as follows:
   a. With a flamed forceps, transfer a sterile absorbent pad to a sterile plastic Petri dish.
   b. Using a 5 ml pipette, transfer 2.0 ml of *m* Endo MF broth to the absorbent pad.
2. Assemble a membrane filtering unit as follows:
   a. *Aseptically* insert the filter holder base into the neck of a 1-liter side-arm flask.
   b. With a flamed forceps, place a sterile membrane filter disk, grid side up, on the filter holder base.
   c. Place the filter funnel on top of the membrane filter disk and secure it to the base with the clamp.
3. Attach the rubber hose to a vacuum source (pump or water aspirator) and pour the appropriate amount of water into the funnel.

   The amount of water used will depend on water quality. No less than 50 ml should be used. Waters with few bacteria and low turbidity permit samples of 200 ml or more. Your instructor will advise you as to the amount of water that you should use. Use a sterile graduate for measuring the water.
4. Rinse the inner sides of the funnel with 20 ml of sterile water.
5. Disconnect the vacuum source, remove the funnel, and carefully transfer the filter disk with sterile forceps to the Petri dish of *m* Endo MF broth. *Keep grid side up.*
6. Incubate at 35° C for 22 to 24 hours. *Don't invert.*
7. After incubation, remove the filter from the dish and dry for 1 hour on absorbent paper.
8. Count the colonies on the disk with low-power magnification, using reflected light. Ignore all colonies that lack the golden metallic sheen. If desired, the disk may be held flat by mounting between two 2″ × 3″ microscope slides after drying. Record your count on the first portion of Laboratory Report 62, 63.

*See Appendix C for special preparation method.

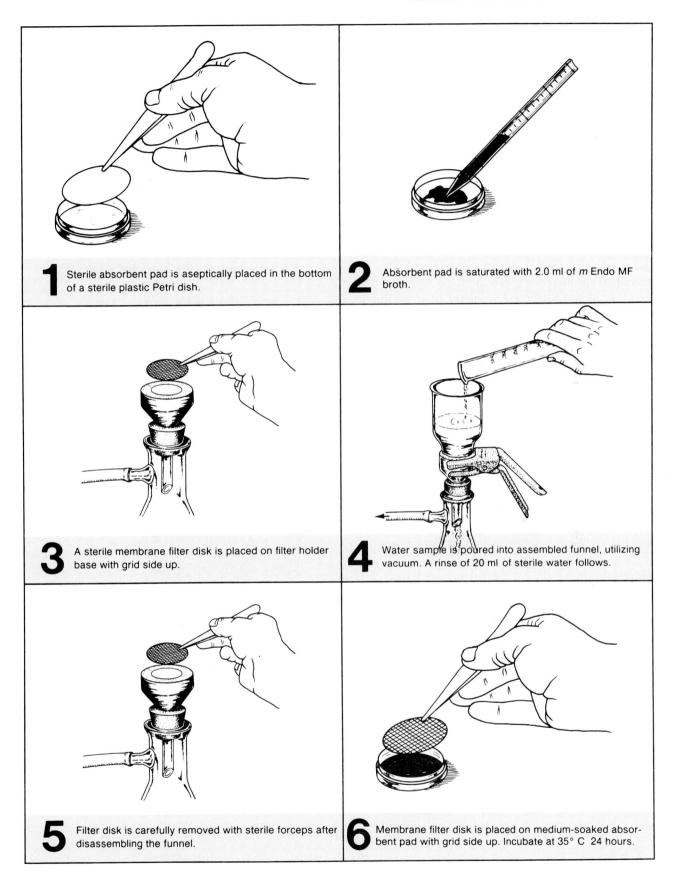

1 Sterile absorbent pad is aseptically placed in the bottom of a sterile plastic Petri dish.

2 Absorbent pad is saturated with 2.0 ml of *m* Endo MF broth.

3 A sterile membrane filter disk is placed on filter holder base with grid side up.

4 Water sample is poured into assembled funnel, utilizing vacuum. A rinse of 20 ml of sterile water follows.

5 Filter disk is carefully removed with sterile forceps after disassembling the funnel.

6 Membrane filter disk is placed on medium-soaked absorbent pad with grid side up. Incubate at 35° C 24 hours.

Figure 62.1 Membrane filter routine

63

Standard Plate Count: A Quantitative Test

In determining the total numbers of bacteria in water, we are faced with the same problems that are encountered with soil. Water organisms have great variability in physiological needs, and no single medium, pH, or temperature is ideal for all types. Despite the fact that only small numbers of organisms in water will grow on nutrient media, the **standard plate count** can perform an important function in water testing. Probably its most important use is to give us a tool to reveal the effectiveness of various stages in the purification of water. Plate counts made of water before and after storage, for example, can tell us how effective holding is in reducing bacterial numbers.

In this exercise, various samples of water will be evaluated by routine standard plate count procedures. Since different dilution procedures are required for different types of water, two methods are given.

Tap Water Procedure

If the water is of low bacterial count, such as in the case of tap water, use the following method.

Materials:
    1.0 ml pipettes
    2 tryptone glucose extract agar pours (TGEA)
    2 sterile Petri plates
    Quebec colony counter and hand counters
    water samples

1. Liquefy two tubes of TGEA and cool to 45° C.
2. After shaking the sample of water 25 times transfer 1 ml of water to each of the two sterile Petri plates.

3. Pour the medium into the dishes, rotate sufficiently to get good mixing of medium and water, and let cool.
4. Incubate at 35° C for 24 hours.
5. Count the colonies of both plates on the Quebec colony counter and record your average count of the two plates on the Laboratory Report.

Surface Water Procedure

If the water is likely to have a high bacterial count, as in the case of surface water, proceed as follows:

Materials:
    1 bottle (75 ml) of tryptone glucose extract
       agar (TGEA)
    6 sterile Petri plates
    2 water blanks (99 ml)
    1.0 ml pipettes

1. Liquefy a bottle of TGEA medium and cool to 45° C.
2. After shaking your water sample 25 times, produce two water blanks with dilutions of 1:100 and 1:1000. See Exercise 23.
3. Distribute aliquots from these blanks to six Petri dishes, which will provide you with two plates each of 1:100, 1:1000, and 1:10,000 dilutions.
4. Pour one-sixth of the TGEA medium into each plate and rotate sufficiently to get even mixing of the water and medium.
5. Incubate at 35° C for 24 hours.
6. Select the pair of plates that has 30 to 300 colonies on each plate and count all the colonies on both plates. Record the average count for the two plates on the second portion of Laboratory Report 62, 63.

PART 12

Microbiology of Milk and Food Products

Milk and food provide excellent growth media for bacteria when suitable temperatures exist. This is in direct contrast to natural waters, which lack the essential nutrients for pathogens. The introduction of a few pathogens into food or milk products becomes a much more serious problem because of the ability of these substances to support tremendous increases in bacterial numbers. Many milk-borne epidemics of human diseases have been spread by contamination of milk by soiled hands of dairy workers, unsanitary utensils, flies, and polluted water supplies. The same thing can be said for improper handling of foods in the home, restaurants, hospitals, and other institutions.

We learned in Part 11 that bacteriological testing of water is primarily qualitative—emphasis being placed on the presence or absence of coliforms as indicators of sewage. Bacteriological testing of milk and food may also be performed in this same manner, using similar media and procedures to detect the presence of coliforms. However, most testing by public health authorities is quantitative. Although the presence of small numbers of bacteria in these substances does not necessarily mean that pathogens are lacking, low counts do reflect better care in handling of food and milk than is true when high counts are present.

Standardized testing procedures for milk products are outlined by the American Public Health Association in *Standard Methods for the Examination of Dairy Products.* The procedures in Exercises 64, 65, and 66 are excerpts from that publication. Copies of the book may be available in the laboratory as well as in the library.

Exercises 67, 68, and 69 pertain to bacterial counts in dried fruit and meats, as well as to spoilage of canned vegetables and meats. Since bacterial counts in foods are performed with some of the techniques you have learned in previous exercises, you will have an opportunity to apply some of those skills here. Exercises 70 and 71 pertain to fermentation methods used in the production of wine and yogurt.

The bacterial count in milk is the most reliable indication we have of its sanitary quality. It is for this reason that the American Public Health Association recognizes the standard plate count as the official method in its *Milk Ordinance and Code.* Although human pathogens may not be present in a high count, it may indicate a diseased udder, unsanitary handling of milk, or unfavorable storage temperatures. In general, therefore, a high count means that there is a greater likelihood of disease transmission. On the other hand, it is necessary to avoid the wrong interpretation of low plate counts, since it is possible to have pathogens such as the brucellosis and tuberculosis organisms when counts are within acceptable numbers. Routine examination and testing of animals act as safeguards against the latter situation.

In this exercise, standard plate counts will be made of two samples of milk: a supposedly good sample and one of known poor quality. *Odd-numbered students will work with the high-quality milk and even-numbered students will test the poor-quality sample.* A modification of the procedures in Exercise 23 will be used.

High-Quality Milk

Materials:
    milk sample
    1 sterile water blank (99 ml)
    4 sterile Petri plates
    1.1 ml dilution pipettes
    1 bottle of TGEA (40 ml)

Quebec colony counter
mechanical hand counter

1. Following the procedures used in Exercise 23, pour four plates with dilutions of 1:1, 1:10, 1:100, and 1:1000. Before starting the dilution procedures, shake the milk sample 25 times in the customary manner.
2. Incubate the plates at 35° C for 24 hours and count the colonies on the plate that has between 30 and 300 colonies.
3. Record your results on the first portion of Laboratory Report 64, 65.

Poor-Quality Milk

Materials:
    milk sample
    3 sterile water blanks (99 ml)
    4 sterile Petri plates
    1.1 ml dilution pipettes
    1 bottle TGEA (50 ml)
    Quebec colony counter
    mechanical hand counter

1. Following the procedures used in Exercise 23, pour four plates with dilutions of 1:10,000, 1:100,000, 1:1,000,000, and 1:10,000,000. Before starting the dilutions, shake the milk sample 25 times in the customary manner.
2. Incubate the plates at 35° C for 24 hours and count the colonies on the plate that has between 30 and 300 colonies.
3. Record your results on the first portion of Laboratory Report 64, 65.

Direct Microscopic Count of Organisms in Milk:
The Breed Count

65

When it is necessary to determine milk quality in a much shorter time than is possible with a standard plate count, one can make a **direct microscopic count** on a slide. This is accomplished by staining a measured amount of milk that has been spread over an area one square centimeter on a slide. The slide is examined under oil and all of the organisms in an entire microscopic field are counted. To increase accuracy, several fields are counted to get average field counts. Before the field counts can be translated into organisms per milliliter, however, it is necessary to calculate the field area.

High-quality milk will have very few organisms per field, necessitating the examination of many fields. A slide made of poor-quality milk, on the other hand, will reveal large numbers of bacteria per field, thus requiring the examination of fewer fields. An experienced technician can determine, usually within 15 minutes, whether or not the milk is of acceptable quality.

In addition to being much faster than the SPC, the direct microscopic count has two other distinct advantages. First of all, it will reveal the presence of bacteria that do not form colonies on an agar plate at 35° C; thermophiles, psychrophiles, and dead bacteria would fall in this category. Secondly, the presence of excessive numbers of leukocytes and pus-forming streptococci on a slide will be evidence that the animal that produced the milk has an udder infection (mastitis).

In view of all these advantages, it is apparent that the direct microscopic count has real value in milk testing. It is widely used for testing raw milk in creamery receiving stations and for diagnosing the types of contamination and growth in pasteurized milk products.

In this exercise, samples of raw whole milk will be examined. Milk that has been separated, blended, homogenized, and pasteurized will lack leukocytes and normal flora.

Slide Preparation

There are several acceptable ways of spreading the milk onto the slide. Figure 65.1 illustrates a method using a guide card. The Breed slide used in fig-

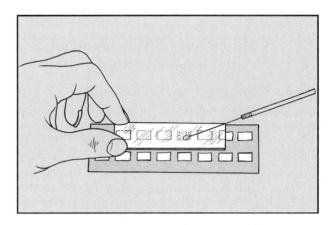

Figure 65.1 Using a guide card to spread milk sample over one square centimeter on a slide

ure 65.2 has five one-centimeter areas that are surrounded by ground glass, obviating the need for a card. Proceed as follows:

Materials:
    Breed slide or guide card
    Breed pipettes (0.01 ml)
    methylene blue, xylol, 95% alcohol
    beaker of water and electric hot plate
    samples of raw milk (poor and high quality)

1. Shake the milk sample 25 times to completely disperse the organisms and break up large clumps of bacteria.

2. Transfer 0.01 ml of milk to one square on the slide. The pipette may be filled by capillary action or by suction, depending on the type of pipette. The instructor will indicate which method to use. *Be sure to wipe off the outside tip of the pipette with tissue before touching the slide* to avoid getting more than 0.01 ml on the slide.

3. Allow the slide to air-dry and then place it over a beaker of boiling water for 5 minutes to steam-fix it.

4. Flood the slide with **xylol** to remove fat globules.

5. Remove the xylol from the slide by flooding the slide with **95% ethyl alcohol.**

6. Gently immerse the slide into a beaker of **distilled water** to remove the alcohol. Do not hold

it under running water; the milk film will wash off.

7. Stain the smear with **methylene blue** for 15 seconds and dip the slide again in water to remove the excess stain.

8. Decolorize the smear to pale blue with 95% alcohol and dip in water to stop decolorization.

9. Allow the slide to completely air-dry before examination.

Calibration of Microscope

(Microscope Factor [MF])

Before counting the organisms in each field it is necessary to know what part of a milliliter of milk is represented in that field. The relationship of the field to a milliliter is the **microscope factor** (MF). To calculate the MF, it is necessary to use a stage micrometer to measure the diameter of the oil immersion field. By applying the formula πr^2 to this measurement, the area is easily determined. With the amount of milk (0.01 ml) and the area of the slide (1 cm^2), it is a simple matter to calculate the MF.

Materials:

stage micrometer

1. Place a stage micrometer on the microscope stage and bring it into focus under oil. Measure the diameter of the field, keeping in mind that each space is equivalent to 0.01 mm.

2. Calculate the area of the field in square millimeters, using the formula πr^2 ($\pi = 3.14$).

3. Convert the area of the field from square millimeters to square centimeters by dividing by 100.

4. Calculate the number of fields in one square centimeter by dividing one square centimeter by the area of the field in square centimeters.

5. To get the part of a milliliter that is represented in a single field (**microscope factor**), multiply the number of fields by 100. The value should be around 500,000. Therefore, a single field represents 1/500,000 of a ml of milk. Record your computations on the Laboratory Report.

Examination of Slide

Two methods of counting the bacteria can be used: individual cells may be tallied or only clumps of bacteria may be counted. In both cases, the number per milliliter will be higher than a standard plate count, but a clump count will be closer to the SPC. Both methods will be used.

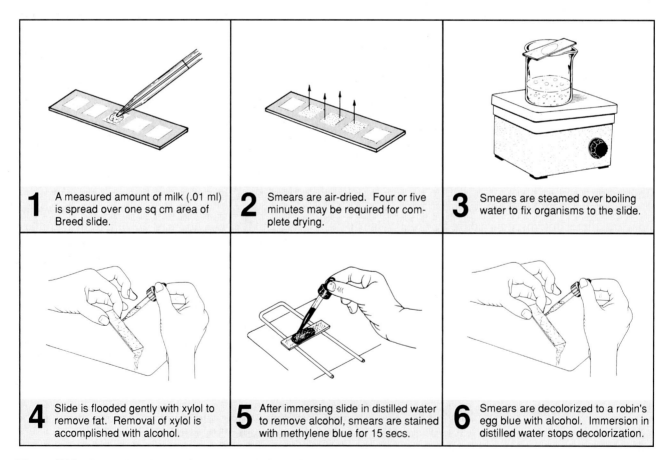

1 A measured amount of milk (.01 ml) is spread over one sq cm area of Breed slide.

2 Smears are air-dried. Four or five minutes may be required for complete drying.

3 Smears are steamed over boiling water to fix organisms to the slide.

4 Slide is flooded gently with xylol to remove fat. Removal of xylol is accomplished with alcohol.

5 After immersing slide in distilled water to remove alcohol, smears are stained with methylene blue for 15 secs.

6 Smears are decolorized to a robin's egg blue with alcohol. Immersion in distilled water stops decolorization.

Figure 65.2   Procedure for making a stained slide of a raw milk sample

1. After the microscope has been calibrated, replace the stage micrometer with the stained slide. Examine it under oil immersion optics.
2. Count the individual cells in five fields and record your results on the Laboratory Report. A field is the entire area encompassed by the oil immersion lens. As you see leukocytes, record their numbers, also.
3. Count only clumps of bacteria in five fields, recording the numbers of leukocytes as well.

Record the totals on the Laboratory Report.

4. Calculate the number of organisms, clumps, and body cells per milliliter using the microscope factor.

Laboratory Report

Complete the last portion of Laboratory Report 64, 65.

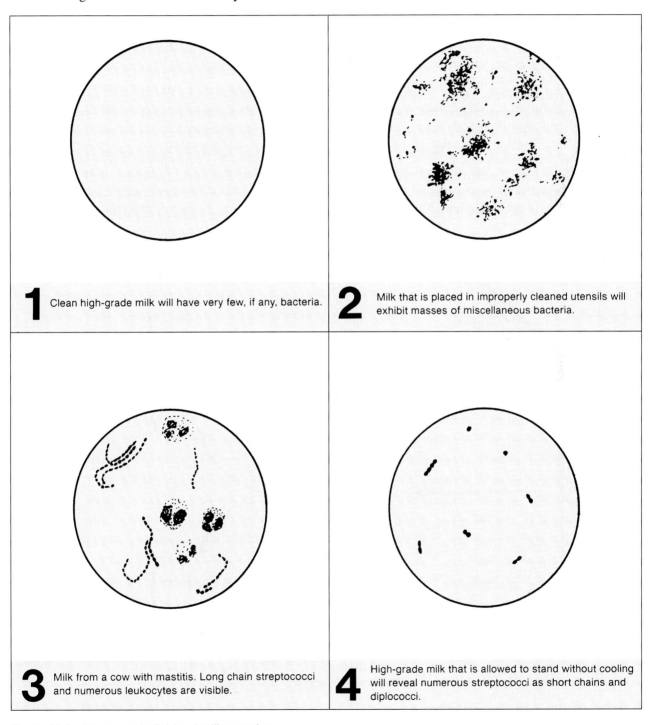

1 Clean high-grade milk will have very few, if any, bacteria.

2 Milk that is placed in improperly cleaned utensils will exhibit masses of miscellaneous bacteria.

3 Milk from a cow with mastitis. Long chain streptococci and numerous leukocytes are visible.

4 High-grade milk that is allowed to stand without cooling will reveal numerous streptococci as short chains and diplococci.

Figure 65.3   Microscopic fields of milk samples

Milk that contains large numbers of actively growing bacteria will have a lowered oxidation-reduction potential due to the exhaustion of dissolved oxygen by microorganisms. The fact that methylene blue loses its color (becomes reduced) in such an environment is the basis for the **reductase test.** In this test, 1 ml of methylene blue (1:25,000) is added to 10 ml of milk. The tube is sealed with a rubber stopper and slowly inverted three times to mix. It is placed in a water bath at 35° C and examined at intervals up to 6 hours. The time it takes for the methylene blue to become colorless is the **methylene blue reduction time** (MBRT). The shorter the MBRT, the lower the quality of milk. An MBRT of 6 hours is very good. Milk with an MBRT of 30 minutes is of very poor quality.

The validity of this test is based on the assumption that all bacteria in milk lower the oxidation-reduction potential at 35° C. Large numbers of psychrophiles, thermophiles, and thermodurics, which do not grow at this temperature, would not produce a positive test. Raw milk, however, will contain primarily *Streptococcus lactis* and *Escherichia coli*, which are strong reducers; thus, this test is suitable

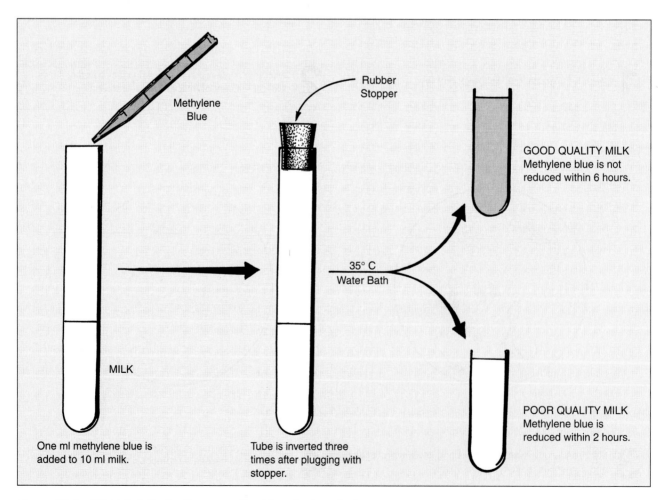

Methylene Blue

Rubber Stopper

GOOD QUALITY MILK
Methylene blue is not reduced within 6 hours.

35° C
Water Bath

MILK

POOR QUALITY MILK
Methylene blue is reduced within 2 hours.

One ml methylene blue is added to 10 ml milk.

Tube is inverted three times after plugging with stopper.

Figure 66.1 Procedure for testing raw milk with reductase test

for screening raw milk at receiving stations. Its principal value is that less technical training of personnel is required for its performance.

In this exercise, samples of low- and high-quality raw milk will be tested.

Materials:
> 2 sterile test tubes with rubber stoppers for each student
> raw milk samples of low- and high-quality (samples A and B)
> water bath set at 35° C
> methylene blue (1:25,000)
> 10 ml pipettes
> 1 ml pipettes
> gummed labels

1. Attach gummed labels with your name and type of milk to two test tubes. Each student will test a good-quality as well as a poor-quality milk.
2. Using separate 10 ml pipettes for each type of milk, transfer 10 ml to each test tube. To the milk in the tubes add 1 ml of methylene blue with a 1 ml pipette. Insert rubber stoppers and gently invert three times to mix. Record your name and the time on the labels and place the tubes in the water bath, which is set at 35° C.
3. After 5 minutes incubation, remove the tubes from the bath and invert once to mix. This is the last time they should be mixed.
4. Carefully remove the tubes from the water bath 30 minutes later and every half hour until the end of the laboratory period. *When at least four-fifths of the tube has turned white,* the end point of reduction has taken place. Record this time on the Laboratory Report. The classification of milk quality is as follows:

Class 1: Excellent, not decolorized in 8 hours.
Class 2: Good, decolorized in less than 8 hours, but not less than 6 hours.
Class 3: Fair, decolorized in less than 6 hours, but not less than 2 hours.
Class 4: Poor, decolorized in less than 2 hours.

Bacterial Counts of Foods

The standard plate count, as well as the multiple tube test, can be used on foods much in the same manner that they are used on milk and water to determine total counts and the presence of coliorms. To get the organisms in suspension, however, a food blender is necessary.

In this exercise, samples of ground meat, dried fruit, and frozen food will be tested for total numbers of bacteria. This will not be a coliform count. The instructor will indicate the specific kinds of foods to be tested and make individual assignments. Figure 67.1 illustrates the general procedure.

Materials:

per student:
  3 Petri plates
  1 bottle (45 ml) of Plate Count agar or
    Standard Methods agar
  1 99 ml sterile water blank
  2 1.1 ml dilution pipettes

per class:
  food blender
  sterile blender jars (one for each type of food)
  sterile weighing paper

180 ml sterile water blanks (one for each type of food)
samples of ground meat, dried fruit, and frozen vegetables, thawed 2 hours

1. Using aseptic techniques, weigh out on sterile weighing paper 20 grams of food to be tested.
2. Add the food and 180 ml of sterile water to a sterile mechanical blender jar. Blend the mixture for 5 minutes. This suspension will provide a 1:10 dilution.
3. With a 1.1 ml dilution pipette dispense from the blender 0.1 ml to plate I and 1.0 ml to the water blank. See figure 67.1.
4. Shake the water blank 25 times in an arc of 1' for 7 seconds with your elbow on the table as done in Exercise 23 (Bacterial Population Counts).
5. Using a fresh pipette, dispense 0.1 ml to plate III and 1.0 ml to plate II.
6. Pour agar (50° C) into the three plates and incubate them at 35° C for 24 hours.
7. Count the colonies on the best plate and record the results on the Laboratory Report.

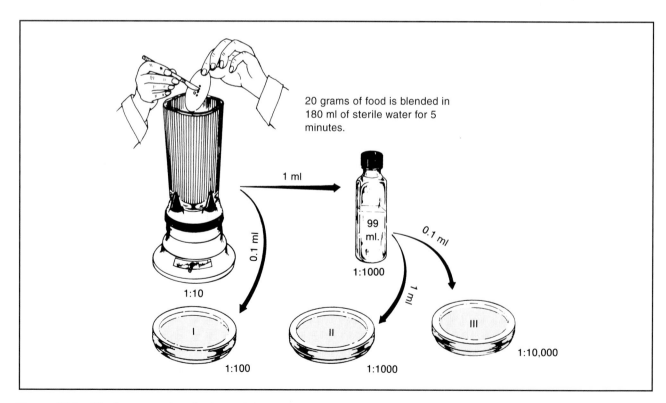

20 grams of food is blended in 180 ml of sterile water for 5 minutes.

1 ml

99 ml.

1:1000

0.1 ml

0.1 ml

1 ml

1:10

I

1:100

II

1:1000

III

1:10,000

Figure 67.1 Dilution procedure for bacterial counts of food

Microbial Spoilage of Canned Food

68

Spoilage of heat-processed, commercially canned foods is confined almost entirely to the action of bacteria that produce heat-resistant endospores. Canning of foods normally involves heat exposure for long periods of time at temperatures that are adequate to kill spores of most bacteria. Particular concern is given to the processing of low-acid foods in which *Clostridium botulinum* can thrive to produce botulism food poisoning.

Spoilage occurs when the heat processing fails to meet accepted standards. This can occur for several reasons: (1) lack of knowledge on the part of the processor (usually the case in home canning); (2) carelessness in handling the raw materials before canning, resulting in an unacceptably high level of contamination that ordinary heat processing may be inadequate to control; (3) equipment malfunction that results in undetected underprocessing; and

(4) defective containers that permit the entrance of organisms after the heat process.

Our concern here will be with the most common types of food spoilage caused by heat-resistant spore-forming bacteria. There are three types: "flat sour," "T.A. spoilage," and "stinker spoilage."

Flat sour pertains to spoilage in which acids are formed with no gas production; result: sour food in cans that have flat ends. **T.A. spoilage** is caused by thermophilic anaerobes that produce acid and gases (CO_2 and H_2, but not H_2S) in low-acid foods. Cans swell to various degrees, sometimes bursting. **Stinker spoilage** is due to spore-formers that produce hydrogen sulfide and blackening of the can and contents. Blackening is due to the reaction of H_2S with the iron in the can to form iron sulfide.

In this experiment you will have an opportunity to become familiar with some of the morphological

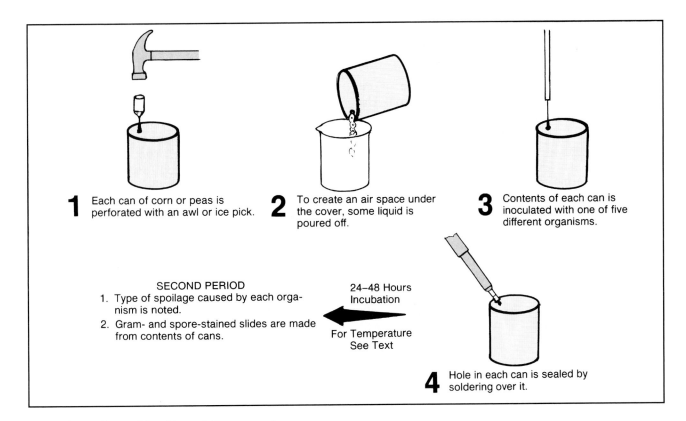

1 Each can of corn or peas is perforated with an awl or ice pick.

2 To create an air space under the cover, some liquid is poured off.

3 Contents of each can is inoculated with one of five different organisms.

SECOND PERIOD
1. Type of spoilage caused by each organism is noted.
2. Gram- and spore-stained slides are made from contents of cans.

24–48 Hours Incubation

For Temperature See Text

4 Hole in each can is sealed by soldering over it.

Figure 68.1   Canned food inoculation procedure

and physiological characteristics of organisms that cause canned food spoilage, including both aerobic and anaerobic endospore formers of *Bacillus* and *Clostridium,* as well as a non-spore-forming bacterium.

Working as a single group, the entire class will inoculate 10 cans of vegetables (corn and peas) with five different organisms. Figure 68.1 illustrates the procedure. Note that the cans will be sealed with solder after inoculation and incubated at different temperatures. After incubation the cans will be opened so that stained microscope slides can be made to determine Gram reaction and presence of endospores. Your instructor will assign individual students or groups of students to inoculate one or more of the 10 cans. One can of corn and one can of peas will be inoculated with each of the organisms. Proceed as follows:

First Period

(Inoculations)

Materials:
5 small cans of corn
5 small cans of peas
cultures of *B. stearothermophilus, B. coagulans, C. sporogenes, C. thermosaccharolyticum,* and *E. coli*
ice picks or awls
hammer
solder and soldering iron
plastic bags
gummed labels & rubber bands

1. Label the can or cans with the name of the organism that has been assigned to you. Use white gummed labels. In addition, place a similar label on one of the plastic bags to be used after sealing of the cans.
2. With an ice pick or awl, punch a small hole through a flat area in the top of each can. This can be done easily with the heel of your hand or a hammer, if available.
3. Pour off a small amount of the liquid from the can to leave an air space under the lid.
4. Use an inoculating needle to inoculate each can of corn or peas with the organism indicated on the label.
5. Take the cans up to the demonstration table where the instructor will seal the hole with solder.

6. After sealing, place each can in two plastic bags. Each bag must be closed separately with rubber bands, and the outer bag must have a label on it.
7. Incubation will be as follows till the next period:
   • **55° C** — *C. thermosaccharolyticum* and *B. stearothermophilus*
   • **37° C** — *C. sporogenes* and *B. coagulans*
   • **30° C** — *E. coli*

Note: If cans begin to swell during incubation they should be placed in refrigerator.

Second Period

(Interpretation)

After incubation place the cans under a hood to open them. The odors of some of the cans will be very strong due to H_2S production.

Materials:
can opener, punch type
small plastic beakers
Parafilm
gram-staining kit
spore-staining kit

1. Open each can carefully with a punch type can opener. If the can is swollen, hold an inverted plastic funnel over the can during perforation to minimize the effects of any explosive release of contents.
2. Remove about 10 ml of the liquid through the opening, pouring it into a small plastic beaker. Cover with Parafilm. This fluid will be used for making stained slides.
3. Return the cans of food to the plastic bags, reclose them, and dispose in a proper trash bin.
4. Prepare gram-stained and endospore-stained slides from your canned food extract as well as from the extracts of all the other cans. Examine under brightfield oil immersion.
5. Record your observations on the report sheet on the demonstration table. It will be duplicated and a copy will be made available to each student.

Laboratory Report

Complete the first portion of Laboratory Report 68, 69.

Microbial Spoilage of Refrigerated Meat

69

Contamination of meats by microbes occurs during and after slaughter. Many contaminants come from the animal itself, others from utensils and equipment. The conditions for rapid microbial growth in freshly cut meats are very favorable, and spoilage can be expected to occur rather quickly unless steps are taken to prevent it. Although immediate refrigeration is essential after slaughter, it will not prevent spoilage indefinitely, or even for a long period of time under certain conditions. In time, cold-tolerant microbes will destroy the meat, even at low refrigerator temperatures.

Microorganisms that grow at temperatures between 5° and 0° C are classified as being either psychrophilic or psychrotrophic. The difference between the two groups is that **psychrophiles** seldom grow at temperatures above 22° C and **psychrotrophs** (psychrotolerants or low-temperature mesophiles) grow well above 25° C. While the optimum growth temperature range for psychrophiles is 15°–18° C, psychrotrophs have an optimum growth temperature range of 25°–30° C. It is the psychrotrophic microorganisms that cause most meat spoilage during refrigeration.

The majority of psychrophiles are gram-negative and include species of *Aeromonas, Alcaligenes, Cytophagia, Flavobacterium, Pseudomonas, Serratia,* and *Vibrio.* Gram-positive psychrophiles include species of *Arthrobacter, Bacillus, Clostridium,* and *Micrococcus.*

Psychrotrophs include a much broader spectrum of gram-positive and gram-negative rods, cocci, vibrios, spore-formers, and non-spore-formers. Typical genera are *Acinetobacter, Chromobacterium, Citrobacter, Corynebacterium, Enterobacter, Escherichia, Klebsiella, Lactobacillus, Moraxella, Staphylococcus,* and *Streptococcus.*

The widespread use of vacuum or modified atmospheric packaging of raw and processed meat has resulted in food spoilage due to facultative and obligate anaerobes, such as *Lactobacillus, Leuconostoc, Pediococcus,* and certain Enterobacteriaceae.

Although most of the previously mentioned psychrotrophic representatives are nonpathogens, there are significant pathogenic psychrotrophs such as *Aeromonas hydrophila, Clostridium botulinum, List-* *eria monocytogenes, Vibrio cholera, Yersinia entercolitica,* and some strains of *E. coli.*

In addition to bacterial spoilage of meat there are many yeasts and molds that are psychrophilic and psychrotrophic. Examples of psychrophilic yeasts are *Cryptococcus, Leucosporidium,* and *Torulopsis.* Psychrotrophic fungi include *Candida, Cryptococcus, Saccharomyces, Alternaria, Aspergillus, Cladosporium, Fusarium, Mucor, Penicillium,* and many more.

Our concern in this experiment will be to test one or more meat samples for the prevalence of psychrophilic-psychrotrophic organisms. To accomplish this, we will liquefy and dilute out a sample of ground meat so that it can be plated out and then incubated in a refrigerator for 2 weeks. After incubation, colony counts will be made to determine the number of organisms of this type that exist in a gram of the sample.

Figure 69.1 illustrates the overall procedure. Work in pairs to perform the experiment.

First Period

Materials:
at demonstration table:
    ground meat and balance
    sterile foil-wrapped scoopula
    1 blank of phosphate buffered water (90 ml)
    blender with sterile blender jar
    sterile Petri dish or sterile filter paper

per pair of students:
    4 large test tubes of sterile phosphate buffered water (9 ml each)
    4 TSA plates
    9 sterile 1 ml pipettes
    L-shaped glass spreading rod
    beaker of 95% ethyl alcohol

At Demonstration Table

1. With a sterile scoopula, weigh 10 gm of ground meat into a sterile Petri plate or onto a sterile piece of filter paper.

2. Pour 90 ml of sterile buffered water from water blank into a sterile blender jar and add the meat.
3. Blend the meat and water at moderate speed for **1 minute.**

Student Pair

1. Label the four water blanks 1 through 4.
2. Label the four Petri plates with their dilutions, as indicated in figure 69.1. Add your initials and date also.
3. Once blender suspension is ready, pipette 1 ml from jar to tube 1.
4. Using a fresh 1 ml pipette, mix the contents in tube 1 and transfer 1 ml to tube 2.
5. Repeat step 4 for tubes 3 and 4, *using fresh pipettes* for each tube.
6. Dispense 0.1 ml from each tube to their respective plates of TSA. Note that by using only 0.1 ml per plate you are increasing the dilution factor by 10 times in each plate.
7. Using a sterile L-shaped glass rod, spread the organisms on the agar surfaces. Sterilize the rod each time by dipping in alcohol and flaming

gently. Be sure to let rod cool completely each time.
8. Incubate the plates for 2 weeks in the back of the refrigerator (away from door-opening) where the temperature will remain between 0° and 5° C.

Second Period

Materials:
   Quebec colony counters
   hand tally counters
   gram-staining kit

1. After incubation, count the colonies on all the plates and calculate the number of psychrophiles and psychrotrophs per gram of meat.
2. Select a colony from one of the plates and prepare a gram-stained slide. Examine under oil immersion and record your observations on the Laboratory Report.

Laboratory Report

Complete the last portion of Laboratory Report 68, 69.

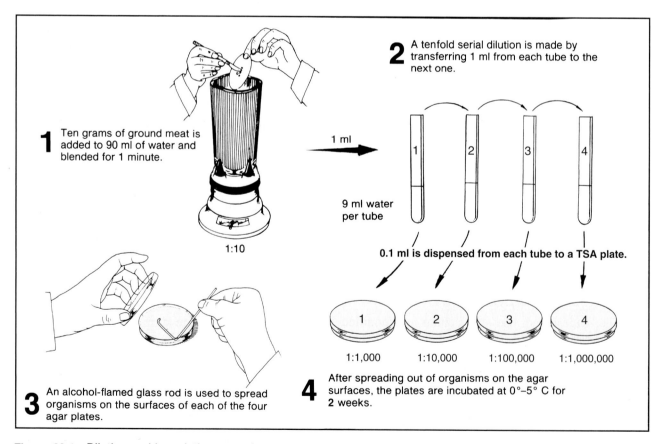

1 Ten grams of ground meat is added to 90 ml of water and blended for 1 minute.

1:10

2 A tenfold serial dilution is made by transferring 1 ml from each tube to the next one.

1 ml

9 ml water per tube

0.1 ml is dispensed from each tube to a TSA plate.

1:1,000   1:10,000   1:100,000   1:1,000,000

3 An alcohol-flamed glass rod is used to spread organisms on the surfaces of each of the four agar plates.

4 After spreading out of organisms on the agar surfaces, the plates are incubated at 0°–5° C for 2 weeks.

Figure 69.1 Dilution and inoculation procedure

Microbiology of Fermented Beverages

70

Fermented food and beverages are as old as civilization. Historical evidence indicates that beer and wine making were well established as long ago as 2000 B.C. An Assyrian tablet states that Noah took beer aboard the ark.

Beer, wine, vinegar, buttermilk, cottage cheese, sauerkraut, pickles, and yogurt are some of the more commonly known products of fermentation. Most of these foods and beverages are produced by different strains of yeasts (*Saccharomyces*) or bacteria (*Lactobacillus, Acetobacter,* etc.).

Fermentation is actually a means of food preservation because the acids formed and the reduced environment (anaerobiasis) hold back the growth of many spoilage microbes.

Wine is essentially fermented fruit juice in which alcoholic fermentation is carried out by *Saccharomyces cerevisiae* var. *ellipsoideus.* Although we usually associate wine with fermented grape juice, it may also be made from various berries, dandelions, rhubarb, etc. Three conditions are necessary: simple sugar, yeast, and anaerobic conditions. The reaction is as follows:

$$C_6H_{12}O_6 \xrightarrow{\text{yeast}} 2C_2H_5OH + 2CO_2$$

Commercially, wine is produced in two forms: red and white. To produce red wines, the distillers use red grapes with the skins left on during the initial stage of the fermentation process. For white wines either red or white grapes can be used, but the skins are discarded. White and red wines are fermented at 13° C (55° F) and 24° C (75° F), respectively.

In this exercise we will set up a grape juice fermentation experiment to learn about some of the characteristics of sugar fermentation to alcohol. Note in figure 70.1 that a balloon will be attached over the mouth of the fermentation flask to exclude oxygen uptake and to trap gases that might be produced. To detect the presence of hydrogen sulfide production we will tape a lead acetate test strip inside the neck

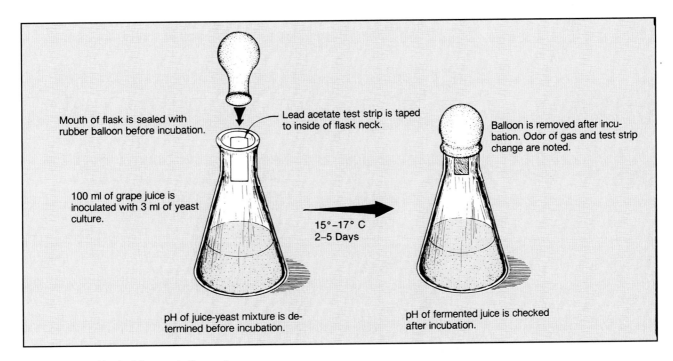

Mouth of flask is sealed with rubber balloon before incubation.

Lead acetate test strip is taped to inside of flask neck.

Balloon is removed after incubation. Odor of gas and test strip change are noted.

100 ml of grape juice is inoculated with 3 ml of yeast culture.

15°–17° C
2–5 Days

pH of juice-yeast mixture is determined before incubation.

pH of fermented juice is checked after incubation.

Figure 70.1 Alcohol fermentation setup

of the flask. The pH of the substrate will also be monitored before and after the reaction to note any changes that occur.

First Period

Materials:
>   100 ml grape juice (no preservative)
>   bottle of juice culture of wine yeast
>   125 ml Erlenmeyer flask
>   1 10 ml pipette
>   balloon
>   hydrogen sulfide (lead acetate) test paper
>   tape
>   pH meter

1. Label an Erlenmeyer flask with your initials and date.
2. Add about 100 ml of grape juice to the flask (fermenter).
3. Determine the pH of the juice with a pH meter and record the pH on the Laboratory Report.
4. Agitate the container of yeast juice culture to suspend the culture, remove 5 ml with a pipette, and add it to the flask.
5. Attach a short strip of tape to a piece of lead acetate test paper (3 cm long), and attach it to the inside surface of the neck of the flask. Make certain that neither the tape nor the test strip protrudes from the flask.
6. Cover the flask opening with a balloon.
7. Incubate at 15°–17° C for 2–5 days.

Second Period

Materials:
>   pH meter

1. Remove the balloon and note the aroma of the flask contents. Describe the odor on the Laboratory Report.
2. Determine the pH and record it on the Laboratory Report.
3. Record any change in color of the lead acetate test strip on the Laboratory Report. If any H_2S is produced, the paper will darken due to the formation of lead sulfide as hydrogen sulfide reacts with the lead acetate.
4. Wash out the flask and return it to the drain rack.

Laboratory Report

Complete the first portion of Laboratory Report 70, 71 by answering all the questions.

71

Microbiology of Fermented Milk Products (Yogurt)

For centuries, people throughout the world have been producing fermented milk products using yeasts and lactic acid–producing bacteria. The **yogurt** of eastern central Europe, the **kefir** of the Cossacks, the **koumiss** of central Asia, and the **leben** of Egypt are just a few examples. In all of these fermented milks lactobacilli act together with some other microorganism to curdle and thicken milk, producing a distinctive flavor desired by the producer. Kefir of the Cossacks is made by charging milk with small cauliflower-like grains that contain *Streptococcus lactis, Saccharomyces delbrueckii,* and *Lactobacillus brevis.* As the grains swell in the milk they release the growing microorganisms to ferment the milk. The usual method for producing yogurt in large-scale production is to add pure cultures of *Streptococcus thermophilus* and *Lactobacillus bulgaricus* to pasteurized milk.

In this exercise you will produce a batch of yogurt from milk by using an inoculum from commercial yogurt. Gram-stained slides will be made from the finished product to determine the types of organisms that control the reaction. If proper safety measures are followed, the sample can be tasted.

Two slightly different ways of performing this experiment are provided here. Your instructor will indicate which method will be followed.

Method A
(First Period)

Figure 71.1 illustrates the procedure for this method. Note that 4 gm of powdered milk is added to 100 ml of whole milk. This mixture is then heated to boiling and cooled to 45° C. After cooling, the milk is inoculated with yogurt and incubated at 45° C for 24 hours. Proceed:

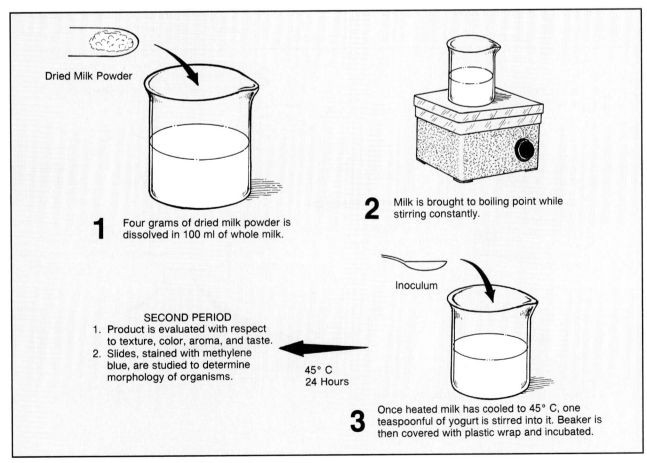

Dried Milk Powder

1 Four grams of dried milk powder is dissolved in 100 ml of whole milk.

2 Milk is brought to boiling point while stirring constantly.

Inoculum

3 Once heated milk has cooled to 45° C, one teaspoonful of yogurt is stirred into it. Beaker is then covered with plastic wrap and incubated.

45° C
24 Hours

SECOND PERIOD
1. Product is evaluated with respect to texture, color, aroma, and taste.
2. Slides, stained with methylene blue, are studied to determine morphology of organisms.

Figure 71.1   **Yogurt production by Method A**

Materials:

dried powdered milk
whole milk
commercial yogurt (with viable organisms)
small beaker, graduate, teaspoon, stirring rod
plastic wrap
filter paper (for weighing)

1. On a piece of filter paper weigh 4 gm of dried powdered milk.
2. To a beaker of 100 ml of whole milk add the powdered milk and stir thoroughly with sterile glass rod to dissolve.
3. Heat to boiling, while stirring constantly.
4. Cool to 45° C and inoculate with one teaspoon of the commercial yogurt. Stir. Be sure to check the label to make certain that product contains a live culture. Cover with plastic wrap.
5. Incubate at 45° C for 24 hours.

Method B

(First Period)

Figure 71.2 illustrates a slightly different method of culturing yogurt, which, due to its simplicity, may be preferred. Note that no whole milk is used and provisions are made for producing a sample for tasting.

Materials:

small beaker, graduate, teaspoon, stirring rod
dried powdered milk

commercial yogurt (with viable organisms)
plastic wrap
filter paper for weighing
paper Dixie cup (5 oz size) and cover
electric hot plate or Bunsen burner and tripod

1. On a piece of filter paper weigh 25 gm of dried powdered milk.
2. Heat 100 ml of water in a beaker to boiling and cool to 45° C.
3. Add the 25 gm of powdered milk and one teaspoonful of yogurt to the beaker of water. Mix the ingredients with a sterile glass rod.
4. Pour some of the mixture into a sterile Dixie cup and cover loosely. Cover the remainder in the beaker with plastic wrap.
5. Incubate at 45° C for 24 hours.

Second Period
(Both Methods)

1. Examine the product and record on the Laboratory Report the color, aroma, texture, and, if desired, the taste.
2. Make slide preparations of the yogurt culture. Fix and stain with methylene blue. Examine under oil immersion and record your results on Laboratory Report 70, 71.

Laboratory Report

Complete the last portion of Laboratory Report 70, 71 by answering all the questions.

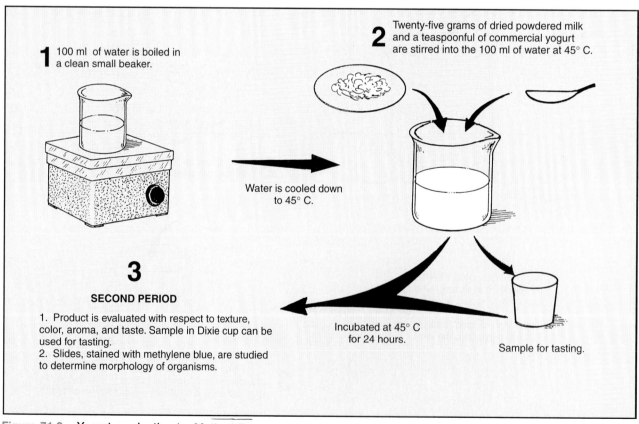

1 100 ml of water is boiled in a clean small beaker.

Water is cooled down to 45° C.

2 Twenty-five grams of dried powdered milk and a teaspoonful of commercial yogurt are stirred into the 100 ml of water at 45° C.

Incubated at 45° C for 24 hours.

Sample for tasting.

3

SECOND PERIOD

1. Product is evaluated with respect to texture, color, aroma, and taste. Sample in Dixie cup can be used for tasting.
2. Slides, stained with methylene blue, are studied to determine morphology of organisms.

Figure 71.2   Yogurt production by Method B

PART 13

Bacterial Genetic Variations

Variations in bacteria that are due to environmental factors and that do not involve restructuring DNA are designated as **temporary variations.** Such variations may be morphological or physiological and disappear as soon as the environmental changes that brought them about disappear. For example, as a culture of *E. coli* becomes old and the nutrients within the tube become depleted, the new cells that form become so short that they appear coccoidal. Reinoculation of the organism into fresh media, however, results in the reappearance of distinct bacilli of characteristic length.

Variations in bacteria that involve alteration of the DNA macromolecule are designated as **permanent variations.** It is because they survive a large number of transfers that they are so named. Such variations are due to mutations. Variations of this type occur spontaneously. They also might be induced by physical and chemical methods. Some permanent variations also are caused by the transfer of DNA from one organism to another, either directly by conjugation or indirectly by phage. It is these permanent genetic variations that the four exercises of this unit represent.

Exercises 72 and 73 of this unit demonstrate how spontaneous mutations are constantly occurring in bacterial populations. The genetic change that occurs in these two exercises pertains to the development of bacterial resistance to streptomycin. In Exercise 74 we will study how chemically induced mutagenicity that causes back mutations is used in the Ames test to determine possible carcinogenicity of chemical compounds. Finally, in Exercise 75 we will study how an organism's genetic structure can be altered by conjugation.

72 Mutant Isolation by Gradient Plate Method

An excellent way to determine the ability of organisms to produce mutants that are resistant to antibiotics is to grow them on a **gradient plate** of a particular antibiotic. Such a plate consists of two different wedgelike layers of media: a bottom layer of plain nutrient agar and a top layer of nutrient agar with the antibiotic. Since the antibiotic is only in the top layer, it tends to diffuse into the lower layer, producing a gradient of antibiotic concentration from low to high.

In this exercise we will make a gradient plate using streptomycin in the medium. *E. coli,* which is normally sensitive to this antibiotic, will be spread over the surface of the plate and incubated for 4 to 7 days. Any colonies that develop in the high concentration area will be streptomycin-resistant mutants.

Plate Preparation

The gradient plate used in this experiment will have a high concentration of 100 mcg of streptomycin per milliliter of medium. This concentration is 10 times the strength used in sensitivity disks in the Kirby-Bauer test method. Prepare a gradient plate as follows:

Materials:
  1 sterile Petri plate
  2 nutrient agar pours (10 ml per tube)
  1 tube of streptomycin solution (1%)
  1 wood spacer (⅛″ × ½″ × 2″)
  1 ml pipette

1. Liquefy two pours of nutrient agar and cool to 50° C.
2. With wood spacer under one edge of Petri plate (see figure 72.1), pour contents of one agar pour into plate. Let stand until solidification has occurred.
3. Remove the wood spacer from under the plate.
4. Pipette 0.1 ml of streptomycin into second agar pour, mix tube between palms, and pour contents over medium of plate that is now resting level on the table.
5. Label the low and high concentration areas on the bottom of the plate.

Inoculation

The inoculation procedure is illustrated in figure 72.2. The technique involves spreading a measured amount of the culture on the surface of the

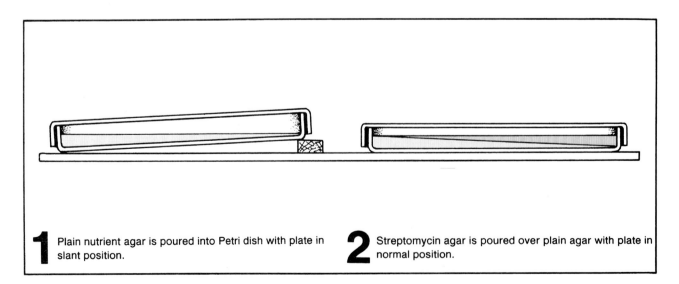

1 Plain nutrient agar is poured into Petri dish with plate in slant position.

2 Streptomycin agar is poured over plain agar with plate in normal position.

Figure 72.1 Procedure for pouring a gradient plate

medium with a glass bent rod to provide optimum distribution.

Materials:

1 beaker of 95% ethanol
1 glass rod spreader
nutrient broth culture of *E. coli*
1 ml pipette

1. Pipette 0.1 ml of *E. coli* suspension onto surface of medium in Petri plate.
2. Sterilize glass spreading rod by dipping it in alcohol first and then passing it quickly through the flame of a Bunsen burner. Cool the rod by placing against sterile medium in plate before contacting organisms.
3. Spread the culture evenly over the surface with the glass rod.
4. Invert and incubate the plate at 37° C for 4 to 7 days in a closed cannister or plastic bag. Unless incubated in this manner, excessive dehydration might occur.

First Evaluation

After 4 to 7 days, look for colonies of *E. coli* in the area of high streptomycin concentration. Count the colonies that appear to be resistant mutants and record your count on the Laboratory Report.

Select a well-isolated colony in the high concentration area and, with a sterile loop, smear the colony over the surface of the medium toward the higher concentration portion of the plate. Do this with two or three colonies. Return the plate to the incubator for another 2 or 3 days.

Final Evaluation

Examine the plate again to note what effect the spreading of the colonies had on their growth. Record your observations on Laboratory Report 72, 73.

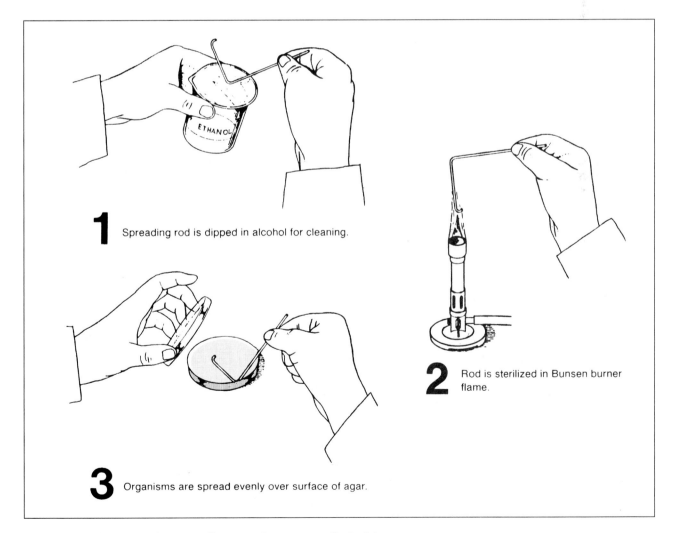

1 Spreading rod is dipped in alcohol for cleaning.

2 Rod is sterilized in Bunsen burner flame.

3 Organisms are spread evenly over surface of agar.

Figure 72.2 Procedure for spreading organisms on gradient plate

73 — Mutant Isolation by Replica Plating

In the last exercise it was observed that *E. coli* could develop mutant strains that are streptomycin-resistant. If we had performed this experiment with other organisms and with other antibiotics, the results would have been quite similar. The question that logically develops in one's mind from this experiment is: What mechanism is involved here? Is a mutation of this sort induced by the antibiotic? Or does the mutation occur spontaneously and independently of the presence of the drug? If we could demonstrate the presence of a streptomycin-resistant mutant occurring on a medium that lacks streptomycin, then we could assume that the mutation occurs spontaneously.

To determine whether or not such a colony exists on a plain agar plate having 500 to 1000 colonies could be a laborious task. One would have to transfer organisms from each colony to a medium containing streptomycin. This is somewhat self-defeating, too, in light of the low incidence of mutations that occur. Many thousands of the transfers might have to be made to find the first mutant. Fortunately, we can resort to replica plating to make all the transfers in one step. Figure 73.1 illustrates the procedure. In this technique a velveteen-covered colony transfer device is used to make the transfers.

Note in figure 73.1 that organisms are first dispersed on nutrient agar with a glass spreading rod. After incubation, all colonies are transferred from the nutrient agar plate to two other plates: first to a nutrient agar plate and second to a streptomycin agar plate. After incubation, streptomycin-resistant strains are looked for on the streptomycin agar.

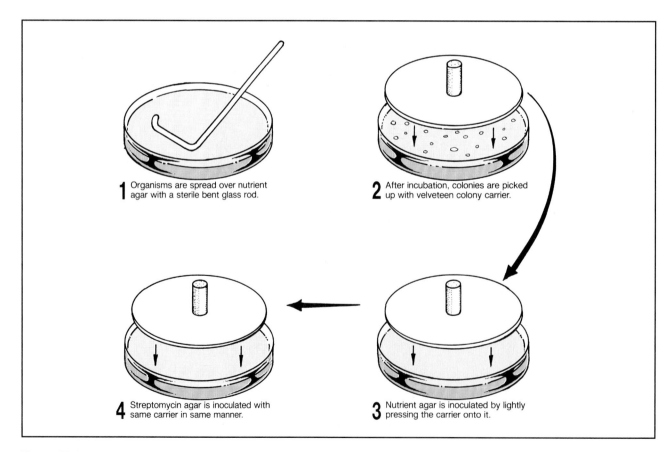

1 Organisms are spread over nutrient agar with a sterile bent glass rod.

2 After incubation, colonies are picked up with velveteen colony carrier.

4 Streptomycin agar is inoculated with same carrier in same manner.

3 Nutrient agar is inoculated by lightly pressing the carrier onto it.

Figure 73.1   Replica plating technique

First Period

Materials:
1 Petri plate of nutrient agar
1 bent glass spreading rod
1 ml serological pipette
beaker of 95% ethanol
Bunsen burner
broth culture of *E. coli*

1. Pipette 0.1 ml of *E. coli* from broth culture to surface of medium in Petri dish.
2. With a sterile bent glass rod, spread the organisms over the plate following the routine shown in figure 73.1.
3. Incubate this plate at 37° C for 24 hours.

Second Period

Materials:
1 Petri plate culture of *E. coli* from previous period
1 Petri plate of nutrient agar per student

1 Petri plate of streptomycin agar (100 micrograms of streptomycin per ml of medium)
1 sterile colony carrier per student

1. Carefully lower the sterile colony carrier onto the colonies of *E. coli* on the plate from the previous period.
2. Inoculate the plate of nutrient agar by lightly pressing the carrier onto the medium.
3. Now *without* returning the carrier to the original culture plate, inoculate the streptomycin agar in the same manner.
4. Incubate both plates at 37° C for 2 to 4 days in an enclosed cannister.

Third Period

Materials:
Quebec colony counter and hand counter

1. Examine both plates and record the information called for on Laboratory Report 72, 73.
2. Tabulate the results of other members of the class.

Bacterial Mutagenicity and Carcinogenesis:
The Ames Test

The fact that carcinogenic compounds induce increased rates of mutation in bacteria has led to the use of bacteria for screening chemical compounds for possible carcinogenesis. The **Ames test,** developed by Bruce Ames at the University of California–Berkeley, has been widely used for this purpose.

The conventional way to determine whether a chemical substance is carcinogenic is to inject the material into animals and look for the development of tumors. If tumors develop, it is presumed that the substance can cause cancer. Although this method works well, it is costly, time-consuming, and cumbersome, especially if it is applied to all the industrial chemicals that have found their way into food and water supplies.

The Ames test serves as a screening test for the detection of carcinogenic compounds by testing the ability of chemical agents to induce bacterial mutations. Although most mutagenic agents are carcinogenic, some are not; however, the correlation between carcinogenesis and mutagenicity is high—around 83%. Once it has been determined that a specific agent is mutagenic, it can be used in animal tests to confirm its carcinogenic capability.

The standard way to test chemicals for mutagenesis has been to measure the rate of *back mutations* in strains of auxotrophic bacteria. In the Ames test a strain of *Salmonella typhimurium,* which is auxotrophic for histidine (unable to grow in the absence of histidine), is exposed to a chemical agent. After chemical exposure and incubation on histidine-deficient medium, the rate of reversion (back mutation) to prototrophy is determined by counting the number of colonies that are seen on the histidine-deficient medium.

Although testing of chemicals for mutagenesis in bacteria has been performed for a long time, two new features are included in the Ames test that make it a powerful tool. The first is that the strain of *S. typhimurium* used here lacks DNA repair enzymes, which prevents the correction of DNA injury. The second feature of the test is the incorporation of mammalian liver enzyme preparations with the chemical agent. The latter is significant because there is evidence that liver enzymes convert many noncarcinogenic chemical agents to carcinogenic ones.

There are two ways to perform the Ames test. The method illustrated in figure 74.1 is a **spot test** that is widely used for screening purposes. The other method is the **plate incorporation test,** which is used for quantitative analysis of the mutagenic effectiveness of compounds. Our concern here will be with the spot test; however, since the concentration of the liver extract is very critical we will omit using it in our test. The test, as performed here, will work well without it.

Success in performing the spot test requires considerable attention to careful measurements and timing. It is for this reason that students will work in pairs to perform the test.

Note in figure 74.1 that 0.1 ml of *S. typhimurium* is first added to a small tube that contains 2 ml of top agar that is held at 45° C. This top agar contains 0.6% agar, 0.5% NaCl, and a *trace* of histidine and biotin. The histidine allows the bacteria to go through several rounds of cell division, which is essential for mutagenesis to occur. Since the histidine deletion extends through the biotin gene, biotin is also needed. This early growth of cells produces a faint background lawn that is barely visible to the naked eye.

Before pouring the top agar over the glucose–minimal salts agar, the tube must be vortexed at slow speed for 3 seconds and poured quickly to get even distribution. The addition of the bacteria, vortexing, and pouring must be accomplished in 20 seconds. Failure to move quickly enough will cause stippling of the top agar.

There are two ways that one can use to apply the chemical agent to the top agar: a filter paper disk may be used, or the chemical can be applied directly to the center of the plate without a disk. The procedure shown in figure 74.1 involves using a disk.

Note the unusual way in which a filter paper disk is impregnated in figure 74.1. To get it to stand on edge it must be put in position with sterile forceps and pressed in slightly to hold it upright. Just the right amount of the chemical agent is then added with a Pasteur pipette to the upper edge of the disk to completely saturate it without making it dripping wet; then the disk is lowered onto the top agar.

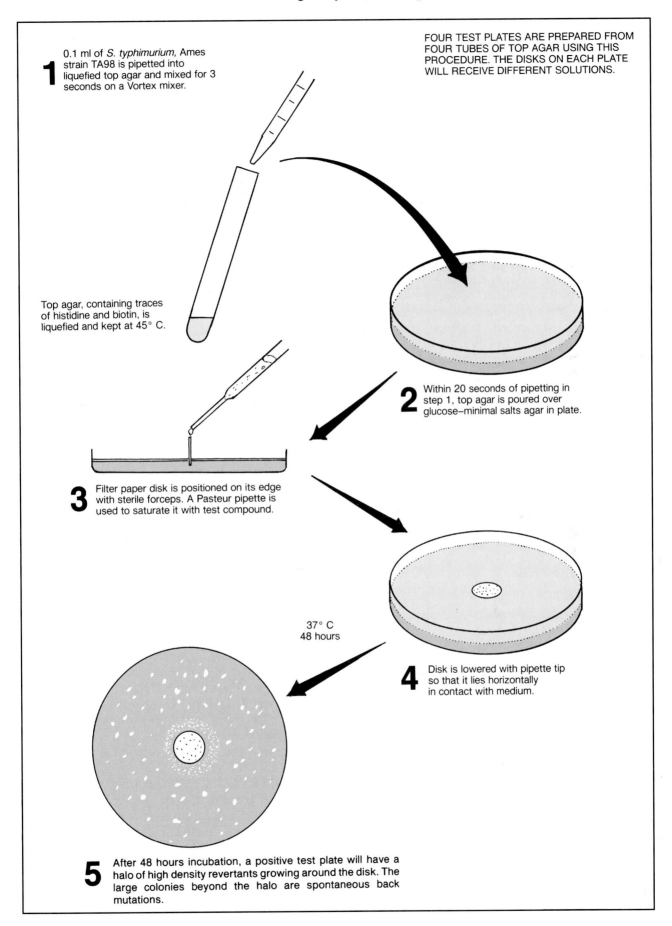

1 0.1 ml of *S. typhimurium,* Ames strain TA98 is pipetted into liquefied top agar and mixed for 3 seconds on a Vortex mixer.

FOUR TEST PLATES ARE PREPARED FROM FOUR TUBES OF TOP AGAR USING THIS PROCEDURE. THE DISKS ON EACH PLATE WILL RECEIVE DIFFERENT SOLUTIONS.

Top agar, containing traces of histidine and biotin, is liquefied and kept at 45° C.

2 Within 20 seconds of pipetting in step 1, top agar is poured over glucose–minimal salts agar in plate.

3 Filter paper disk is positioned on its edge with sterile forceps. A Pasteur pipette is used to saturate it with test compound.

37° C
48 hours

4 Disk is lowered with pipette tip so that it lies horizontally in contact with medium.

5 After 48 hours incubation, a positive test plate will have a halo of high density revertants growing around the disk. The large colonies beyond the halo are spontaneous back mutations.

Figure 74.1   Procedure for performing a modified Ames test

Once the test reagent is deposited on the top agar, the plate is incubated at 37° C for 48 hours. If the agent is mutagenic, a halo of densely packed revertant colonies will be seen around the disk. Scattered larger colonies will show up beyond the halo that represent spontaneous back mutations, not related to the test reagent.

You will be issued an unknown chemical agent to test and you will have an opportunity to test some other substance you have brought to the laboratory. In addition to these two tests you will be inoculating positive and negative test controls: thus, each pair of students will be responsible for four plates.

Keep in mind as you perform this experiment that there is a lot more to the Ames test than revealed here. While we are using only one tester strain of *S. typhimurium*, there are several others that are used in routine testing. The additional strains are needed to accommodate different kinds of chemical compounds. While one chemical agent may be mutagenic on one tester strain, it may produce a negative result on another strain. Also, keep in mind that we are not taking advantage of using the liver extract.

First Period

(Inoculations)

Materials:
per pair of students:

4 plates of glucose–minimal salts agar (30 ml per plate)
4 tubes of top agar (2 ml per tube)
tube of sterile water
Vortex mixer

sterile Pasteur pipettes, forceps
serological pipettes (1 ml size)
filter paper disks, sterile in Petri dish
test reagents:

4-NOPD (10 µg/ml) solution*
tube of unknown possible carcinogen
substance from home for testing

culture of *S. typhimurium,* Ames strain,
TA98 in trypticase soy broth
*4-nitro-o-phenylenediamine

1. Working with your laboratory partner, label the bottoms of four glucose–minimal salts agar plates as follows: POSITIVE CONTROL, NEGATIVE CONTROL, UNKNOWN, and OPTIONAL.
2. Liquefy four tubes of top agar and cool to 45° C.

3. With a 1 ml serological pipette, inoculate a tube of top agar with 0.1 ml of *S. typhimurium.*
4. Thoroughly mix the organisms into the top agar by vortexing (slow speed) for 3 seconds, or rolling the tube between the palms of both hands. Pour the contents onto the positive control plate of glucose–minimal salts agar. The agar plate must be at room temperature. **Work rapidly to achieve pipetting, mixing, and spreading in 20 seconds.**
5. Repeat steps 3 and 4 for each of the other three tubes of top agar.
6. With sterile forceps place a disk on its edge near the center of the **positive control plate.** Sterilize the forceps by dipping in alcohol and flaming.
7. With a sterile Pasteur pipette, deposit just enough 4-NOPD on the upper edge of the disk to saturate it; then, push over the disk with the pipette tip onto the agar so that it lies flat.
8. Insert a sterile disk on the **negative control plate** in the same manner as above. Moisten this disk with sterile water, and reposition it flat on the agar surface. Be sure to use a fresh Pasteur pipette.
9. Place a disk on the **unknown plate,** and, using the same procedures, infiltrate it with your unknown, and position it flat on the agar.
10. On the fourth plate (**optional**) deposit a drop of your unknown from home. If the test substance from home is crystalline, place a few crystals directly on the top agar of the optional plate in its center. Liquid substances should be handled in same manner as above.
11. Incubate all four plates for 48 hours at 37° C.

Second Period

(Evaluation)

Examine all four plates. You should have a pronounced halo of revertant colonies around the disk on the positive control plate, and no, or very few, revertants on the negative control plate. The presence of a few scattered revertants on the negative control plate is due to spontaneous back mutations, which always occur. Examine the areas beyond the halo to see if you can detect a faint lawn of bacterial growth.

CAUTION: Since much of the glassware in this experiment contains carcinogens, do not dispose of any of it in the usual manner. Your instructor will indicate how this glassware is to be handled.

Record your results on the Laboratory Report and answer all the questions.

Bacterial Conjugation

Although bacteria do not possess the type of sexual reproduction characteristic of eukaryons, some of them do exhibit a form of sexuality called conjugation. **Conjugation** may be defined as unidirectional transfer of genetic material (DNA) by one organism to another. To enable the DNA to pass from the donor (male) to the recipient (female), a "conjugation tube" forms between the two organisms. Only a fractional segment of the donor's chromosome passes into the recipient. As indicated in figure 75.1, the DNA acquired by the recipient is incorporated into its single, circular chromosome. With this new acquisition of genes, the recipient becomes a different organism.

In this experiment we will attempt to demonstrate the existence of this phenomenon by using two auxotrophs of *E. coli*. An **auxotroph** is a bacterial mutant that requires one or more growth factors that the wild strain, or **prototroph,** can synthesize. Auxotrophs are produced in the laboratory by subjecting prototrophs to mutagenic agents, such as ultraviolet irradiation or mitomycin C.

One of the auxotrophs used in this experiment is able to synthesize methionine, but not threonine, from a minimal medium containing only glucose, ammonia, and inorganic salts. In other words, this auxotroph has lost the gene that exists in the prototroph that would enable it to synthesize threonine in the presence of a minimal medium. Thus, it cannot grow on such a medium. Its genetic composition is designated as **thr−met+.**

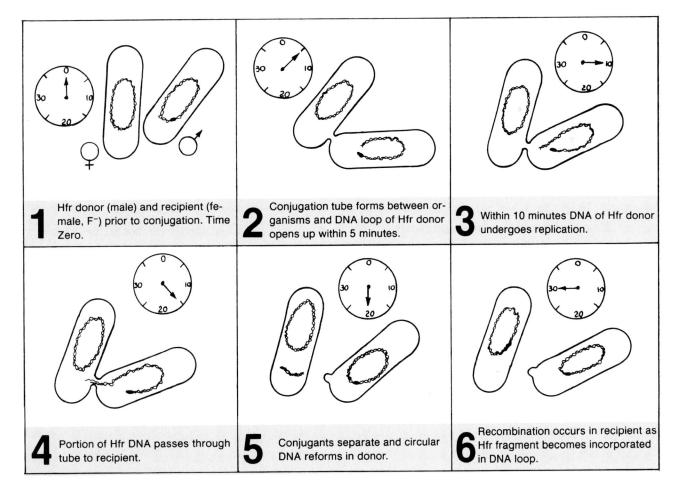

1 Hfr donor (male) and recipient (female, F⁻) prior to conjugation. Time Zero.

2 Conjugation tube forms between organisms and DNA loop of Hfr donor opens up within 5 minutes.

3 Within 10 minutes DNA of Hfr donor undergoes replication.

4 Portion of Hfr DNA passes through tube to recipient.

5 Conjugants separate and circular DNA reforms in donor.

6 Recombination occurs in recipient as Hfr fragment becomes incorporated in DNA loop.

Figure 75.1 Stages in bacterial conjugation

The other auxotroph in this experiment is able to synthesize threonine, but not methionine, and also is unable to grow on the minimal medium. It is designated as **thr+met−.**

If these two auxotrophs are able to conjugate to produce a strain that is **thr+met+,** then you have a genetic combination, like the original prototroph, that will grow on the minimal medium. It is the purpose of this experiment to demonstrate that conjugation can occur to bring about such a genetic recombination of genes. The donor in this experiment is **thr+met−.** The recipient is **thr−met+.**

Materials:

1 sterile serological tube (13 × 100 mm)
3 tubes of sterile distilled water (9 ml per tube)
2 Petri plates of trypticase soy agar (TSA)
5 Petri plates of minimal agar (MA)
3 sterile 1 ml pipettes
1 nutrient broth culture of *E. coli* (10^8 cells per ml), auxotroph (*thr+met−*), male (M)
1 nutrient broth culture of *E. coli* (10^8 cells per ml), auxotroph (*thr−met+*), female (F)
beaker of 95% alcohol
1 bent glass rod

Control Inoculations

To establish that the two auxotrophic strains of *E. coli* grow well on a complete medium (TSA) but not at all on minimal agar (MA), inoculate two plates of each of these media as follows:

1. Label one TSA plate "TSA-Male" and the other "TSA-Female." Label one MA plate "MA-Male" and the other "MA-Female."
2. With a good isolation technique, streak all four plates with the appropriate organisms.

Recombination Inoculations

The procedure in figure 75.2 will be performed to bring about conjugation of the two auxotrophs of *E. coli*. Note that one milliliter of each strain is placed in an empty test tube and is allowed to stand for 30 minutes. After this period, when conjugation has presumably occurred, the conjugants are diluted out onto MA plates in dilutions of 1:100, 1:1000, and 1:10,000. Any colonies that appear on this medium

will be organisms that can synthesize both threonine and methionine (*thr+met+*). By counting the colonies on these plates, it is possible to determine the frequency of recombination.

1. Label a sterile serological tube "M × F." This will be used for conjugation of the two strains of *E. coli.*
2. With separate pipettes, transfer 1 ml from each culture into the conjugation tube. Let stand for **30 minutes.**
3. While conjugation is taking place, label the three water blanks: I, II, and III. Also, label the three plates of minimal agar: I-1:100, II-1:1000, and III-1:10,000.
4. With a sterile pipette, transfer 1 ml from the conjugation tube into water blank I. Mix the contents in tube I by drawing the dilution up into the pipette three times and discharging slowly into the tube.
5. With the same pipette, transfer 0.1 ml from tube I to plate I. Do not spread at this time.
6. Still using the same pipette, transfer 1.0 ml from tube I to tube II and mix contents in tube II as above. Now transfer 0.1 ml from tube II to plate II. Do not spread at this time.
7. Still using the same pipette, transfer 1.0 ml from tube II to tube III and repeat all above procedures to inoculate plate III from tube III with 0.1 ml of diluted organisms.
8. With a sterile bent glass rod, spread the organisms over each plate following the routine shown in figure 75.2. Be sure to sterilize the rod before spreading organisms on each plate.
9. Incubate the two TSA and five MA plates, inverted, for 2 to 4 days at 30° C.

Evaluation of Results

Examine the control streak plates. Did both strains grow on the TSA plates? Did they grow on the MA plates?

Now examine the I, II, and III MA plates made in the second part of the experiment. Do you see any colonies here? Consult the Laboratory Report to determine the *number of recombinants per ml* and the *percentage of recombinants* in the M × F mixture.

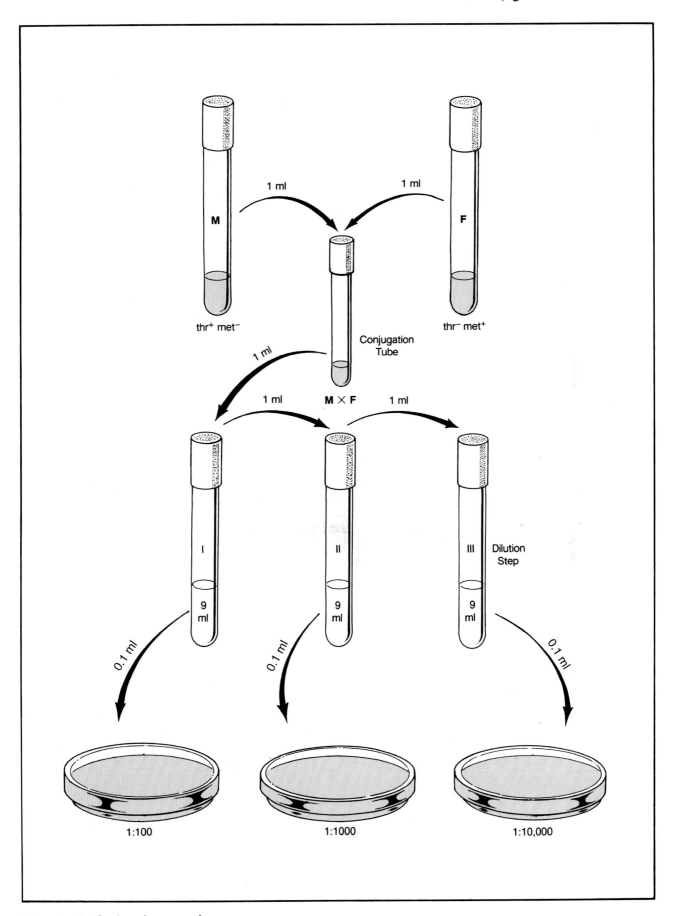

Figure 75.2 Conjugation procedure

PART 14 Medical Microbiology and Immunology

Although many of the exercises up to this point in this manual pertain in some way to medical microbiology, they also have applications that are nonmedical. The exercises of this unit, however, are primarily medical or dental in nature.

Medical (clinical) microbiology is primarily concerned with the isolation and identification of pathogenic organisms. Naturally, the techniques for studying each type of organism are different. A complete coverage of this field of microbiology is very extensive, encompassing the Mycobacteriaceae, Brucellaceae, Enterobacteriaceae, Corynebacteriaceae, Micrococcaceae, *ad infinitum.* It is not possible to explore all of these groups in such a short period of time; however, this course would be incomplete if it did not include some of the routine procedures that are used in the identification of some of the more common pathogens.

Exercise 76 in this unit differs from the other 13 exercises in that it pertains to the spread of disease (epidemiology) rather than to specific microorganisms. Its primary function is to provide an understanding of some of the tools used by public health epidemiologists to determine the sources of infection in the disease transmission cycle.

Since the most frequently encountered pathogenic bacteria are the gram-positive pyogenic cocci and the intestinal organisms, Exercises 77, 78, and 79 have been devoted to the study of those bacteria. The exercise that provides the greatest amount of depth is Exercise 78 (The Streptococci). To provide assistance in the identification of streptococci, it has been necessary to provide supplementary information in Appendix E.

Four exercises (81, 82, 83, and 84) are related to various applications of the agglutination reaction to serological testing. Two of these exercises pertain to slide tests and two of them are tube tests. It is anticipated that the instructor will select those tests from this group that fit time and budget limitations.

Exercises 86, 87, and 88 cover some of the basic hematological tests that might be included in a microbiology laboratory. The last exercise (89) pertains to an old test that has been revived pertaining to caries susceptibility.

A disease caused by microorganisms that enter the body and multiply in the tissues at the expense of the host is said to be an **infectious disease.** Infectious diseases that are transmissible to other persons are considered to be **communicable.** The transfer of communicable infectious agents between individuals can be accomplished by direct contact, such as in handshaking, kissing, and sexual intercourse, or they can be spread indirectly through food, water, objects, animals, and so on.

Epidemiology is the study of how, when, where, what, and who are involved in the spread and distribution of diseases in human populations. An epidemiologist is, in a sense, a medical detective who searches out the sources of infection so that the transmission cycle can be broken.

Whether an epidemic actually exists is determined by the epidemiologist by comparing the number of new cases with previous records. If the number of newly reported cases in a given period of time in a specific area is excessive, an **epidemic** is considered to be in progress. If the disease spreads to one or more continents, a **pandemic** is occurring.

In this experiment we will have an opportunity to approximate, in several ways, the work of the epidemiologist. Each member of the class will take part in the spread of a "synthetic infection." The mode of transmission will be handshaking. For obvious safety reasons, the agent of transmission will not be a pathogen.

Two different approaches to this experiment are given: procedures A and B. In procedure A a white powder is used. In Procedure B two non-pathogens (*Micrococcus luteus* and *Serratia marcescens*) will be used. The advantage of procedure A is that it can be completed in one laboratory session. Procedure B, on the other hand, is more realistic in that viable organisms are used; however, it involves two periods. Your instructor will indicate which procedure is to be followed.

Procedure A

In this experiment each student will be given a numbered container of white powder. Only one member in the class will be given a powder that is to be considered the infectious agent. The other members will be issued a transmissible agent that is considered noninfectious. After each student has spread the powder on his or her hands, all members of the class will engage in two rounds of handshaking, directed by the instructor. A record of the handshaking contacts will be recorded on a chart similar to the one on the Laboratory Report. After each round of handshaking, the hands will be rubbed on blotting paper so that a chemical test can be applied to it to determine the presence or absence of the infections agent.

Once all the data are compiled, an attempt will be made to determine two things: (1) the original source of the infection, and (2) who the carriers are. The type of data analysis used in this experiment is similar to the procedure that an epidemiologist would employ. Proceed as follows:

Materials:
    1 numbered container of white powder*
    1 piece of white blotting paper
    spray bottles of "developer solution"*

Preliminaries

1. After assembling your materials, write your name and unknown number at the top of your sheet of blotting paper. In addition, draw a line down the middle, top to bottom, and label the left side ROUND 1 and the right side ROUND 2.
2. Wash and dry your hands thoroughly.
3. Moisten the right hand with water and prepare it with the agent by thoroughly coating it with the white powder, especially on the palm surface. This step is similar to the contamination that would occur to one's hand if it were sneezed into during a cold.

 IMPORTANT: Once the hand has been prepared do not rest it on the tabletop or allow it to touch any other object.

---

*Instructor: To prevent students from preguessing the outcome of this experiment, the compositions of the powders and developer solution are known only to the instructor. The *Instructor's Handbook* provides this information.

Round 1

1. On the cue of the instructor, we will begin the first round of handshaking. Your instructor will inform you when it is your turn to shake hands with someone. You may shake with anyone, but it is best not to shake your neighbor's hand. *Be sure to use only your treated hand, and avoid extracurricular glad-handing.*

2. In each round of handshaking you will be selected by the instructor *only once* for handshaking; however, due to the randomness of selection by the handshakers, it is possible that you may be selected as the "shakee" several times.

3. After every member of the class has shaken someone's hand, we need to assess just who might have picked up the "microbe." To accomplish this, wipe your fingers and palm of the contaminated hand on the left side of your blotting paper. Press fairly hard, but don't tear the surface.
   IMPORTANT: Don't allow your hand to touch any other object. A second round of handshaking follows.

Round 2

1. On the cue of your instructor, proceed to shake hands with another person. Avoid contact with any other objects.

2. Once the second handshaking episode is finished, rub the fingers and palm of the contaminated hand on the right side of the blotting paper.
   CAUTION: Keep your contaminated hand off the left side of the blotting paper.

Chemical Identification

1. To determine who has been "infected" we will now spray the developer solution on the handprints of both rounds. One at a time, each student, with the help of the instructor, will spray his or her blotting paper with developer solution.

2. Color interpretation is as follows:

   Blue: positive for infectious agent
   Brown or yellow: negative

Tabulation of Results

1. Tabulate the results on the chalkboard, using a table similar to the one on the Laboratory Report.

2. Once all results have been recorded, proceed to determine the originator of the epidemic. The easiest way to determine this is to put together a flow chart of shaking.

3. Identify those persons that test positive. You will be working backward with the kind of information an epidemiologist has to work with (contacts and infections). Eventually, a pattern will emerge that shows which person started the epidemic.

4. Complete the Laboratory Report.

Procedure B

In this experiment each student will be given a piece of hard candy that has had a drop of *Micrococcus luteus* or *Serratia marcescens* applied to it. Only one person in the class will receive candy with *S. marcescens,* the presumed pathogen. All others will receive *M. luteus.*

After each student has handled the piece of candy with a glove-covered right hand, he or she will shake hands (glove to glove) with another student as directed by the instructor. A record will be kept of who takes part in each contact. Two rounds of handshaking will take place. After each round a plate of trypticase soy agar will be streaked.

After incubating the plates a tabulation will be made for the presence or absence of *S. marcescens* on the plates. From the data collected an attempt will be made to determine two things: (1) the original source of the infection, and (2) who the carriers are. The type of data analysis used in this experiment is similar to the procedure that an epidemiologist would employ. Proceed as follows:

Materials:
   sterile rubber surgical gloves (1 per student)
   hard candy contaminated with *M. luteus*
   hard candy contaminated with *S. marcescens*
   sterile swabs (2 per student)
   TSA plates (1 per student)

Preliminaries

1. Draw a line down the middle of the bottom of a TSA plate, dividing it into two halves. Label one half ROUND 1 and the other ROUND 2. Be sure to put your initials on the plate.

2. Put a sterile rubber glove on your right hand. Avoid contaminating the palm surface.

3. Grasp the piece of candy in your gloved hand, rolling it around the surface of your palm. Discard the candy into a beaker of disinfectant set aside for disposal. You are now ready to do the first-round handshake.

Round 1

1. *On the cue of your instructor,* select someone to shake hands with. You may shake with anyone, but it is best not to shake hands with your neighbor.
2. In each round of handshaking you will be selected by the instructor *only once* for handshaking; however, due to the randomness of selection by the handshakers, it is possible that you may be selected as the "shakee" several times. The instructor or a recorder will record the initials of the shaker and shakee each time.
3. After you have shaken someone's hand, swab the surface of your palm and transfer the organisms to the side of your plate designated as ROUND 1. Discard this swab into the appropriate container for disposal.

Round 2

1. Again, on the cue of your instructor, select someone at random to shake hands with. Be sure not to contaminate your gloved hand by touching something else.
2. With a fresh swab, swab the palm of your hand and transfer the organisms to the side of your plate designated as ROUND 2. Make sure that your initials and the initials of the shakee are recorded by the instructor or recorder.
3. Incubate the TSA plate at room temperature for 48 hours.

Tabulation and Analysis

1. After 48 hours incubation look for typical red *S. marcescens* colonies on your Petri plate. If such colonies are present, record them as positive on your Laboratory Report chart and on the chart on the chalkboard.
2. Fill out the chart on your Laboratory Report with all the information from the chart on the chalkboard.
3. Identify those persons that test positive. You will be working backwards with the kind of information an epidemiologist has to work with (contacts and infections). Eventually a pattern will emerge that shows which person started the epidemic.

Laboratory Report

Complete the Laboratory Report for this exercise.

The Staphylococci:
Isolation and Identification

77

Often in conjunction with streptococci, the staphylococci cause abscesses, boils, carbuncles, osteomyelitis, and fatal septicemias. Collectively, the staphylococci and streptococci are referred to as the pyogenic (pus-forming) gram-positive cocci. Originally isolated from pus in wounds, the staphylococci were subsequently demonstrated to be normal inhabitants of the nasal membranes, the hair follicles, the skin, and the perineum of healthy individuals. The fact that 90% of hospital personnel are carriers of staphylococci portends serious epidemiological problems, especially since most strains are penicillin-resistant.

The **staphylococci** are gram-positive spherical bacteria that divide in more than one plane to form irregular clusters of cells. They are listed in section 12, volume 2, of *Bergey's Manual of Systematic Bacteriology*. The genus *Staphylococcus* is grouped with three other genera in family Micrococcaceae:

SECTION 12 GRAM-POSITIVE COCCI

Family I   Micrococcaceae
  Genus I   *Micrococcus*
  Genus II   *Stomatococcus*
  Genus III   *Planococcus*
  Genus IV   **Staphylococcus**

Family II   Deinococcaceae
  Genus I   *Deinococcus*
  Genus II   **Streptococcus**

Although the staphylococci make up a coherent phylogenetic group, they have very little in common with the streptococci except for their basic similarities of being gram-positive, non-spore-forming cocci. Note that *Bergey's Manual* puts these two genera into separate families due to their inherent differences.

Of the 19 species of staphylococci listed in *Bergey's Manual*, the most important ones are *S. aureus, S. epidermidis,* and *S. saprophyticus.* The single most significant characteristic that separates these species is the ability or inability of these organisms to coagulate plasma: only *S. aureus* has this ability; the other two are coagulase-negative.

Although *S. aureus* has, historically, been considered to be the only significant pathogen of the three, the others do cause infections. Some cerebrospinal fluid infections (2), prosthetic joint infec-

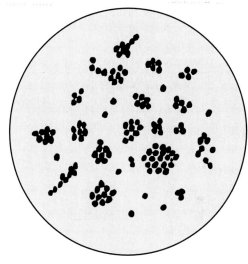

Figure 77.1   Staphylococci
└ commonly found in wound infections

tions (3), and vascular graft infections (1) have been shown to be due to coagulase-negative staphylococci. Numbers in parentheses designate references at the end of this exercise.

Our concern in this exercise will pertain exclusively to the differentiation of only three species of staphylococci. If other species are encountered, the student may wish to use the API Staph-Ident miniaturized test strip system (Exercise 55).

In this experiment we will attempt to isolate staphylococci from (1) the nose, (2) a fomite, and (3) an "unknown-control." The unknown-control will be a mixture containing staphylococci, streptococci, and some other contaminants. If the nasal membranes and fomite prove to be negative, the unknown-control will yield positive results, providing all inoculations and tests are performed correctly.

Since *S. aureus* is by far the most significant pathogen in this group, most of our concern here will be with this organism. It is for this reason that the characteristics of only this pathogen will be outlined next.

Staphylococcus aureus cells are 0.8 to 1.0 μm in diameter and may occur singly, in pairs, or as clusters. Colonies of *S. aureus* on trypticase soy agar or blood agar are opaque, 1 to 3 mm in diameter, and yellow, orange, or white. They are salt-tolerant, growing well on media containing 10% sodium chloride. Virtually all strains are coagulase-positive. Mannitol is fermented aerobically to produce acid. Alpha toxin is produced that causes a wide zone of clear (beta-type) hemolysis on blood agar; in rabbits it causes local necrosis and death.

The other two species lack alpha toxin (do not exhibit hemolysis) and are coagulase-negative. Mannitol is fermented to produce acid (no gas) by all strains of *S. aureus* and most strains of *S. saprophyticus.* Table 77.1 lists the principal characteristics that differentiate these three species of staphylococcus.

Table 77.1 Differentiation of three species of staphylococci

	S. Aureus	S. Epidermidis	S. Saprophyticus
Alpha toxin	+	–	–
Mannitol (acid only)	+	–	(+)
Coagulase	+	–	–
Biotin for growth	–	+	NS
Novobiocin	S	S	R

Note: NS = not significant; S = sensitive; R = resistant; (+) = mostly positive

To determine the incidence of carriers in our classroom, as well as the incidence of the organism on common fomites, we will follow the procedure illustrated in figure 77.2. Results of class findings will be tabulated on the chalkboard so that all members of the class can record data required on the Laboratory Report. The characteristics we will look for in our isolates will be (1) beta-type hemolysis (alpha toxin), (2) mannitol fermentation, and (3) coagulase production. Organisms found to be positive for these three characteristics will be *presumed* to be *S. aureus*. Final confirmation will be made with additional tests. Proceed as follows:

First Period

(Specimen Collection)

Note in figure 77.2 that swabs that have been applied to the nasal membranes and fomites will be placed in tubes of enrichment medium containing 10% NaCl (*m*-staphylococcus broth). Since your unknown-control will lack a swab, initial inoculations from this culture will have to be done with a loop.

Materials:
   1 tube containing numbered unknown-control
   3 tubes of *m*-staphylococcus broth
   2 sterile cotton swabs

1. Label the three tubes of *m*-staphylococcus broth NOSE, FOMITE, and the number of your unknown- control.
2. Inoculate the appropriate tube of *m*-staphylococcus broth with one or two loopfuls of your unknown-control.
3. After moistening one of the swabs by immersing partially into the "nose" tube of broth, swab the nasal membrane just inside your nostril. A small amount of moisture on the swab will enhance the pickup of organisms. Place this swab into the "nose" tube.
4. Swab the surface of a fomite with the other swab that has been similarly moistened and deposit this swab in the "fomite" tube.

   The fomite you select may be a coin, drinking glass, telephone mouthpiece, or any other item that you might think of.
5. Incubate these tubes of broth for 4 to 24 hours at 37° C.

Second Period

(Primary Isolation Procedure)

Two kinds of media will be streaked for primary isolation: mannitol salt agar and staphylococcus medium 110. *(selective plate)*

Mannitol salt agar (MSA) contains mannitol, 7.5% sodium chloride, and phenol red indicator. The NaCl inhibits organisms other than staphylococci. If the mannitol is fermented to produce acid, the phenol red in the medium changes color from red to yellow.

Staphylococcus medium 110 (SM110) also contains NaCl and mannitol, but it lacks phenol red. Its advantage over MSA is that it favors colony pigmentation by different strains of *S. aureus*. Since this medium lacks phenol red, no color change takes place as mannitol is fermented. *(will be yellow – becoz it ferments mannitol)*

Materials:
   3 culture tubes from last period
   2 Petri plates of MSA
   2 Petri plates of SM110

1. Label the bottoms of the MSA and SM110 plates as shown in figure 77.2. Note that to minimize the number of plates required, it will be necessary to make half-plate inoculations for the nose and fomite. The unknown-control will be inoculated on separate plates.
2. Quadrant streak the MSA and SM110 plates with the unknown control.
3. Inoculate a portion of the nose side of each plate with the swab from the nose tube; then, with a

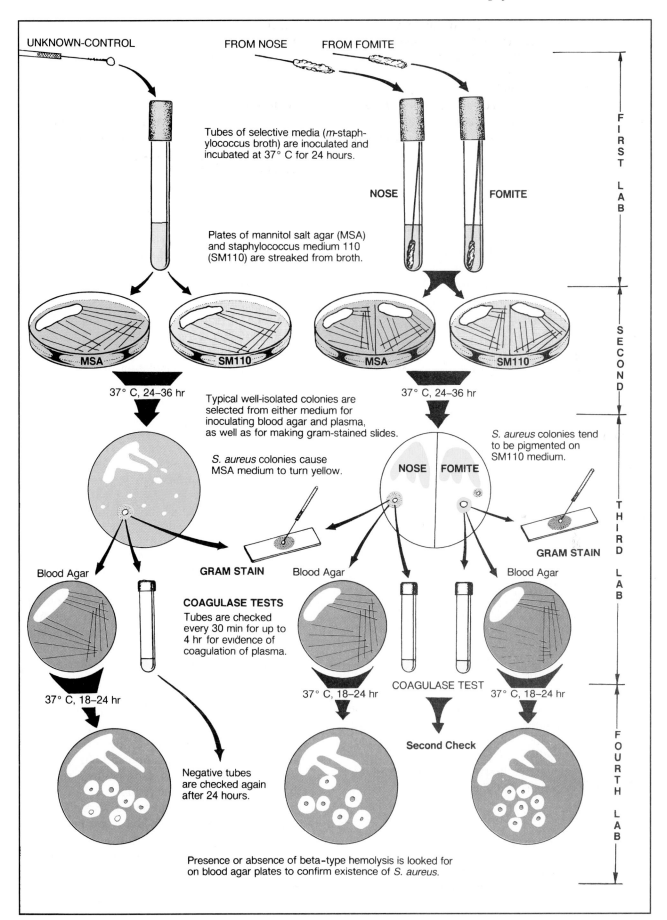

Figure 77.2 Procedure for presumptive identification of staphylococci

sterile loop, streak out the organisms on the remainder of the agar on that half of each plate. The swabbed areas will provide massive growth; the streaked-out areas should yield good colony isolation.

4. Repeat step 3 to inoculate the other half of each agar plate with the swab from the fomite tube.
5. Incubate the plates aerobically at 37° C for 24 to 36 hours.

Third Period

(Plate Evaluations and Coagulase/DNase Tests)

During this period we will perform the following tasks: (1) evaluate the plates from the previous period, (2) inoculate blood agar plates, (3) make gram-stained slides, and (4) perform coagulase and/or DNase tests on organisms from selected colonies. Proceed as follows:

Materials:
    MSA and SM110 plates from previous period
    2 blood agar plates
    serological tubes containing 0.5 ml of 1:4
        saline dilution of rabbit or human plasma
        (one tube for each isolate)
    Petri plates of DNase agar
    gram-staining kit

Evaluation of Plates

1. Examine the mannitol salt agar plates. Has the phenol red in the medium surrounding any of the colonies turned yellow?

    If this color change exists, it can be presumed that you have isolated a strain of *S. aureus.* Record your results on the Laboratory Report and chalkboard. (Your instructor may wish to substitute a copy of the chart from the Laboratory Report to be filled out at the demonstration table.)

2. Examine the plates of SM110. The presence of growth here indicates that the organisms are salt-tolerant. Note color of the colonies (white, yellow, or orange).

3. Record your observations of these plates on the Laboratory Report and chalkboard.

Blood Agar Inoculations

1. Label the bottom of one blood agar plate with your unknown-control number, and streak out the organisms from a staph-like colony.

2. Select staphylococcus-like colonies from the MSA and SM110 plates from the nose and fomites for streaking out on another blood agar plate. Use half-plate streaking methods, if necessary.

3. Incubate the blood agar plates at 37° C for 18 to 24 hours. *Don't leave plates in incubator longer than 24 hours.* Overincubation will cause blood degeneration.

Coagulase Tests

The fact that 97% of the strains of *S. aureus* have proven to be coagulase-positive and that the other two species are *always* coagulase-negative makes the coagulase test an excellent definitive test for confirming identification of *S. aureus.*

The procedure is simple. It involves inoculating a small tube of plasma with several loopfuls of the organism and incubating it in a 37° C water bath for several hours. If the plasma coagulates, the organism is coagulase-positive. Coagulation may occur in 30 minutes or several hours later. *Any degree of coagulation, from a loose clot suspended in plasma to a solid immovable clot, is considered to be a positive result, even if it takes 24 hours to occur.*

It should be emphasized that this test is valid only for gram-positive, staphylococcus-like bacteria, because some gram-negative rods, such as *Pseudomonas,* can cause a false-positive reaction. The mechanism of clotting in such organisms is not due to coagulase. Proceed as follows:

1. Label the plasma tubes NOSE, FOMITE, or UNKNOWN, depending on which of your plates have staph-like colonies.

2. With a wire loop, inoculate the appropriate tube of plasma with organisms from one or more colonies on SM110 or MSA. Use several loopfuls. Success is more rapid with a heavy inoculation. If positive colonies are present on both nose and fomite sides, be sure to inoculate a separate tube for each side.

3. Place the tubes in a 37° C water bath.

4. Check for solidification of the plasma every 30 minutes for the remainder of the period. Note in figure 77.3 that solidification may be complete, as in the lower tube, or show up as a semisolid ball, as seen in the middle tube.

    Any cultures that are negative at the end of the period will be left in the water bath. At 24 hours your instructor will remove them from the water bath and place them in the refrigerator, so that you can evaluate them in the next laboratory period.

5. Record your results on the Laboratory Report.

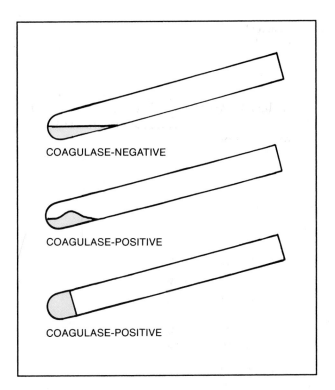

Figure 77.3 Coagulase test results: one negative and two positive tests

DNase Test

The fact that coagulase-positive bacteria are also able to hydrolyze DNA makes the DNase test a reliable means of confirming *S. aureus* identification. The following procedure can be used to determine if a staph-like organism can hydrolyze DNA.

1. Heavily streak the organism on a plate of DNase test agar. One plate can be used for several test cultures by making short streaks about 1 inch long.
2. Incubate for 18–24 hours at 35° C.

Gram-Stained Slides

While your tubes of plasma are incubating in the water bath, prepare gram-stained slides from the same colonies that were used for the blood agar plates and coagulase tests.

Examine the slides under oil immersion lens and draw the organisms in the appropriate areas on the Laboratory Report.

Fourth Period

(Confirmation)

During this period we will make final assessment of all tests and perform any other confirmatory tests that might be available to us.

Materials:

coagulase tubes from previous tests
blood agar plates from previous period
DNase test agar plates from previous period
0.1N HCl

1. Examine any coagulase tubes that were carried over from the last laboratory period that were negative at the end of that period. Record your results on the Laboratory Report.
2. Examine the colonies on your blood agar plates. Look for clear (beta-type) hemolysis around the colonies. The presence of alpha toxin is a definitive characteristic of *S. aureus*. Record your results on the Laboratory Report.
3. Look for zones of clearing near the streaks on the DNase agar plate. If none is seen, develop by flooding the plate with 0.1N HCl. The acid will render the hydrolyzed areas somewhat opaque.
4. Record your results on the chart on the chalkboard or chart on demonstration table. If an instructor-supplied tabulation chart is used, the instructor will have copies made of it to be supplied to each student.

Further Testing

In addition to using the API Staph-Ident miniaturized test strip system (Exercise 55) to confirm your identification of staphylococci, you may wish to use the latex agglutination slide test described in Exercise 82. Your instructor will inform you as to the availability of these materials and the desirability of proceeding further.

Laboratory Report

After recording your results on the chalkboard (or on chart on demonstration table), complete the chart on the Laboratory Report and answer all the questions.

Literature Cited

1. Liekweg, W. G., Jr., and L. T. Greenfield. 1977. Vascular prosthetic infection: Collected experience and results of treatment. *Surgery* 81: 355–400.
2. Schoenbaum, S. C., P. Gardner, and J. Shillito. 1975. Infections in cerebrospinal shunts: Epidemiology, clinical manifestations, and therapy. *J. Infect. Dis.* 131:543–52.
3. Wilson, P. D., Jr., E. A. Salvati, P. Aglietti, and L. J. Kutner. 1973. The problem of infection in endoprosthetic surgery of the hip joint. *Clin. Orthop. Relat. Res.* 96:213–21.

78

3 types of Hemolysis
1) Beta - β - area
2) alpha - α
3) gamma - δ

you use the blood plate
to see hemolysis

(complete) Beta hemolysis - area surrounding colony
is clear
(particial) alpha - see a greenish/grayish area
surrounding colony

(no) gamma - no hemolysis around colonies

The Streptococci:
Isolation and Identification

The streptococci differ from the staphylococci in that they are arranged primarily in chains rather than in clusters. In addition to causing many mixed infections with staphylococci, the streptococci can also, separately, cause diseases such as pneumonia, meningitis, endocarditis, pharyngitis, erysipelas, and glomerulonephritis.

Several species of streptococci are normal inhabitants of the pharynx. They can also be isolated from surfaces of the teeth, the saliva, skin, colon, rectum, and vagina.

The streptococci of greatest medical significance are *S. pyogenes, S. agalactiae,* and *S. pneumoniae.* Of lesser importance are *S. faecalis, S. faecium,* and *S. bovis.* Appendix E describes in greater detail the characteristics and significance of these and other streptococcal species.

The purpose of this exercise is twofold: (1) to learn about standard procedures for isolating streptococci from the pharynx and (2) to learn how to differentiate between the most significant medically important streptococci.

Figure 78.2 illustrates the overall procedure to be followed in the pursuit of the above two goals. Note that blood agar is used to separate the streptococci into two groups on the basis of the type of hemolysis they produce on blood agar. Those organisms that produce alpha hemolysis on blood agar can be differentiated by four tests. Those that produce beta-type hemolysis can be differentiated with the CAMP test and three other tests. The procedure outlined here is, primarily, designed to achieve *presumptive identification* of seven groups of streptococci. A few extra tests are usually required to confirm identification.

To broaden the application of these tests you may be given two or three unknown cultures of streptococci to be identified along with the pharyngeal isolates. *If unknowns are to be used, they will not be issued until physiological media are to be inoculated.*

First Period

(Making a Streak-Stab Agar Plate)

During this period a plate of blood agar is swabbed and streaked in a special way to determine the type of hemolytic bacteria that are present in the pharynx.

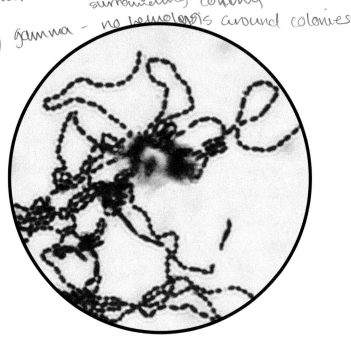

Figure 78.1 Streptococci

Before making such a streak plate, however, clinicians prefer to use a tube of enrichment broth (TSB) or a selective medium of TSB with a little crystal violet added to it (TSBCV). Media of this type are usually incubated at 37° C for 24 hours. This is particularly useful if the number of organisms might be low or if the swab cannot be applied to blood agar immediately. *Although this enrichment/selective step has been omitted here, it should be understood that the procedure is routine.*

Since swabbing one's own throat properly can be difficult, it will be necessary for you to work with your laboratory partner to swab each other's throats. Once your throat has been swabbed, you will proceed to use the swab to streak and stab your own agar plate according to a special procedure shown in figure 78.3.

Materials:
1 tongue depressor
1 sterile cotton swab
inoculating loop
1 blood agar plate

1. With the subject's head tilted back and the tongue held down with the tongue depressor, rub

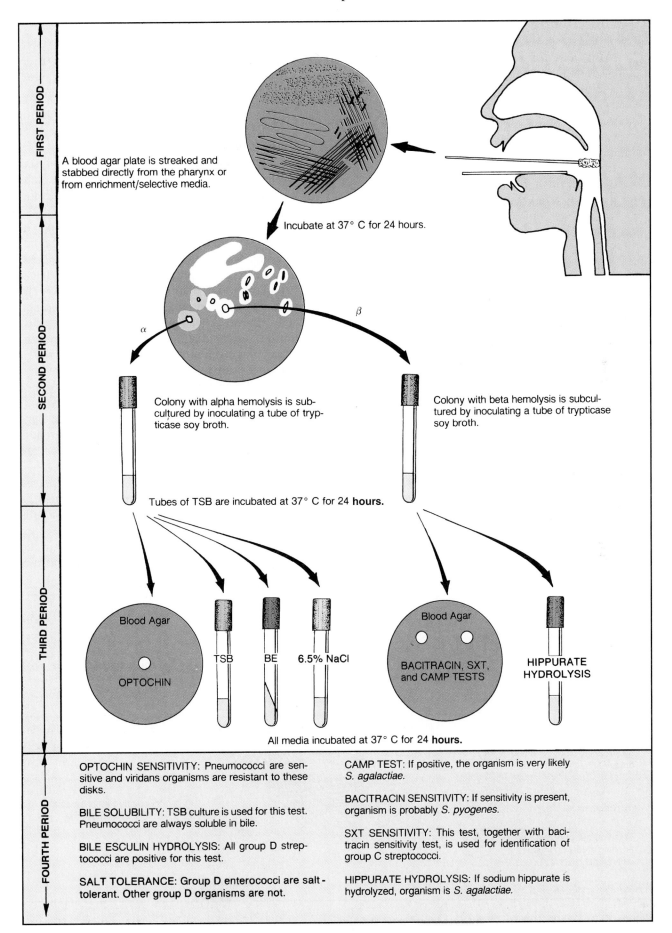

A blood agar plate is streaked and stabbed directly from the pharynx or from enrichment/selective media.

Incubate at 37° C for 24 hours.

α

β

Colony with alpha hemolysis is subcultured by inoculating a tube of trypticase soy broth.

Colony with beta hemolysis is subcultured by inoculating a tube of trypticase soy broth.

Tubes of TSB are incubated at 37° C for 24 **hours.**

Blood Agar

OPTOCHIN

TSB   BE   6.5% NaCl

Blood Agar

BACITRACIN, SXT, and CAMP TESTS

HIPPURATE HYDROLYSIS

All media incubated at 37° C for 24 **hours.**

FIRST PERIOD

SECOND PERIOD

THIRD PERIOD

FOURTH PERIOD

OPTOCHIN SENSITIVITY: Pneumococci are sensitive and viridans organisms are resistant to these disks.

BILE SOLUBILITY: TSB culture is used for this test. Pneumococci are always soluble in bile.

BILE ESCULIN HYDROLYSIS: All group D streptococci are positive for this test.

SALT TOLERANCE: Group D enterococci are salt-tolerant. Other group D organisms are not.

CAMP TEST: If positive, the organism is very likely *S. agalactiae.*

BACITRACIN SENSITIVITY: If sensitivity is present, organism is probably *S. pyogenes.*

SXT SENSITIVITY: This test, together with bacitracin sensitivity test, is used for identification of group C streptococci.

HIPPURATE HYDROLYSIS: If sodium hippurate is hydrolyzed, organism is *S. agalactiae.*

Figure 78.2   Media inoculations for the presumptive identification of streptococci

253

the back surface of the pharynx up and down with the sterile swab.

Also, *look for white patches* in the tonsillar area. Avoid touching the cheeks and tongue.

2. Since streptococcal hemolysis is most accurately analyzed when the colonies develop anaerobically beneath the surface of the agar, it will be necessary to use a streak-stab technique as shown in figure 78.3. The essential steps are as follows:

- Roll the swab over an area approximating one-fifth of the surface. The entire surface of the swab should contact the agar.
- With a wire loop, streak out three areas as shown to thin out the organisms.
- Stab the loop into the agar to the bottom of the plate at an angle perpendicular to the surface to make a clean cut without ragged edges.
- Be sure to make one set of stabs in an unstreaked area so that streptococcal hemolysis will be easier to interpret with a microscope.

3. Incubate the plate aerobically at 37° C for 24 hours. *Do not incubate longer than 24 hours.*

Second Period

(Analysis and Subculturing)

During this period, two things must be accomplished: first, the type of hemolysis must be correctly determined and, second, well-isolated colonies must be selected for making subcultures. The importance of proper subculturing cannot be overemphasized: without a pure culture, future tests are certain to fail. Proceed as follows:

Materials:
blood agar plate from previous period
tubes of TSB (one for each different type of colony)
dissecting microscope

1. Look for isolated colonies that have alpha or beta hemolysis surrounding them. Streptococcal colonies are characteristically very small.

2. Do any of the stabs appear to exhibit hemolysis? Examine these hemolytic zones near the stabs under 60× magnification with a dissecting microscope.

3. Consult figure 78.4 to analyze the type of hemolysis. Note that the illustrations on the left side indicate what the colonies would look like if they were submerged under a layer of blood agar (two-layer pour plate). The illustrations on the right indicate the nature of hemolysis around stabs on streak-stab plates. Although this illustration is very diagrammatic, it reveals the microscopic differences between three kinds of hemolysis: alpha, alpha-prime, and beta.

Only those stabs that are completely free of red blood cells in the hemolytic area are considered to be **beta hemolytic.** The chance of isolating a colony of this type from your own throat is very slim, for the beta hemolytic streptococci are the most serious pathogens.

If some red blood cells are seen dispersed throughout the hemolytic zone, the organism is said to be **alpha hemolytic.** *S. pneumoniae* falls into this category.

If you see some RBCs near the growth in the stab, and none in the clear areas, the organism is

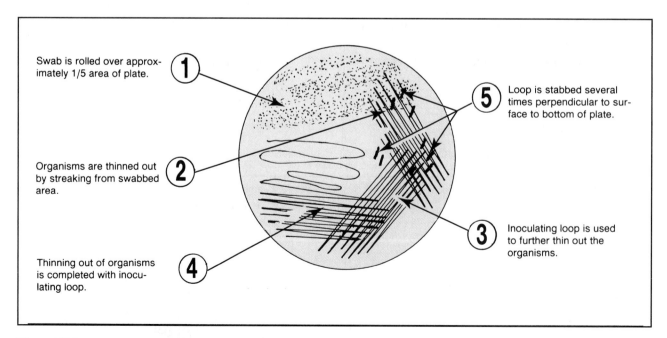

Swab is rolled over approximately 1/5 area of plate.

Organisms are thinned out by streaking from swabbed area.

Thinning out of organisms is completed with inoculating loop.

Loop is stabbed several times perpendicular to surface to bottom of plate.

Inoculating loop is used to further thin out the organisms.

Figure 78.3 Streak-stab procedure for blood agar inoculations

classified as **alpha-prime hemolytic.** Viridans streptococci often fall in this category.

4. Record your observations on the Laboratory Report.

5. Select well-isolated colonies that exhibit hemolysis (alpha, beta, or both) for inoculating tubes of TSB. Be sure to label the tubes ALPHA or BETA. Whether or not the organism is alpha or beta is crucial in identification.

   Since the chances of isolating beta hemolytic streptococci from the pharynx are usually quite slim, notify your instructor if you think you have isolated one.

6. Incubate the tubes at 37° C for 24 hours.

7. **Important:** At some time prior to the next laboratory session, review the material in Appendix E that pertains to this exercise.

Third Period

(Inoculations for Physiological Tests)

Presumptive identification of the various groups of streptococci is based on seven or eight physiological tests. Table 78.1 on page 257 reveals how they perform on these tests. Note that groups A, B, and C are all beta hemolytic; a few enterococci are also beta hemolytic. The remainder are all alpha hemolytic or nonhemolytic.

Since each of the physiological tests is specific for differentiating only two or three groups, it is not desirable to do all the tests on all unknowns. For economy and preciseness, only four tests that are mentioned for the third period in figure 78.2 should be performed on an isolate or unknown.

Before any inoculations are made, however, it is desirable to do a purity check on each TSB culture from the previous period. To accomplish this it will be necessary to make a gram-stained slide of each of the cultures.

If unknowns are to be issued, they will be given to you at this time. They will be tested along with your pharyngeal isolates. The only information that will be given to you about each unknown is its hemolytic type so that you will be able to determine what physiological tests to perform on each one. Proceed as follows:

Gram-Stained Slides (Purity Check)

Materials:
   TSB cultures from previous period
   gram-staining kit

1. Make a gram-stained slide from each of the pharyngeal isolates and examine them under oil

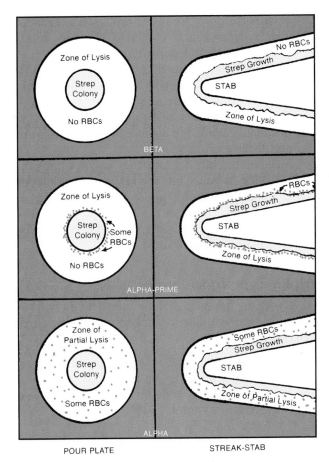

Figure 78.4 Comparison of hemolysis types as seen on pour plates and streak-stab plates

immersion lens. Do they appear to be pure cultures?

2. Draw the organisms in the appropriate circles on the Laboratory Report.

Beta-Type Inoculations

Use the following procedure to perform tests on each isolate that has beta-type hemolysis:

Materials:
for each isolate:
   1 blood agar plate
   1 tube of sodium hippurate broth
   1 bacitracin differential disk
   1 SXT sensitivity disk
   1 broth culture of *S. aureus*
   dispenser or forceps for transferring disks

1. Label a blood agar plate and a tube of sodium hippurate broth with proper identification information of each isolate and unknown to be tested.

2. Follow the procedure outlined in figure 78.5 to inoculate each blood agar plate with the isolate (or unknown) and *S. aureus.*

Note that a streak of the unknown is brought down perpendicular to the *S. aureus* streak, keeping the two organisms about 1 cm apart.

3. With forceps or dispenser, place one bacitracin differential disk and one SXT disk on the heavily streaked area at points shown in figure 78.5. Press down on each disk slightly.

4. Inoculate one tube of sodium hippurate broth for each isolate or unknown.

5. Incubate the blood agar plates at 37° C, aerobically, for 24 hours, and the hippurate broth tubes at 35° C, aerobically, for 24 hours. If the hippurate broths prove to be negative or weakly positive at 24 hours, they should be given more time to see if they change.

Alpha-Type Inoculations

As shown in figure 78.2, four inoculations will be made for each isolate or unknown that is alpha hemolytic.

Materials:

1 blood agar plate (for up to 4 unknowns)
1 6.5% sodium chloride broth
1 trypticase soy broth (TSB)
1 bile esculin (BE) slant
1 optochin (Taxo P) disk
candle jar setup or CO_2 incubator

1. Mark the bottom of a blood agar plate to divide it into halves, thirds, or quarters, depending on the number of alpha hemolytic organisms to be tested. Label each space with the code number of each test organism.

2. Completely streak over each area of the blood agar plate with the appropriate test organism,

and place one optochin (Taxo P) disk in the center of each area. Press down slightly on each disk to secure them to the medium.

3. Inoculate one tube each of TSB, BE, and 6.5% NaCl broth with each test organism.

4. Incubate all media at 35°–37° C as follows:

> Blood agar plates: 24 hours in a candle jar
> 6.5% NaCl broths: 24, 48, and 72 hours
> Bile esculin slants: 48 hours
> Trypticase soy broths: 24 hours

Note: While the blood agar plates should be incubated in a candle jar or CO_2 incubator, the remaining cultures can be incubated aerobically.

Fourth Period
(Evaluation of Physiological Tests)

Once all of the inoculated media have been incubated for 24 hours, you are ready to examine the plates and tubes and add test reagents to some of the cultures. Some of the tests will also have to be checked at 48 and 72 hours.

After you have assembled all the plates and tubes from the last period examine the blood agar plates first that were double-streaked with the unknowns and *S. aureus*. Note that the second, third, and fourth tests listed in table 78.1 can be read from these plates. Proceed as follows:

CAMP Reaction

If you have an unknown that produces an enlarged arrowhead-shaped hemolytic zone at the juncture where the unknown meets the *S. aureus* streak, as

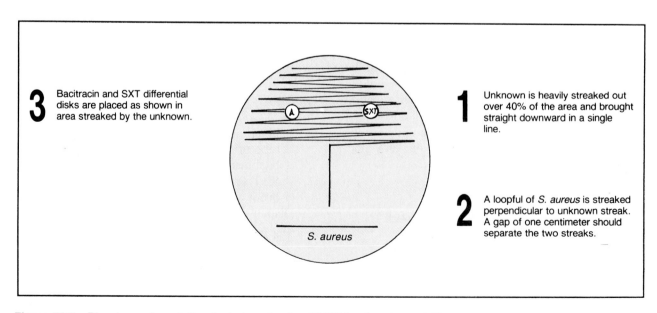

3 Bacitracin and SXT differential disks are placed as shown in area streaked by the unknown.

S. aureus

1 Unknown is heavily streaked out over 40% of the area and brought straight downward in a single line.

2 A loopful of *S. aureus* is streaked perpendicular to unknown streak. A gap of one centimeter should separate the two streaks.

Figure 78.5 Blood agar inoculation technique for the CAMP, bacitracin, and SXT tests

seen in figure 78.6, the organism is *S. agalactiae*. This phenomenon is due to what is called the *CAMP factor.* The only problem that can arise from this test is that if the plate is incubated anaerobically, a positive CAMP reaction can occur on *S. pyogenes* inoculated plates.

Record the CAMP reactions for each of your isolates or unknowns on the Laboratory Report.

Bacitracin Susceptibility

Any size zone of inhibition seen around the bacitracin disks should be considered to be a positive test result. Note in table 78.1 that *S. pyogenes* is positive for this characteristic.

S. py – susceptible
S. ag – resistant

This test has two limitations: (1) the disks must be of the *differential type,* not sensitivity type, and (2) the test should not be applied to alpha hemolytic streptococci. Reasons: Sensitivity disks have too high a concentration of the antibiotic, and many alpha hemolytic streptococci are sensitive to these disks.

Record the results of this test on table under D of your Laboratory Report.

SXT Sensitivity Test

The disks used in this test contain 1.25 mg of trimethoprim and 27.75 mg of sulfamethoxazole (SXT). The purpose of this test is to distinguish

Table 78.1 Physiological Tests for Streptococcal Differentiation

GROUP	Type of Hemolysis	Bacitracin Susceptibility	CAMP Reaction or Hippurate Hydrolysis	SXT Sensitivity	Bile Esculin Hydrolysis	Tolerance to 6.5% NaCl	Optochin Susceptibility	Bile Solubility
Group A *S. pyogenes*	beta	+	–	R	–	–	–	–
Group B *S. agalactiae*	beta	–*	+	R	–	±	–	–
Group C *S. equi* *S. equisimilis* *S. zooepidemicus*	beta	–*	–	S	–	–	–	–
Group D** (enterococci) *S. faecalis* *S. faecium* etc.	alpha beta none	–	–	R	+	+	–	–
Group D** (nonenterococci) *S. bovis* etc.	alpha none	–	–	R/S	+	–	–	–
Viridans *S. mitis* *S. salivarius* *S. mutans* etc.	alpha none	–*	–*	S	–	–	–	–
Pneumococci *S. pneumoniae*	alpha	±	–	–	–	+	+	

Note: R = resistant; S = sensitive; blank = not significant.
*Exceptions occur occasionally.
**See comments on pp. 443 and 444 concerning correct genus.

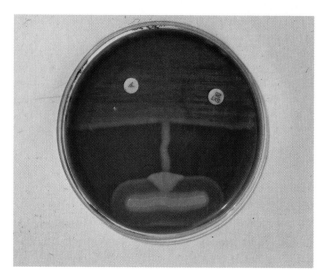

Figure 78.6 Note positive SXT disk on right, negative bacitracin disk on left, and positive CAMP reaction (arrowhead). Organism: *S. agalactiae*.

groups A and B from other beta hemolytic streptococci. Note in table 78.1 that both groups A and B are uniformly resistant to SXT.

If a beta hemolytic streptococcus proves to be bacitracin-resistant and SXT-susceptible, it is classified as being a **non-group-A or -B beta hemolytic streptococcus.** This means that the organism is probably a species within group C. *Keep in mind that an occasional group A streptococcal strain is susceptible to both bacitracin and SXT disks.* One must always remember that exceptions to most tests do occur; that is why this identification procedure leads us only to *presumptive* conclusions.

Record any zone of inhibition (resistance) as positive for this test.

Hippurate Hydrolysis

Note in table 78.1 that hippurate hydrolysis and the CAMP test are grouped together as positive tests for *S. agalactiae*. If an organism is positive for both tests, or either one, one can assume with almost 100% certainty that the organism is *S. agalactiae*.

Proceed as follows to determine which of your isolates are able to hydrolyze sodium hippurate:

Materials:
>   serological test tubes
>   serological pipettes (1 ml size)
>   ferric chloride reagent
>   centrifuge

1. Centrifuge the culture for 3 to 5 minutes.
2. Pipette 0.2 ml of the supernatant and 0.8 ml of ferric chloride reagent into an empty serological test tube. Mix well.

3. Look for a **heavy precipitate** to form. If the precipitate forms and persists for 10 minutes or longer, the test is positive. If the culture proves to be weakly positive, incubate the culture for another 24 hours and repeat the test.
4. Record your results on the Laboratory Report.

Bile Esculin Hydrolysis

This is the best physiological test that we have for the identification of group D streptococci. Both enterococcal and nonenterococcal species of group D are able to hydrolyze esculin in the agar slant, causing the slant to blacken.

A positive BE test tells us that we have a group D streptococcus; differentiation of the two types of group D streptococci depends on the salt-tolerance test.

Examine the BE agar slants, looking for **blackening of the slant,** as illustrated in figure 78.7. If less than half of the slant is blackened, or if no blackening occurs within 24 to 48 hours, the test is negative.

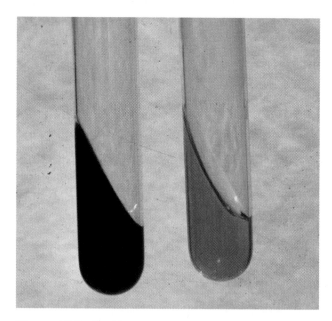

Figure 78.7 Positive bile esculin hydrolysis on left; negative on the right

Salt Tolerance (6.5% NaCl)

All enterococci of group D produce heavy growth in 6.5% NaCl broth. As indicated in table 78.1, none of the nonenterococci, group D, grow in this medium. This test, then, provides us with a good method for differentiating the two types of group D streptococci.

A positive result shows up as turbidity within 72 hours. A color change of **purple to yellow** may also be present. If the tube is negative at 24 hours, incubate it and check it again at 48 and 72 hours. *If* *the organism is salt-tolerant and BE-positive, it is considered to be an enterococcus. Parenthetically, it should be added here that approximately 80% of group B streptococci will grow in this medium.*

Optochin

Optochin susceptibility is used for differentiation of the alpha hemolytic viridans streptococci from the pneumococci. The pneumococci are sensitive to these disks; the viridans organisms are resistant.

Materials:
blood agar plates with optochin disks
plastic metric ruler

1. Measure the diameters of zones of inhibition that surround each disk, evaluating whether the zones are large enough to be considered positive. The standards are as follows:
   - For 6 mm diameter disks, the zone must be at least 14 mm diameter to be considered positive.
   - For 10 mm diameter disks, the zone must be at least 16 mm diameter to be considered positive.
2. Record your results on the Laboratory Report.

Bile Solubility

If an alpha hemolytic streptococcal organism is soluble in bile and positive on the optochin test, presumptive evidence indicates that the isolate is *S. pneumoniae*. Perform the bile solubility test on each of your alpha hemolytic isolates as follows:

Materials:
2 empty serological tubes (per test)

dropping bottle of phenol red indicator
dropping bottle of 0.05N NaOH
TSB culture of unknown
2% bile solution (sodium desoxycholate)
bottle of normal saline solution
2 serological pipettes (1 ml size)
water bath (37° C)

1. Mark one empty serological tube BILE and the other SALINE. Into their respective tubes, pipette 0.5 ml of 2% bile and 0.5 ml of saline.
2. Shake the TSB unknown culture to suspend the organisms and pipette 0.5 ml of the culture into each tube.
3. Add one or two drops of phenol red indicator to each tube and adjust the pH to 7.0 by adding drops of 0.05N NaOH.
4. Place both tubes in a 37° C water bath and examine periodically for 2 hours. If the turbidity clears in the bile tube, it indicates that the cells have disintegrated and the organism is *S. pneumoniae*. Compare the tubes side by side.
5. Record your results on the Laboratory Report.

Final Confirmation

All the laboratory procedures performed so far lead us to presumptive identification. To confirm these conclusions it is necessary to perform serological tests on each of the unknowns. If commercial kits are available for such tests, they should be used to complete the identification procedures.

Laboratory Report

Complete the Laboratory Report for this exercise.

Gram-Negative Intestinal Pathogens

The enteric pathogens of prime medical concern are the **salmonella** and **shigella.** They cause enteric fevers, food poisoning, and bacillary dysentery. *Salmonella typhi,* which causes typhoid fever, is by far the most significant pathogen of the salmonella group. In addition to the typhoid organism, there are 10 other distinct salmonella species and over 2200 serotypes. The shigella, which are the prime causes of human dysentery, comprise four species and many serotypes. *Serotypes* within genera are organisms of similar biochemical characteristics that can most easily be differentiated by serological typing.

Routine testing for the presence of these pathogens is a function of public health laboratories at various governmental levels. The isolation of these pathogenic enterics from feces is complicated by the fact that the colon contains a diverse population of bacteria. Species of such genera as *Escherichia, Proteus, Enterobacter, Pseudomonas,* and *Clostridium* exist in large numbers: hence it is necessary to use media that are differential and selective to favor the growth of the pathogens.

Figure 79.1 is a separation outline that is the basis for the series of tests that are used to demonstrate the presence of salmonella or shigella in a patient's blood, urine, or feces. Note that lactose fermentation separates the salmonella and shigella from most of the other Enterobacteriaceae. Final differentiation of the two enteric pathogens from *Proteus* relies on motility, hydrogen sulfide production, and urea hydrolysis. The differentiation information of the positive lactose fermenters on the left side of the separation outline is provided here mainly for comparative references that can be used for the identification of other unknown enterics.

The procedural diagram in figure 79.2 on the opposite page reveals how we will apply these facts in the identification of an unknown salmonella or shigella. The entire process will involve four laboratory periods.

In this experiment you will be given a mixed culture containing a coliform, *Proteus,* and a salmonella or shigella. The pathogens will be of the less dangerous types, but their presence will, naturally, demand utmost caution in handling. Your problem will be to isolate the pathogen from the mixed culture and make a genus identification. There are five steps that are used to prove the presence of these

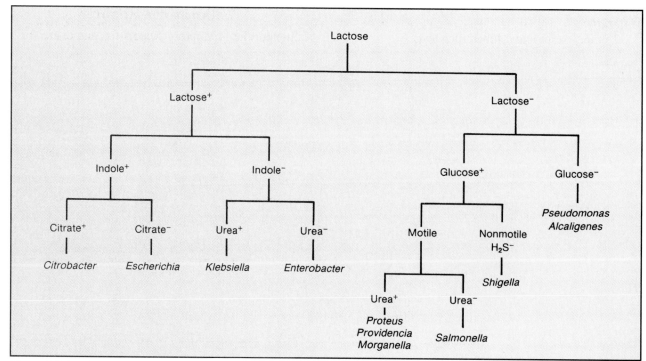

Figure 79.1 Separation outline of Enterobacteriaceae

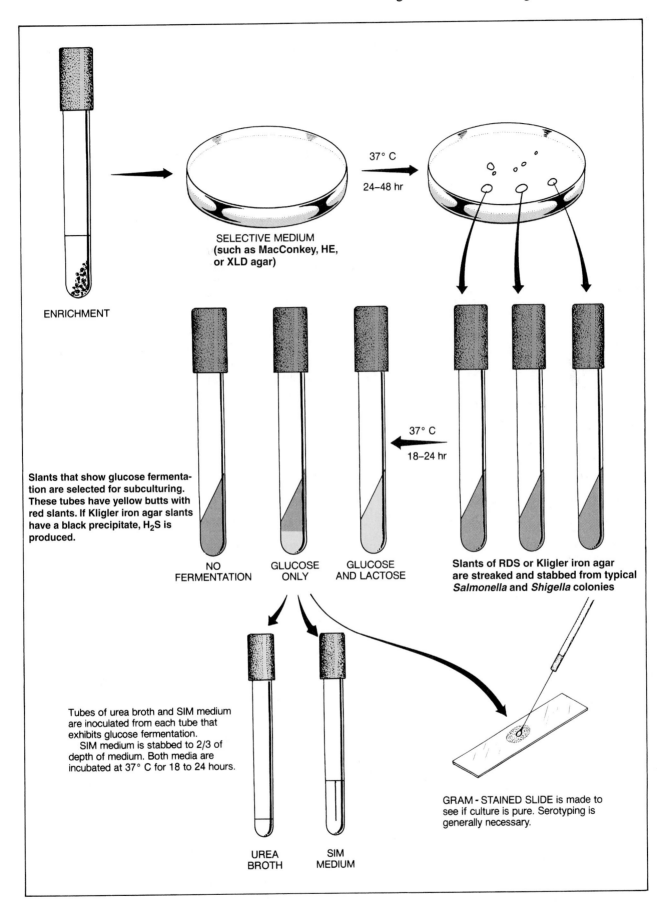

ENRICHMENT

SELECTIVE MEDIUM
(such as MacConkey, HE,
or XLD agar)

37° C
24–48 hr

Slants that show glucose fermentation are selected for subculturing.
These tubes have yellow butts with
red slants. If Kligler iron agar slants
have a black precipitate, H₂S is
produced.

37° C
18–24 hr

NO
FERMENTATION

GLUCOSE
ONLY

GLUCOSE
AND LACTOSE

Slants of RDS or Kligler iron agar
are streaked and stabbed from typical
Salmonella and *Shigella* colonies

Tubes of urea broth and SIM medium
are inoculated from each tube that
exhibits glucose fermentation.
    SIM medium is stabbed to 2/3 of
depth of medium. Both media are
incubated at 37° C for 18 to 24 hours.

GRAM - STAINED SLIDE is made to
see if culture is pure. Serotyping is
generally necessary.

UREA
BROTH

SIM
MEDIUM

Figure 79.2   Isolation and presumptive identification of *Salmonella* and *Shigella* pathogens in a stool sample: (1) enrichment, (2) isolation, (3) fermentation tests, (4) final physiological tests, and (5) serotyping.

Enrichment

There are two enrichment media that are most frequently used to inhibit the nonpathogens and favor the growth of pathogenic enterics. They are selenite F and gram-negative (GN) broths. While most salmonella grow unrestricted in these two media, some of the shigella are inhibited to some extent in selenite F broth; thus, for shigella isolation, GN broth is preferred. In many cases, stool samples are plated directly on isolation media.

In actual practice, 1 to 5 grams of feces are placed in 10 ml of enrichment broth. In addition, plates of various kinds of selective media are inoculated directly. The broths are usually incubated for 4 to 6 hours.

Since we are not using stool samples in this exercise, the enrichment procedure is omitted. Instead, you will streak the isolation media directly from the unknown broth.

First Period

(Isolation)

There are several excellent selective differential media that have been developed for the isolation of these pathogens. Various inhibiting agents such as brilliant green, bismuth sulfite, sodium desoxycholate, and sodium citrate are included in them. For *Salmonella typhi,* bismuth sulfite agar appears to be the best medium. Colonies of *S. typhi* on this medium appear black due to the reduction of sulfite to sulfide.

Other widely used media are MacConkey agar, Hektoen Enteric agar (HE), and Xylose Lysine Desoxycholate (XLD) agar. These media may contain bile salts and/or sodium desoxycholate to inhibit gram-positive bacteria. To inhibit coliforms and other nonenterics, they may contain a citrate. All of them contain lactose and a dye so that if an organism is a lactose fermenter, its colony will take on a color characteristic of the dye present.

Since the enrichment procedure is being omitted here, you will be issued an unknown broth culture with a pathogenic enteric. Your instructor will indicate which selective media will be used. Proceed as follows to inoculate the selective media with your unknown mixture:

Materials:
    unknown culture (mixture of a coliform,
       Proteus, and a salmonella or shigella)
    1 or more Petri plates of different selective
       media: MacConkey, Hektoen Enteric (HE),
       or Xylose Lysine Desoxycholate (XLD) agar

1. Label each plate with your name and unknown number.
2. With a loop, streak each plate with your unknown in a manner that will produce good isolation.
3. Incubate the plates at 37° C for 24 to 48 hours.

Second Period

(Fermentation Tests)

As stated above, the fermentation characteristic that separates the SS pathogens from the coliforms is their *inability to ferment lactose.* Once we have isolated colonies on differential media that look like salmonella or shigella, the next step is to determine whether the isolates can ferment lactose. All media for this purpose contain at least two sugars, glucose and lactose. Some contain a third sugar, sucrose. They also contain phenol red to indicate when fermentation occurs. Russell Double Sugar (RDS) agar is one of the simpler media that works well. Kligler iron agar may also be used. It is similar to RDS with the addition of iron salts for detection of H_2S. Your instructor will indicate which one will be used.

Proceed as follows to inoculate three slants from colonies on the selective media that look like either salmonella or shigella. The reason for using three slants is that the you may have difficulty distinguishing *Proteus* from the SS pathogens. By inoculating three tubes from different colonies, you will be increasing your chances of success.

Materials:
    3 agar slants (RDS or Kligler iron)
    streak plates from first period

1. Label the three slants with your name and the number of your unknown.
2. Look for isolated colonies that look like salmonella or shigella organisms. The characteristics to look for on each medium are as follows:
   - **MacConkey agar**—*Salmonella, Shigella* and other non-lactose-fermenting species produce smooth, colorless colonies. Coliforms that ferment lactose produce reddish, mucoid, or dark-centered colonies.
   - **Hektoen Enteric (HE) agar**—*Salmonella* and *Shigella* colonies are greenish-blue. Some species of *Salmonella* will have greenish-blue colonies with black centers due to H_2S production. Coliform colonies are salmon to orange and may have a bile precipitate.
   - **Xylose Lysine Desoxycholate (XLD) agar**— although most *Salmonella* produce red colonies with black centers, a few may produce red colonies that lack black centers. *Shigella* colonies are red. Coliform colonies are yellow.

Some *Pseudomonas* produce false-positive red colonies.

3. With a straight wire, inoculate the three agar slants from separate SS-appearing colonies. Use the streak-stab technique. When streaking the surface of the slant before stabbing, move the wire over the entire surface for good coverage.

4. Incubate the slants at 37° C for 18 to 24 hours. Longer incubation time may cause alkaline reversion. Even refrigeration beyond this time may cause reversion.

Alkaline reversion is a condition in which the medium turns yellow during the first part of the incubation period and then changes to red later due to increased alkalinity.

Third Period

(Slant Evaluations and Final Inoculations)

During this period you will inoculate tubes of SIM medium and urea broth with organisms from the slants of the previous period. Examination of the separation outline in figure 79.1 reveals that the final step in the differentiation of the SS pathogens is to determine whether a non-lactose-fermenter can do three things: (1) exhibit motility, (2) produce hydrogen sulfide, and (3) produce urease. You will also be making a gram-stained slide to perform a purity check. If miniaturized multitest media are available, they can also be inoculated at this time.

Materials:

    RDS or Kligler agar slants from previous period
    1 tube of SIM medium for each positive slant
    1 tube of urea broth for each positive slant
    miniaturized multitest media such as API 20E or Enterotube II (optional)

1. Examine the slants from the previous period and **select those tubes that have a yellow butt with a red slant.** These tubes contain organisms that ferment only glucose (non-lactose-fermenters). If you used Kligler's iron agar, a black precipitate in the medium will indicate that the organism is a producer of H_2S.

    Note in figure 79.1 that slants that are completely yellow are able to ferment lactose, as well as glucose. Tubes that are completely red are either nonfermenters or examples of alkaline reversion. Ignore those tubes.

2. With a loop, inoculate one tube of urea broth from each slant that has a yellow butt and red slant (non-lactose-fermenter).

3. With a straight wire, stab one tube of SIM medium from each of the same agar slants. Stab in the center to two-thirds of depth of medium.

4. Incubate these tubes at 37° C for 18 to 24 hours.

5. Make gram-stained slides from the same slants and confirm the presence of gram-negative rods.

6. If miniaturized multitest media are available, such as API 20E or Enterotube II, inoculate and incubate for evaluation in the next period. Consult Exercises 52 and 53 for instructions.

7. Refrigerate the positive RDS and Kligler iron slants for future use, if needed.

Fourth Period

(Final Evaluation)

During this last period the tubes of SIM medium, urea broth, and any miniaturized multitest media from the last period will be evaluated. Serotyping can also be performed, if desired.

Materials:

    tubes of urea broth and SIM medium from previous period
    Kovacs' reagent and chloroform
    5 ml pipettes
    miniaturized multitest media from previous period
    serological testing materials (optional)

1. Examine the tubes of SIM medium, checking for motility and H_2S production. If you see **cloudiness** spreading from the point of inoculation, the organism is motile. A **black precipitate** will be evidence of H_2S production.

2. Test for indole production by pipetting 2 ml of chloroform into each SIM tube and then adding 2 ml of Kovacs' reagent. A **pink to deep red color** will form in the chloroform layer if indole is produced.

    Salmonella are negative. Some *Shigella* may be positive. *Citrobacter* and *Escherichia* are positive.

3. Examine the urea broth tubes. If the medium has changed from yellow to **red** or **cerise color,** the organism is urease-positive.

4. If a miniaturized multitest media was inoculated in the last period, complete them now.

5. If time and materials are available, confirm the identification of your unknown with serological typing. Refer to Exercise 81.

Laboratory Report

Record the identity of your unknown on the Laboratory Report and answer all the questions.

Chronic or acute infections of the urinary tract may involve the kidneys, ureters, bladder, or urethra. Such infections may cause high blood pressure, kidney damage, uremia, or death. Some infections are inapparent and may go unnoticed for some time. Most infections of this tract enter by way of the urethra; very few originate in the blood.

A multitude of organisms can cause urinary infections. The most common cause of such infections in women of childbearing age is *Escherichia coli.* In order of frequency after *E. coli,* are other members of the Enterobacteriaceae, *Psuedomonas aeruginosa* and *Staphylococcus* species.

The importance of performing microbial analyses of urine on patients with urinary infections cannot be overemphasized. Some physicians tend to treat patients with antimicrobics and watch for symptomatic improvement without performing follow-up urinary tests, but this practice is not reliable. Clinical testing of urine 48 to 78 hours after the start of chemotherapy should be performed to evaluate the effectiveness of the therapy. If the antimicrobics are effective, the urine will be free of bacteria at this time.

A thorough microbial analysis of urine from a patient with urinary distress should be both quantitative and qualitative. The steps are as follows:

- Collect a urine sample as aseptically as possible.
- Do a plate count to determine the presence or absence of infection.
- Isolate the pathogen, if an infection is known to be present.
- Make a presumptive identification of the pathogen.
- Do an antimicrobic sensitivity test.

Except for antimicrobic testing, all of the above steps will be addressed in this laboratory exercise. Note that there are two parts to this experiment. The first portion pertains to doing a **plate count.** Note, also, that in figures 80.1 and 80.2 two different methods are available for making the inoculations. Your instructor will indicate which method will be used.

The second portion of this exercise is concerned with the protocol that one can follow to identify the genus of a pathogen that might be causing a urinary infection. Figure 80.3 depicts the routine that we will use to make this determination. Completion of this second portion of the exercise will yield only a **presumptive identification** of a pathogen. Further physiological tests, which are not included in this exercise, would be necessary to make species identification.

Aseptic Collection of Urine

Since the urethra in all individuals contains some bacteria, especially near its external orifice, the mere presence of bacteria in urine does not necessarily indicate that a urinary infection exists. One might assume that aseptic collection can be achieved with a catheter. However, collection by catheterization is, generally, not desirable because bacteria may be dislodged in the urethra, and there is the danger of causing an infection with this procedure.

Meaningful results, however, can be obtained with *midstream voided specimens* collected in sterile containers. For best results, the external genitalia should be cleansed with liquid soap containing chlorhexidine. Even with midstream samples, however, one can expect to find low counts of the following contaminants in normal urine: coagulase-negative streptococci, diphtheroid bacilli, enterococci, *Proteus,* hemolytic streptococci, yeasts, and aerobic gram-positive spore-forming rods.

Specimens are most reliable when plated out immediately after collection. If bacterial tests cannot be performed immediately, refrigeration is mandatory. It must be kept in mind that urine is an ideal growth medium for many bacteria. Specimens not properly refrigerated should be considered unsatisfactory for study.

The Plate Count

Before attempting to isolate pathogens from a urine sample, one should determine, first of all, that an infection actually exists. Since normal urine always contains some bacteria, yeasts, and other organisms, we need to know if there is *an excess number of organisms* present due to an infection somewhere in the urinary tract. The best way to make this evaluation is to do a plate count to determine the number of organisms per ml that are present.

Generally speaking, if a urine sample contains 100,000 or more organisms per ml, one may assume that *significant bacteriuria* exists. In some instances, however, urine from completely normal individuals may exceed these numbers. It is also possible for counts between 1000 and 100,000 to be significant. Thus, it is apparent that precise evaluation of plate counts must take into consideration other factors. The clinician, aware of the effects of certain variables, will subjectively evaluate the results. Our purpose here in this experiment is not to interpret but simply to become familiar with the basic procedures.

Proceed to inoculate two plates of trypticase soy agar with inocula from a urine sample. After 24 hours incubation, the colonies will be counted on the best plate. Your instructor will indicate whether Method A or Method B will be used.

Method A:
Using Calibrated Inoculating Loops

Note in figure 80.1 that calibrated loops are used to inoculate tubes of TSA. After the poured plates have been incubated for 24 hours, counts will be made. Proceed as follows:

Materials:
first period:
    urine sample
    1 sterile empty shake bottle
    2 sterile Petri plates
    2 trypticase soy agar pours
    calibrated wire loops (0.01 μm and 0.001 μm)

second period:
    plates from previous period
    Quebec colony counter
    mechanical hand counter

1. Liquefy two TSA pours and cool to 50° C.
2. Label one Petri plate 1:100 and the other 1:1000.
3. Pour the urine into an empty sterile shake bottle, cap it tightly, and shake 25 times, as in figure 23.3, page 91.
4. With a 0.01 calibrated sterile loop transfer one loopful to one of the pours, mix by rolling the tube between both palms, and pour into the 1:100 plate. Be sure to flame the neck of the tube before pouring into the plate.
5. Repeat step 4 using the 0.001 calibrated loop and the other TSA pour. Pour into the 1:1000 plate.
6. Incubate the plates at 37° C for 24 hours.
7. After incubation, select the plate that contains between 30 and 300 colonies. Count all colonies on a Quebec colony counter, using a mechanical hand counter to tally.
8. Record your count and number of organisms per ml on the Laboratory Report.

Method B: Using Pipettes

This method differs from Method A in that pipettes are used instead of calibrated loops for making the dilutions. Proceed as follows:

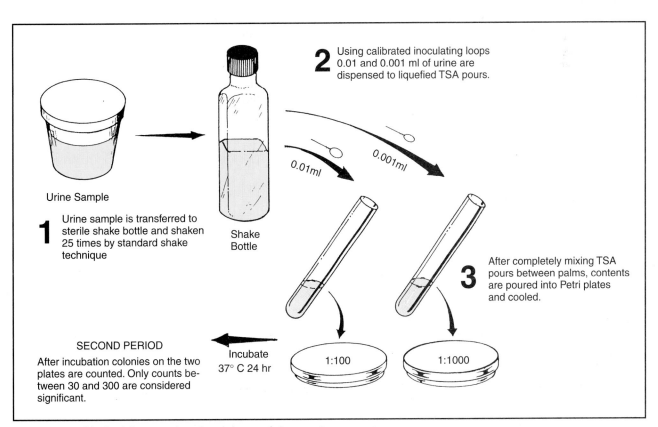

2 Using calibrated inoculating loops 0.01 and 0.001 ml of urine are dispensed to liquefied TSA pours.

0.01ml

0.001ml

Urine Sample

1 Urine sample is transferred to sterile shake bottle and shaken 25 times by standard shake technique

Shake Bottle

3 After completely mixing TSA pours between palms, contents are poured into Petri plates and cooled.

SECOND PERIOD
After incubation colonies on the two plates are counted. Only counts between 30 and 300 are considered significant.

Incubate 37° C 24 hr

1:100

1:1000

Figure 80.1 Method A procedure for doing a plate count

Materials:

first period:

    urine sample
    1 99 ml sterile water blank
    1 sterile empty shake bottle
    2 sterile Petri plates
    2 trypticase soy agar pours
    2 1.1 ml dilution pipettes
    mechanical pipetting device

second period:

    plates from previous period
    Quebec colony counter
    mechanical hand counter

1. Liquefy two TSA pours and cool to 50° C.
2. Label one Petri plate 1:100 and the other 1:1000.
3. Pour the urine into an empty sterile shake bottle, cap it tightly, and shake 25 times, as in figure 23.3, page 91.
4. Transfer 1 ml of the mixed urine to a 99 ml sterile water blank. Use a mechanical delivery device with the pipette.
5. Mix the water blank with 25 shakes and, with a fresh pipette, transfer 0.1 ml to the 1:1000 plate and 1.0 ml to the 1:100 plate.
6. Empty the tubes of TSA into the plates, swirl them, and let stand to cool.
7. Incubate the plates at 37° C for 24 hours.
8. After incubation, select the plate that contains between 30 and 300 colonies. Count all colonies on a Quebec colony counter, using a mechanical hand counter to tally.

9. Record your count and number of organisms per ml on the Laboratory Report.

Presumptive Identification

Once it is established that an infection exists, the next step is to isolate the pathogen and identify it. Figure 80.3 illustrates the overall procedure for making a presumptive identification of the genus of a urinary pathogen.

The minimum number of laboratory periods required to arrive at a presumptive identification is two; to be more explicit in identifying the unknown, a total of three or four periods will be required.

First Period

Note in figure 80.3 that two things will be done during this period with a concentrated sample of urine: (1) two microscope slides will be made for direct examination, and (2) four kinds of media will be inoculated. Proceed as follows:

Materials:

    1 sterile centrifuge tube (with screw cap)
    1 tube of thioglycollate medium (BBL135C)
    1 plate of blood agar (TSA base)
    1 plate of desoxycholate lactose agar (DLA)
    1 plate of phenylethyl alcohol medium with blood (PEA-B)

1. Shake the sample to resuspend the organisms and decant 10 ml into a centrifuge tube. Keep the tube capped.

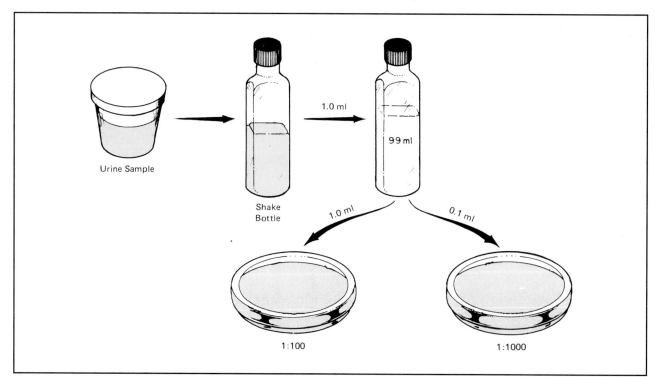

Figure 80.2   Method B procedure for doing a plate count

2. Centrifuge for 10 minutes at 2000 rpm.
3. Decant all but 0.5 ml from the tube and resuspend the sediment with a sterile wire loop.
4. Inoculate a tube of thioglycollate medium with a loopful of the sediment.
5. Streak out a loopful of the sediment on each of the three agar plates (blood agar, DLA, and PEA-B). Use a good isolation technique.
6. Incubate the thioglycollate tube and three plates at 37° C for 18 to 24 hours.
7. Make a wet mount slide from material in the bottom of the centrifuge tube and examine under high-dry, preferably with phase optics. Look for casts, pus cells, and other elements.

   Refer to figure 80.4 for help in identifying objects that are present. Normal urine will contain an occasional leukocyte, some epithelial cells, mucus, bacteria, and crystals of various kinds.
8. Make a gram-stained slide and examine it under oil immersion. Determine the morphology and staining reaction of the predominant organism. Record your observations on the Laboratory Report.

Second Period

After the plates have been incubated for 18 to 24 hours lay them out and evaluate them according to the characteristics of each medium.

Thioglycollate Medium This medium is inoculated to promote the growth of organisms that are not present in large numbers or are too fastidious to grow readily in nutrient broth. In the event that none of the plates produce colonies from the urine of a patient known to have a urinary infection, this tube can be used for reinoculation or to provide information pertaining to growth characteristics.

Blood Agar Practically all pathogens of the urinary tract will grow on this medium. This includes

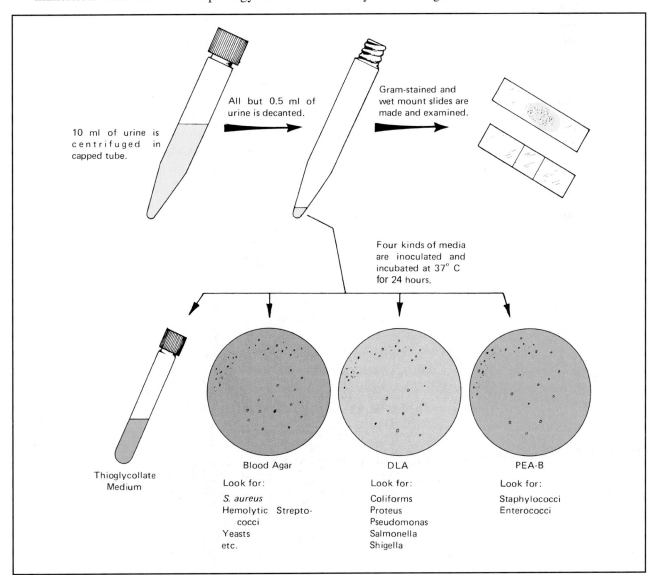

Figure 80.3 Procedure for presumptive identification of urinary pathogens

the coliforms, *Proteus, Pseudomonas, Candida,* staphylocci, streptococci, and so on.

Subculturing from this plate to trypticase soy broth can provide pure cultures of the pathogen for physiological testing or antimicrobic sensitivity testing.

The presence or absence of hemolytic activity can also be determined at this time. If the pathogen appears to be a hemolytic, gram-positive coccus, one should follow the procedures outlined in Exercises 77 and 78 for identification.

Desoxycholate Lactose Agar The presence of sodium desoxycholate and sodium citrate in this medium is inhibitory to gram-positive bacteria. Coliforms and other gram-negative bacteria grow well on it.

If the predominant organism is gram-negative, some differentiation may be made at this point. If the colonies on this medium are flat and rose-red in color, the organism is *E. coli. Pseudomonas* and *Proteus,* which do not ferment the lactose in the medium, produce white colonies. *Proteus* can be confirmed with the urease test, being positive for urease production.

Pseudomonas species give a positive reaction with Taxo N disks. Fermentation and additional physiological testing may be necessary for species identification. Exercises 48, 49, and 79 should be consulted for further testing.

Phenylethyl Alcohol Medium This medium, to which blood has been added, is highly inhibitory to gram-negative organisms. *Proteus,* in particular, is prevented from growing on it. If considerable growth occurs on the DLA plate, and very little or no growth here, then one can assume that the disease is due to a gram-negative organism. This, of course, would be confirmed by the findings on the gram-stained slide.

The findings on this plate should be correlated with those on blood agar. If enterococci (*S. faecalis*) are suspected, a plate of Mead agar should be streaked and incubated at 37° C. Enterococci produce pink colonies on this medium.

Laboratory Report

Record all findings on the Laboratory Report.

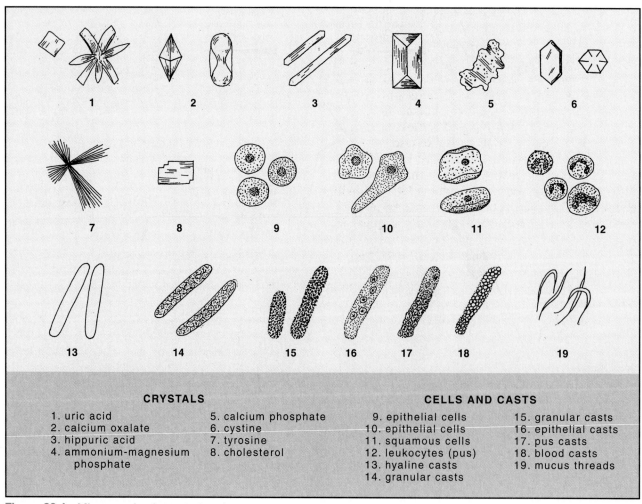

CRYSTALS

1. uric acid	5. calcium phosphate
2. calcium oxalate	6. cystine
3. hippuric acid	7. tyrosine
4. ammonium-magnesium phosphate	8. cholesterol

CELLS AND CASTS

9. epithelial cells	15. granular casts
10. epithelial cells	16. epithelial casts
11. squamous cells	17. pus casts
12. leukocytes (pus)	18. blood casts
13. hyaline casts	19. mucus threads
14. granular casts	

Figure 80.4 Microscopic elements in urine

Slide Agglutination Test:
Serological Typing

81

Organisms of different species not only differ morphologically and physiologically, but they also differ in protein makeup. The different proteins of bacterial cells that are able to stimulate antibody production when injected into an animal are **antigens.** The antigenic structure of each species of bacteria is unique to that species and, like the fingerprint of an individual, can be used to identify the organism. Many closely related microorganisms that are identical physiologically can be differentiated only by determining their antigenic nature.

The method of determining the presence of specific antigens in a microorganism is called **serological typing** (serotyping). It consists of adding a suspension of the organisms to **antiserum,** which contains antibodies that are specific for the known antigens. If the antigens are present, the antibodies in the antiserum will combine with the antigens, causing **agglutination,** or clumping, of the bacterial cells. Serotyping is particularly useful in the identification of various organisms that cause salmonella and shigella infections. In the identification of the various serotypes of these two genera, the use of antisera is generally performed after basic biochemical tests have been utilized as in Exercise 79.

In this exercise you will be issued two unknown organisms, one of which is a salmonella. By following the procedure shown in figure 81.1, you will determine which one of the unknowns is salmonella. Note that you will use two test controls. A **negative test control** will be set up in depression A on the slide to see what the absence of agglutination looks like. The negative control is a mixture of antigen and saline (antibody is lacking). A **positive test control** will be performed in depression C with standardized antigen and antiserum to give you a typical reaction of agglutination.

Materials:
2 numbered unknowns per student (slant
    cultures of a salmonella and a coliform)
salmonella O antigen, group B
    (Difco #2840-56)
salmonella O antiserum, poly A-I
    (Difco #2264-47)
depression slides or spot plates

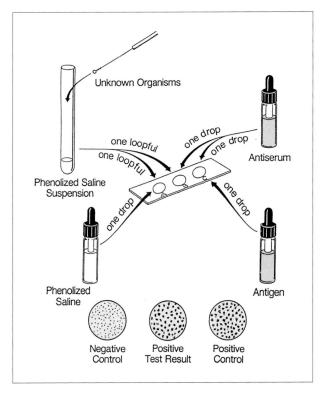

Figure 81.1   **Slide agglutination technique**

dropping bottle of phenolized saline solution
    (0.85% sodium chloride, 0.5% phenol)
2 serological tubes per student
1 ml pipettes

1. Label three depressions on a spot plate or depression slide **A, B,** and **C,** as shown in figure 81.1.
2. Make a phenolized saline suspension of each unknown in separate serological tubes by suspending one or more loopfuls of organisms in 1 ml of phenolized saline. Mix the organisms sufficiently to ensure complete dispersion of clumps of bacteria. The mixture should be very turbid.
3. Transfer one loopful (0.05 ml) from the phenolized saline suspension of one tube to depressions A and B.
4. To depressions B and C add one drop of salmonella O polyvalent antiserum. To depression A,

add one drop of phenolized saline, and to depression C, add one drop of salmonella O antigen, group B.

5. Mix the organisms in each depression with a clean wire loop. Do not go from one depression to the other without washing the loop first.

6. Compare the three mixtures. Agglutination should occur in depression C (positive control), but not in depression A (negative control). If agglutination occurs in depression B, the organism is salmonella.

7. Repeat this process on another slide for the other organism.

8. **Caution:** Deposit all slides and serological tubes in the container of disinfectant provided by the instructor.

Laboratory Report

Record your results on the first portion of Laboratory Report 81, 82.

Slide Agglutination (Latex) Test:
For *S. aureus* Identification

82

Many manufacturers of reagents for slide agglutination tests utilize polystyrene latex particles as carriers for the antibody particles. By adsorbing reactive antibody units to these particles, an agglutination reaction results that occurs rapidly and is much easier to see than ordinary precipitin type reactions that might be used to demonstrate the presence of a soluble antigen.

In this exercise we will use reagents manufactured by Difco Laboratories to determine if a suspected staphylococcus organism produces coagulase and/or protein A. The test reagent (*Difco Staph Reagent*) is a suspension of yellow latex particles sensitized with antibodies for coagulase and protein A. Reagents are also included to provide positive and negative controls in the test. Instead of using depression slides or spot plates, Difco provides disposable cards with eight black circles printed on them for performing the test. As indicated in figure 82.1, only three circles are used when performing the test on one unknown. The additional circles are provided for testing five additional unknowns at the same time. The black background of the cards facilitates rapid interpretation by providing good contrast for the yellow clumps that form.

There are two versions of this test: direct and indirect. The procedure for the direct method is illustrated in figure 82.1. The indirect method differs in that saline is used to suspend the organism being tested.

It should be pointed out that the reliability correlation between this test for coagulase and the tube test (page 250) is very high. Studies reveal that a reliability correlation of over 97% exists. Proceed as follows to perform this test.

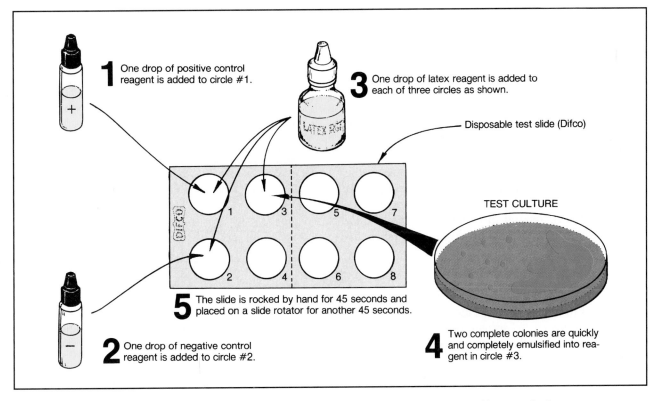

1 One drop of positive control reagent is added to circle #1.

3 One drop of latex reagent is added to each of three circles as shown.

Disposable test slide (Difco)

TEST CULTURE

5 The slide is rocked by hand for 45 seconds and placed on a slide rotator for another 45 seconds.

2 One drop of negative control reagent is added to circle #2.

4 Two complete colonies are quickly and completely emulsified into reagent in circle #3.

Figure 82.1 Slide agglutination test (direct method) for the presence of coagulase and/or protein A

Materials:

> plate culture of staphylococcus-like organism (trypticase soy agar plus blood)
> Difco Staph Latex Test kit #3850-32-7, which consists of:
> > bottle of Bacto Staph Latex Reagent
> > bottle of Bacto Staph Positive Control
> > bottle of Bacto Staph Negative Control
> > bottle of Bacto Normal Saline Reagent
> > disposable test slides (black circle cards)
> > mixing sticks (minimum of 3)
> slide rotator

Direct Method

If the direct method is to be used, as illustrated in figure 82.1, follow this procedure:

1. Place one drop of Bacto Staph Positive Control reagent onto circle #1.
2. Place one drop of Bacto Staph Negative Control reagent on circle #2.
3. Place one drop of Bacto Staph Latex Reagent onto circles #1, #2, and #3.
4. Using a sterile inoculating needle or loop, quickly and completely emulsify *two isolated colonies* from the culture to be tested into the drop of Staph Latex Reagent in circle #3.

   Also, emulsify the Staph Latex Reagent in the positive and negative controls in circles #1 and #2 using separate mixing sticks supplied in the kit.

   All mixing in these three circles should be done quickly to minimize drying of the latex on the slide and to avoid extended reaction times for the first cultures emulsified.
5. Rock the slide by hand for 45 seconds.
6. Place the slide on a slide rotator capable of providing 110 to 120 rpm and rotate it for another 45 seconds.
7. Read the results immediately, according to the descriptions provided in the table at right. If agglutination occurs before 45 seconds, the results may be read at that time. *The slide should be read at normal reading distance under ambient light.*

Indirect Method

The only differences between the direct and indirect methods pertain to the amount of inoculum and the use of saline to emulsify the unknown being tested. Proceed as follows:

1. Place one drop of Bacto Staph Positive Control reagent onto test circle #1.
2. Place one drop of Bacto Staph Negative Control onto circle #2.
3. Place one drop of Bacto Normal Saline Reagent onto circle #3.
4. Using a sterile inoculating needle or loop, completely emulsify *four isolated colonies* from the culture to be tested into the circle containing the drop of saline (circle #3).
5. Add one drop of Bacto Staph Latex Reagent to each of the three circles.
6. Quickly mix the contents of each circle, using individual mixing sticks.
7. Rock the slide by hand for 45 seconds.
8. Place the slide on a slide rotator capable of providing 110 to 120 rpm and rotate it for another 45 seconds.
9. Read the results immediately according to the descriptions provided in the table below. If agglutination occurs before 45 seconds, the results may be read at that time. *The slide should be read at normal reading distance under ambient light.*

POSITIVE REACTIONS	
4 +	Large to small clumps of aggregated yellow latex beads; clear background
3 +	Large to small clumps of aggregated yellow latex beads; slightly cloudy background
2 +	Medium to small but clearly visible clumps of aggregated yellow latex beads; moderately cloudy background
1 +	Fine clumps of aggregated yellow latex beads; cloudy background
NEGATIVE REACTIONS	
+	Smooth cloudy suspension; particulate grainy appearance that cannot be identified as agglutination
−	Smooth, cloudy suspension; free of agglutination or particles

Laboratory Report

Record your results on the last portion of Laboratory Report 81, 82.

Tube Agglutination Test: The Heterophile Antibody Test

83

Infectious mononucleosis (IM) is a benign disease, occurring principally in individuals in the 13 to 25 year age group. It is caused by the Epstein-Barr virus (EBV), a herpesvirus, that is one of the most ubiquitous viruses in humans. Studies have shown that the virus can be isolated from saliva of patients with IM, as well as from some healthy, asymptomatic individuals. Between 80% and 90% of all adults possess antibodies for EBV.

The disease is characterized by a sudden onset of fever, a sore throat, and pronounced enlargement of the cervical lymph nodes. There is also moderate leukocytosis with a marked increase in the number of lymphocytes (50% to 90%).

The serological test for IM takes advantage of an unusual property: the antibodies produced against the EBV coincidentally agglutinate sheep red blood cells. This is an example of a **heterophile antigen**—a substance isolated from a living form that stimulates the production of antibodies capable of reacting with tissues of other organisms. The antibodies are referred to as **heterophile antibodies.**

This test is performed by adding a suspension of sheep red blood cells to dilutions of inactivated patient's serum and incubating the tubes overnight in the refrigerator. Figure 83.1 illustrates the overall procedure. Agglutination titers of 320 or higher are considered significant. Titers of 40,960 have been obtained.

Proceed as follows to perform this test on a sample of test serum:

First Period

Materials:

    test-tube rack (Wassermann type) with 10 clean serological tubes

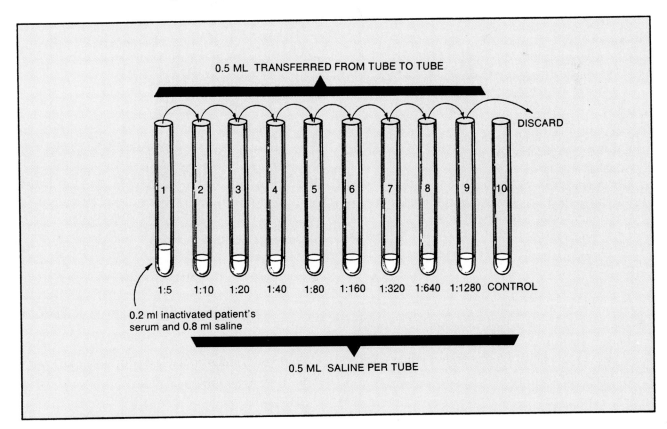

Figure 83.1 Procedure for setting up heterophile antibody test

bottle of saline solution (0.85% NaCl), clear or
filtered
1 ml pipettes
5 ml pipettes
2% suspension of sheep red blood cells
patient's serum (known to be positive)

1. Place the test serum in a 56° C water bath for 30 minutes to inactivate the complement.
2. Set up a row of 10 serological tubes in the front row of a test-tube rack and number them from 1 to 10 (left to right) with a marking pencil.
3. Into tube 1, pipette 0.8 ml of physiological saline.
4. Dispense 0.5 ml of physiological saline to tubes 2 through 10. Use a 5 ml pipette.
5. With a 1 ml pipette add 0.2 ml of the inactivated serum to tube 1. Mix the contents of this tube by drawing into the pipette and expelling about 5 times.
6. Transfer 0.5 ml from tube 1 to tube 2, mix five times, and transfer 0.5 ml from tube 2 to tube 3, etc., through the ninth tube. *Discard 0.5 ml from the ninth tube after mixing.* Tube 10 is the **control.**

7. Add 0.2 ml of 2% sheep red blood cells to all tubes (1 through 10) and shake the tubes. Final dilutions of the serum are shown in figure 83.1.
8. Allow the rack of tubes to stand at room temperature for 1 hour, then transfer the tubes to a small wire basket, and place in a refrigerator to remain overnight.

Second Period

Set up the tubes in a tube rack in order of dilution and compare each tube with the control by holding the tubes overhead and looking up at the bottoms of the tubes. Nonagglutinated cells will tumble to the bottom of the tube and form a small button (as in control tube). Agglutinated cells will form a more-amorphous "blanket."

The **titer** should be recorded as the reciprocal of the last tube in the series that shows positive agglutination.

Laboratory Report

Complete the first portion of Laboratory Report 83, 84.

Tube Agglutination Test: The Widal Test

84

A tube test for determining the quantity of agglutinating antibodies, or **agglutinins,** in the serum of a patient with typhoid fever was described by Grunbaum and Widal in 1896. This technique is still in use today and has been adapted to many other diseases as well.

The procedure involves adding a suspension of dead typhoid bacterial cells to a series of tubes containing the patient's serum, which has been diluted out to various concentrations. After the tubes have been incubated for 30 minutes at 37° C, they are centrifuged and examined to note the amount of agglutination that has occurred.

The reciprocal of the highest dilution at which agglutination is seen is designated as the **antibody titer** of the patient's serum. For example, if the highest dilution at which agglutination occurs is 1:320, the titer is 320 antibody units per milliliter of serum. Naturally, the higher the titer, the greater is the antibody response of the individual to the disease.

This technique can be used clinically to determine whether a patient with typhoidlike symptoms actually has the disease. If successive daily tests on a patient's serum reveal no antibody titer, or a low titer that does not increase from day to day, it can be assumed that some other disease is present. On the other hand, if one sees a daily increase in the titer, it can be assumed that a typhoid infection does exist. Since the treatment of typhoid fever requires powerful antibiotics that are not widely used on other similar diseases, it is very important to diagnose this disease early to begin the proper form of chemotherapy as soon as possible.

In this exercise you will be given a sample of blood serum that is known to contain antibodies for the typhoid organism. By using the Widal tube agglutination method, you will determine the antibody titer.

Materials:

test-tube rack (Wassermann type) with 10 clean
   serological tubes
bottle of saline solution (0.85%), clear or
   filtered

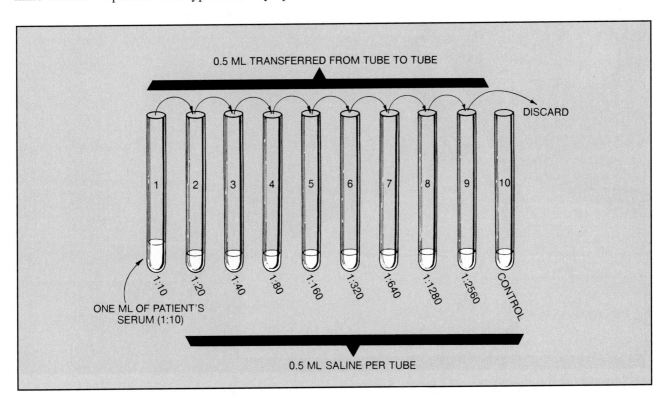

Figure 84.1   Procedure for dilution of serum

1 ml pipettes
5 ml pipettes
water bath at 37° C
centrifuges
antigen (1:10 dilution) *Salmonella typhi* "O"
patient's serum (1:10 dilution), known positive
  "O"

1. Dilute the patient's serum as shown in figure 84.1. Follow this routine:

    a. Set up 10 clean serological test tubes in the front row of a test-tube rack and number them from 1 to 10 (left to right) with a marking pencil.

    b. Into tube 1 pipette 1 ml of the patient's serum (1:10 dilution). *For convenience, the instructor may wish to dispense this material to each student.*

    c. With a 5 ml pipette, dispense 0.5 ml of saline to each of the remaining 9 tubes.

    d. With a 1 ml pipette, transfer 0.5 ml of the serum from tube 1 to tube 2. Mix the serum and saline in tube 2 by carefully drawing the liquid up into the pipette and discharging it slowly back down into the tube a minimum of three times.

    e. Repeat this process by transferring 0.5 ml from tube 2 to 3, tube 3 to 4, 4 to 5, etc. *When you get to tube 9, discard the 0.5 ml drawn from it instead of adding it to tube 10;* thus, tube 10 will contain only saline and can be used as a **negative test control** for comparing with the other tubes.

2. With a fresh 5 ml pipette, transfer 0.5 ml of antigen to each tube. Shake the rack to completely mix the antigen and diluted serum.

3. Place the rack in a water bath at 37° C for 30 minutes.

4. Centrifuge all tubes for 3 minutes at 2000 rpm. (If time permits, 7 minutes centrifugation is preferable.)

5. Examine each tube for agglutination and record the titer as the reciprocal of the highest dilution in which agglutination is seen.

    When examining each tube, jar it first by rapping the side of the tube with a snap of the finger to suspend the clumps of agglutinated cells. Hold it up against the light of a desk lamp in the manner shown in figure 84.2. Do not look directly into the light. The reflection of the light off the particles is best seen against a dark background. Compare each tube with tube 10, which is your negative test control.

6. Record your results on the last portion of Laboratory Report 83, 84.

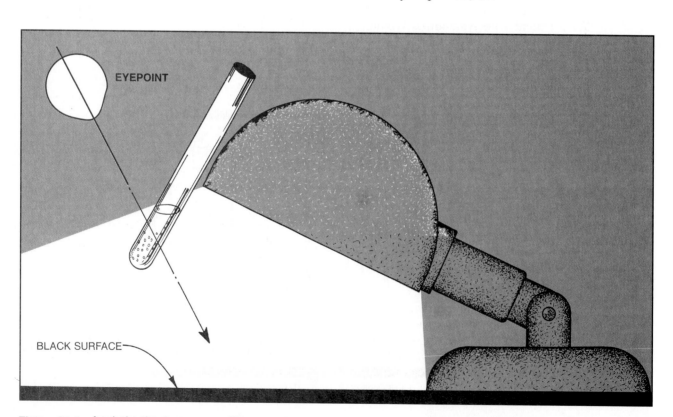

EYEPOINT

BLACK SURFACE

Figure 84.2 Agglutination is more readily seen when the tube is examined against a black surface

Phage Typing

85

The host specificity of bacteriophage is such that it is possible to delineate different strains of individual species of bacteria on the basis of their susceptibility to various kinds of bacteriophage. In epidemiological studies, where it is important to discover the source of a specific infection, determining the phage type of the causative organism can be an important tool in solving the riddle. For example, if it can be shown that the phage type of *S. typhi* in a patient with typhoid fever is the same as the phage type of an isolate from a suspected carrier, chances are excellent that the two cases are epidemiologically related. Since all bacteria are probably parasitized by bacteriophages, it is theoretically possible, through research, to classify each species into strains or groups according to their phage type susceptibility. This has been done for *Staphylococcus aureus, Salmonella typhi,* and several other pathogens. The following table illustrates the lytic groups of *S. aureus* as proposed by M. T. Parker.

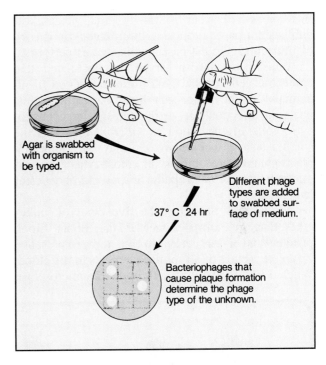

Figure 85.1 Bacteriophage typing

Lytic Group	Phages in Group
I	29 52 52A 79 80
II	3A 3B 3C 55 71
III	6 7 42E 47 53 54 75 77 83A
IV	42D
not allotted	81 187

In bacteriophage typing, a suspension of the organism to be typed is swabbed over an agar surface. The bottom of the plate is marked off in squares and labeled to indicate which phage types are going to be used. To the organisms on the surface, a small drop of each phage type is added to their respective squares. After incubation, the plate is examined to see which phages were able to lyse the organisms. This is the procedure to be used in this exercise. See figure 85.1.

Materials:
1 Petri plate of tryptone yeast extract agar
bacteriophage cultures (available types)
nutrient broth cultures of *S. aureus* with swabs

1. Mark the bottom of a plate of tryptone yeast extract agar with as many squares as there are phage types to be used. Label each square with the phage type numbers.
2. Swab the entire surface of the agar with the organisms.
3. Deposit one drop of each phage in its respective square.
4. Incubate the plate at 37° C for 24 hours and record the lytic group and phage type of the culture.
5. Record your results on the Laboratory Report.

86

White Blood Cell Study:
The Differential WBC Count

In 1883, at the Pasteur Institute in Paris, Metchnikoff published a paper proposing the **phagocytic theory of immunity.** On the basis of his studies performed on transparent starfish larvae, he postulated that amoeboid cells in the tissue fluid and blood of all animals are the major guardians of health against bacterial infection. He designated the large phagocytic cells of the blood as *macrophages* and the smaller ones as *microphages.* Today, Metchnikoff's macrophages are known as monocytes and his microphages as neutrophils or polymorphonuclear leukocytes.

Figure 86.1 illustrates the five types of leukocytes that are normally seen in the blood. Blood platelets and erythrocytes also are shown to present a complete picture of all formed elements in the blood. When observed as living cells under the microscope,

they appear as refractile, colorless structures. As shown here, however, they reflect the dyes that are imparted by Wright's stain.

In this exercise we will do a study of the white blood cells in human blood. This study may be made from a prepared stained microscope slide or from a slide made from your own blood. By scanning an entire slide and counting the various types, you will have an opportunity to encounter most, if not all, types. The erythrocytes and blood platelets will be ignored.

Figures 86.1 and 86.2 will be used to identify the various types of cells. Figure 86.3 illustrates the procedure for preparing a slide stained with Wright's stain. The relative percentages of each type will be determined after a total of 100 white blood cells have been identified. This method of white blood

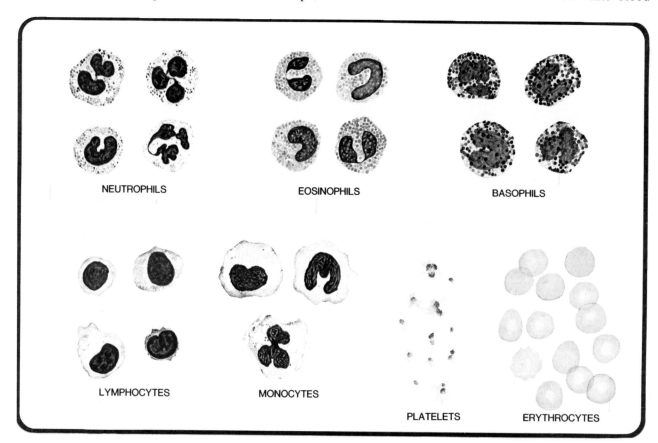

K.P. Talaro

Figure 86.1   Formed elements of blood

278

cell enumeration is called a **differential WBC count.**

As you proceed with this count, it will become obvious that the neutrophils are most abundant (50%–70%). The next most prominent cells are the lymphocytes (20%–30%). Monocytes constitute about 2%–6%; eosinphils, 1%–5%; and basophils, less than 1%.

A normal white blood cell count is between 5000 and 10,000 white cells per cubic millimeter. Elevated white blood cell counts are referred to as *leukocytosis;* counts of 30,000 or 40,000 represent marked leukocytosis. When counts fall considerably below 5000, *leukopenia* is said to exist. Both conditions can have grave implications.

The value of a differential count is immeasurable in the diagnosis of infectious diseases. High neutro-

phil counts, or *neutrophilia,* often signal localized infections, such as appendicitis or abscesses in some other part of the body. *Neutropenia,* a condition in which there is a marked decrease in the numbers of neutrophils, occurs in typhoid fever, undulant fever, and influenza. *Eosinophilia* may indicate allergic conditions or invasions by parasitic roundworms such as *Trichinella spiralis,* the "pork worm." Counts of eosinophils may rise to as high as 50% in cases of trichinosis. High lymphocyte counts, or *lymphocytosis,* are present in whooping cough and some viral infections. Increased numbers of monocytes, or *monocytosis,* may indicate the presence of the Epstein-Barr virus, which causes infectious mononucleosis.

Note in the materials list that items needed for making a slide (option B) are listed separately. If a

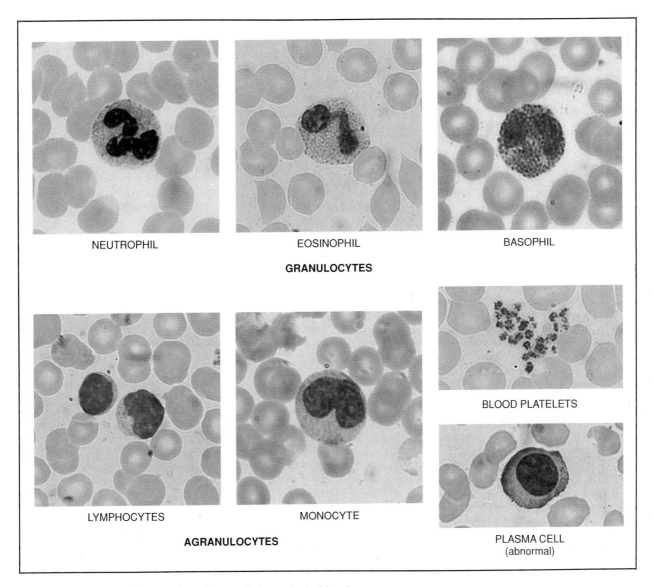

NEUTROPHIL EOSINOPHIL BASOPHIL

GRANULOCYTES

BLOOD PLATELETS

LYMPHOCYTES MONOCYTE

AGRANULOCYTES

PLASMA CELL
(abnormal)

Figure 86.2 Photomicrographs of formed elements in blood

prepared slide (option A) is to be used, ignore the instructions under the heading "Preparation of Slide" and proceed to the heading "Performing the Cell Count." Your instructor will indicate which option will be used. Proceed as follows:

PRECAUTIONS

Although it may be assumed that you are always practicing strict aseptic techniques in this laboratory, a reminder is in order here as you approach an experiment dealing with fresh blood. A review of all the sanitary procedures listed in the Introduction on page ix may be necessary.

Materials:

prepared blood slide (option A):
    stained with Wright's or Giemsa's stains

for staining a blood smear (option B):
    2 or 3 clean microscope slides (should have
        polished edges)
    sterile disposable lancets
    sterile absorbent cotton, 70% alcohol
    Wright's stain, wax pencil, bibulous paper
    distilled water in dropping bottle

Preparation of Slide

Figure 86.3 illustrates the procedure that will be used to make a stained slide of a blood smear. The most difficult step in making such a slide is getting a good spread of the blood, which is thick at one end and thin at the other end. If done properly, the smear will have a gradient of cellular density that will make it possible to choose an area that is ideal for study. The angle at which the spreading slide is held in making the smear will determine the thickness of the smear. It may be necessary for you to make more than one slide to get an ideal one.

1. Clean three or four slides with soap and water. Handle them with care to avoid getting their flat surfaces soiled by your fingers. Although only two slides may be used, it is often necessary to repeat the spreading process, thus the extra slides.

2. Scrub the middle finger with 70% alcohol and stick it with a lancet. Put a drop of blood on the slide $\frac{3}{4}''$ from one end and spread with another slide in the manner illustrated in figure 86.3.

    Note that the blood is dragged over the slide, not pushed. Do not pull the slide over the smear a second time. If you don't get an even

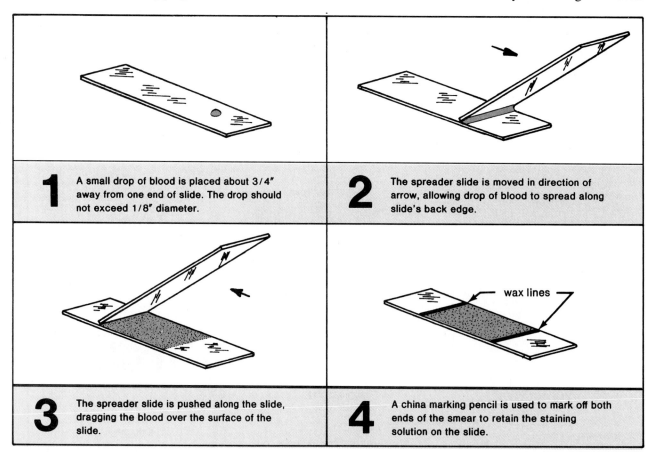

Figure 86.3 Smear preparation technique for making a stained blood slide

smear the first time, repeat the process on a fresh clean slide. To get a smear that will be the proper thickness, hold the spreading slide at an angle somewhat greater than 45°.

3. Draw a line on each side of the smear with a wax pencil to confine the stain that is to be added. (Note: This step is helpful for beginners, and usually omitted by professionals.)

4. Cover the film with Wright's stain, *counting the drops* as you add them. Stain for **4 minutes** and then add the same number of drops of distilled water to the stain and let stand for another **10 minutes.** Blow gently on the mixture every few minutes to keep the solutions mixed.

5. Gently wash off the slide under running water for 30 seconds and shake off the excess. Blot dry with bibulous paper.

Performing the Cell Count

Whether you are using a prepared slide or one that you have just stained, the procedure is essentially the same. Although the high-dry objective can be used for the count, the oil immersion lens is much better. Differentiation of some cells is difficult with high-dry optics. Proceed as follows:

1. Scan the slide with the low-power objective to find an area where cell distribution is best. A good area is one in which the cells are not jammed together or scattered too far apart.

2. Systematically scan the slide, following the pathway indicated below in figure 86.4. As each leukocyte is encountered, identify it, using figures 86.1 and 86.2 for reference.

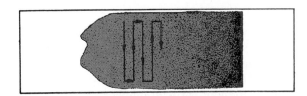

Figure 86.4   Path to follow when seeking cells

3. Tabulate your count on the Laboratory Report sheet according to the instructions there. It is best to remove the Lab Report sheet from the back of the manual for this identification and tabulation procedure.

Laboratory Report

Complete the first portion of Laboratory Report 86–88.

87

Total WBC Count

Although the differential white blood cell count provides us with the relative percentages of leukocytes, it alone cannot reveal the true picture of the extent of an infection. For a more complete picture, one must also know the total number of WBCs per cubic millimeter of blood.

Although the number of leukocytes may vary with the time of day, exercise, and other factors, a range of 5000 to 9000 WBCs per cubic millimeter is considered normal. If an individual were to have an abnormally high neutrophil percentage, and a total count of, say, 17,000 WBCs, the presence of an infection of some sort would be highly probable.

To determine the number of leukocytes in a cubic millimeter of blood, one must dilute the blood and count the WBCs on a specialized slide called a **hemacytometer.** Figures 87.1 and 87.2 reveal various steps in preparation for this count.

Note in figure 87.1 that blood is drawn up into a special pipette and then diluted in the pipette with a weak acid solution. After shaking the pipette to mix the acid and blood, a small amount of diluted blood is allowed to flow under the cover glass of the hemacytometer. The count of white blood cells is made with the low-power microscope objective. Proceed as follows.

Preparation of Hemacytometer

Working with your laboratory partner, assist each other to prepare a "charged" hemacytometer as follows:

Materials:
  hemacytometer and cover glass
  WBC diluting pipette and rubber tubing
  WBC diluting fluid
  mechanical hand counter
  pipette cleaning solutions
  cotton, alcohol, lancets, clean cloth

1. Wash the hemacytometer and cover glass with soap and water, rinse well, and dry with a clean cloth or Kimwipes.
2. Produce a free flow of blood, wipe away the first drop and draw the blood up into the diluting pipette to the 0.5 mark. See illustration 1 of figure 87.1. If the blood happens to go a little above the mark, the volume can be reduced to the mark by placing the pipette tip on a piece of blotting paper.

    If the blood goes substantially past the 0.5 mark, discharge the blood, wash the pipette in the four cleansing solutions (illustration 3, figure 87.2), and start over. The ideal way is to draw up

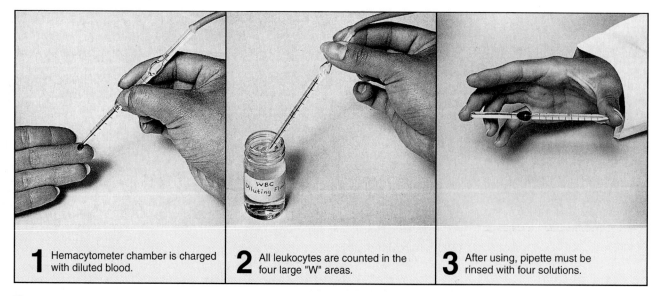

1 Hemacytometer chamber is charged with diluted blood.

2 All leukocytes are counted in the four large "W" areas.

3 After using, pipette must be rinsed with four solutions.

Figure 87.1 Dilution and mixing procedures for WBC blood count

the blood exactly to the mark on the first attempt. To clean the pipette, rinse it first with acid, then water, alcohol, and finally with acetone.

3. As shown in illustration 2, figure 87.1, draw the WBC diluting fluid up into the pipette until it reaches the 11.0 mark.

4. Place your thumb over the tip of the pipette, slip off the tubing, and place your third finger over the other end (illustration 3, figure 87.1).

5. Mix the blood and diluting fluid in the pipette for **2–3 minutes** by holding it as shown in illustration 3, figure 87.1. The pipette should be held *parallel to the tabletop* and moved through a 90° arc, with the wrist held rigidly.

6. Discharge one-third of the bulb fluid from the pipette by allowing it to drop onto a piece of paper toweling.

7. While holding the pipette as shown in illustration 1, figure 87.2, deposit a tiny drop on the polished surface of the counting chamber next to the edge of the cover glass. **Do not let the tip of the pipette touch the polished surface for more than an instant.** If it is left there too long, the chamber will overfill.

A properly filled chamber will have diluted blood filling only the space between the cover glass and the counting chamber. No fluid should run down into the moat.

8. Charge the other side if the first side was overfilled.

Performing the Count

Place the hemacytometer on the microscope stage and bring the grid lines into focus under the **low-power** (10×) objective. Use the coarse adjustment knob and reduce the lighting somewhat to make both the cells and lines visible.

Locate one of the large "W" (white) areas shown in illustration 2, figure 87.2. One of these areas should fill up the entire field. Since the diluting fluid contains acid, all red blood cells have been destroyed; only the leukocytes will show up as very small dots.

Do the cells seem to be evenly distributed? If not, charge the other half of the counting chamber after further mixing. If the other chamber had been previously charged unsuccessfully by overflooding, wash off the hemacytometer and cover glass, shake the pipette for 2–3 minutes, and recharge it.

Count all the cells in the four "W" areas, using a mechanical hand counter. To avoid overcounting of cells at the boundaries, **count the cells that touch the lines on the left and top sides only.** Cells that touch the boundary lines on the right and bottom sides should be ignored. This applies to the boundaries of each entire "W" area.

Discharge the contents of the pipette and rinse it out by sequentially flushing with the following fluids: acid, water, alcohol, and acetone. See illustration 3, figure 87.2.

Calculations

To determine the number of leukocytes per cubic millimeter, multiply the total number of cells counted in the four "W" areas by 50. The factor of 50 is the product of the volume correction factor and dilution factor, or

$$2.5 \times 20 = 50$$

Laboratory Report

Answer the questions that pertain to this experiment on Laboratory Report 86–88.

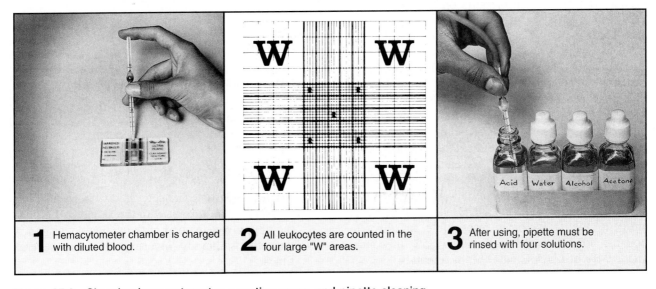

1 Hemacytometer chamber is charged with diluted blood.

2 All leukocytes are counted in the four large "W" areas.

3 After using, pipette must be rinsed with four solutions.

Figure 87.2 Charging hemacytometer, counting areas, and pipette cleaning

Exercises 81 through 84 illustrate three uses of agglutination tests as related to (1) the identification of serological types, (2) species identification (*S. aureus*), and (3) disease identification (infectious mononucleosis and typhoid fever). The typing of blood is another example of a medical procedure that relies on this useful phenomenon.

The procedure for blood typing was developed by Karl Landsteiner around 1900. He is credited with having discovered that human blood types can be separated into four groups on the basis of two antigens that are present on the surface of red blood cells. These antigens are designated as A and B. The four groups (types) are A, B, AB, and O. The last group type O, which is characterized by the absence of A or B antigens is the most common type in the United States (45% of the population). Type A is next in frequency, found in 39% of the population. The incidences of types B and AB are 12% and 4%, respectively.

Blood typing is performed with antisera containing high titers of anti-A and anti-B antibodies. The test may be performed by either slide or tube methods. In both instances, a drop of each kind of antiserum is added to separate samples of saline suspension of red blood cells. Figure 88.1 illustrates the slide technique. If agglutination occurs only in the suspension to which the anti-A serum was added, the blood is type A. If agglutination occurs only in the anti-B mixture, the blood is type B. Agglutination in both samples indicates that the blood is type AB. The absence of agglutination indicates that the blood is type O.

Between 1900 and 1940 a great deal of research was done to uncover the presence of other antigens in human red blood cells. Finally, in 1940, Landsteiner and Wiener reported that rabbit sera containing antibodies against the red blood cells of the rhesus monkey would agglutinate the red blood cells of 5% of white humans. This antigen in humans, which was first designated as the **Rh factor** (in due respect to the rhesus monkey), was later found to exist as six antigens: C, c, D, d, E, and e. Of these six antigens, the D factor is responsible for the Rh-positive condition and is found in 85% of caucasians, 94% of blacks, and 99% of orientals.

Typing blood for the Rh factor can also be performed by both tube and slide methods, but certain differences in the techniques are involved. First of all, the antibodies in the typing sera are of the incomplete albumin variety, which *will not agglutinate human red cells when they are diluted with saline.* Therefore, it is necessary to use whole blood or dilute the cells with plasma. Another difference is that the test *must be performed at higher temperatures:* 37° C for tube test, 45° C for the slide test.

In this exercise, two separate slide methods are presented for typing blood. If only the Landsteiner ABO groups are to be determined, the first method may be preferable. If Rh typing is to be included, the second method, which utilizes a slide warmer, will be followed. The availability of materials will determine which method is to be used.

ABO Blood Typing

Materials:

small vial (10 mm dia × 50 mm long)
disposable lancets (B-D Microlance, Serasharp, etc.)
70% alcohol and cotton
china marking pencil
microscope slides
typing sera (anti-A and anti-B)
applicators or toothpicks
saline solution (0.85% NaCl)
1 ml pipettes

1. Mark a slide down the middle with a marking pencil, dividing the slide into two halves as shown in figure 88.1. Write "anti-A" on the left side and "anti-B" on the right side.
2. Pipette 1 ml of saline solution into a small vial or test tube.
3. Scrub the middle finger with a piece of cotton saturated with 70% alcohol and pierce it with a sterile disposable lancet. Allow two or three drops of blood to mix with the saline by holding the finger over the end of the vial and washing it with the saline by inverting the tube several times.

4. Place a drop of this red cell suspension on each side of the slide.
5. Add a drop of anti-A serum to the left side of the slide and a drop of anti-B serum to the right side. **Do not contaminate the tips of the serum pipettes with the material on the slide.**

6. After mixing each side of the slide with separate applicators or toothpicks, look for agglutination. The slide should be held about 6″ above an illuminated white background and rocked gently for 2 or 3 minutes. Record your results on the Laboratory Report as of 3 minutes.

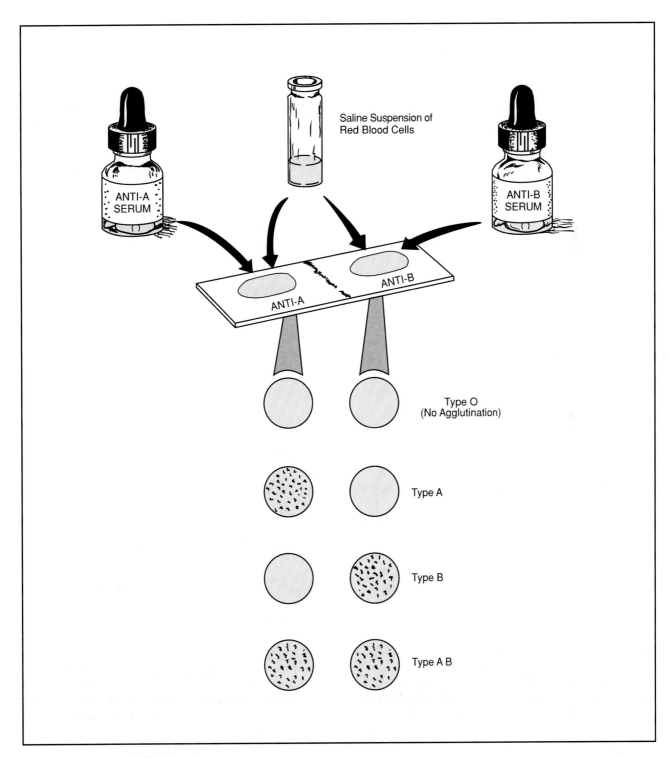

Figure 88.1 **Typing of ABO blood groups**

Combined ABO and Rh Typing

As stated, Rh typing must be performed with heat on blood that has not been diluted with saline. A warming box such as the one in figure 88.2 is essential in this procedure. In performing this test, two factors are of considerable importance: first, only a small amount of blood must be used (a drop of about 3 mm diameter on the slide) and, second, proper agitation must be executed. The agglutination that occurs in this antibody-antigen reaction results in finer clumps; therefore, closer examination is essential. If the agitation is not properly performed, agglutination may not be as apparent as it should be.

In this combined method we will use whole blood for the ABO typing as well as for the Rh typing. Although this method works satisfactorily as a classroom demonstration for the ABO groups, it is *not as reliable* as the previous method in which saline and room temperature are used. *This method is not recommended for clinical situations.*

Materials:

slide warming box with a special marked slide
anti-A, anti-B, and anti-D typing sera
applicators or toothpicks
70% alcohol and cotton
disposable sterile lancets (B-D Microlances, Serasharp, etc.)

1. Scrub the middle finger with a piece of cotton saturated with 70% alcohol and pierce it with a sterile disposable lancet. Place a small drop in each of three squares on the marked slides on the warming box.

   To get the proper proportion of serum to blood, do not use a drop larger than 3 mm diameter on the slide.

2. Add a drop of anti-D serum to the blood in the anti-D square, mix with a toothpick, and note the time. **Only 2 minutes should be allowed for agglutination.**

3. Add a drop of anti-B serum to the anti-B square and a drop of anti-A serum to the anti-A square. Mix the sera and blood in both squares with *separate* fresh toothpicks.

4. Agitate the mixtures on the slide by slowly rocking the box back and forth on its pivot. At the end of 2 minutes, examine the anti-D square carefully for agglutination. If no agglutination is apparent, consider the blood to be Rh-negative. By this time the ABO type can also be determined.

5. Record your results on Laboratory Report 86–88.

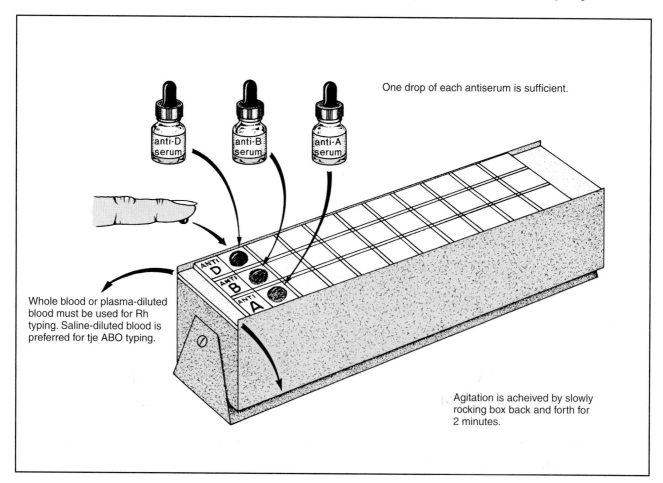

One drop of each antiserum is sufficient.

anti-D serum   anti-B serum   anti-A serum

Whole blood or plasma-diluted blood must be used for Rh typing. Saline-diluted blood is preferred for tje ABO typing.

Agitation is acheived by slowly rocking box back and forth for 2 minutes.

Figure 88.2 Blood typing with warming box

The Snyder Caries Susceptibility Test

89

The degradation of enamel and dentin in the formation of tooth decay (dental caries) occurs as a result of the production of lactic acid by bacteria (*Streptococcus mutans* and others) in the presence of high levels of sucrose. Of the various methods that have been devised to determine one's susceptibility to tooth decay, M. L. Snyder's caries susceptibility test is a relatively simple test that has been shown to have a fairly high reliability correlation.

This method relies on the rapidity of organisms in saliva to lower the pH in a medium that contains 2% dextrose (Snyder test agar). Since decalcification of enamel begins at a pH of 5.5, and progresses rapidly as the pH is lowered to 4.4 and less, the demonstration of pH lowering becomes evidence of susceptibility to caries.

To indicate the presence of acid production in the medium, the indicator bromcresol green is incorporated in it. This indicator is green at pH 4.8 and becomes yellow at pH 4.4, remaining yellow below 4.4.

Figure 89.1 illustrates the procedure that is used in the Snyder caries susceptibility test. Note that 0.2 ml of saliva is added to a tube of liquefied Snyder test agar (50° C) and mixed well by rotating the tube between the palms of both hands. After the medium has solidified, the tube is incubated at 37° C for a period of 24–72 hours. If the medium turns yellow in 24–48 hours, the individual is said to be susceptible to caries.

Although we will be performing this test only once, it should be noted that test reliability is enhanced by performing the test on three consecutive days at the same time each day. If the test is performed right after toothbrushing, it is not as reliable as if two or three hours have elapsed after brushing. Proceed as follows to perform this test:

Materials:

1 tube of Snyder test agar (5 ml in 15 mm dia tube)
1 30 ml sterile beaker
1 piece of paraffin ($\frac{1}{4}'' \times \frac{1}{4}'' \times \frac{1}{8}''$)
1 ml pipette
1 gummed label

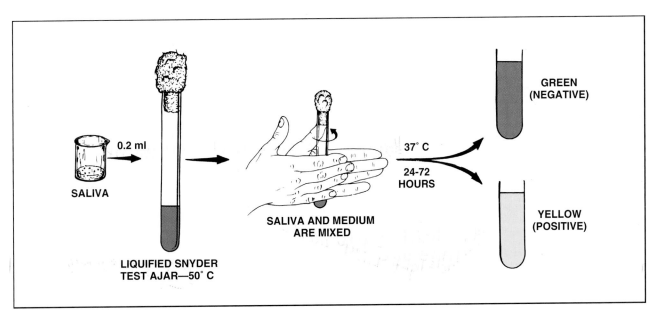

Figure 89.1 Snyder caries susceptibility test

1. Liquefy a tube of Snyder test agar and cool it to 50° C.
2. After allowing a piece of paraffin to soften under the tongue for a few minutes, start chewing it. Chew it for **3 minutes,** moving it from one side of the mouth to the other. *Do not swallow the saliva.* As it accumulates, deposit it in the small sterile beaker.
3. Vigorously shake the sample in the beaker from side to side for 30 seconds to disperse the organisms.
4. With a 1 ml pipette transfer 0.2 ml of saliva to the tube of agar. Do not allow the pipette to touch the side of the tube or agar.
5. Before the medium solidifies, mix the contents of the tube by rotating the tube vigorously between the palms of the hands.
6. Write your name on a gummed label and attach it to the tube.
7. Incubate the tube at 37° C. Examine the tube every 24 hours to see if the bromcresol green indicator has changed to yellow. If it has, the test is positive. The degree of caries susceptibility is determined from the table below.
8. Record your results on the Laboratory Report.

Caries Susceptibility	Medium turns yellow in:		
	24 hours	48 hours	72 hours
Marked	Positive		
Moderate	Negative	Positive	
Slight	Negative	Negative	Positive
Negative	Negative	Negative	Negative

LABORATORY REPORT

6

Student: _____

Desk No.: _____ Section: _____

Protozoa, Algae, and Cyanobacteria

A. Tabulation of Observations

In this study of freshwater microorganisms, record your observations in the following tables. The number of organisms to be identified will depend on the availability of time and materials. Your instructor will indicate the number of each type that should be recorded.

Record the genus of each identifiable type. Also, indicate the phylum or division to which the organism belongs. Microorganisms that you are unable to identify should be sketched in the space provided. It is not necessary to draw those that are identified.

PROTOZOA

GENUS	PHYLUM	BOTTLE NO.	SKETCHES OF UNIDENTIFIED

ALGAE

GENUS	DIVISION	BOTTLE NO.	SKETCHES OF UNIDENTIFIED

CYANOBACTERIA

GENUS	BOTTLE NO.	SKETCHES OF UNIDENTIFIED

B. General Questions

Record the answers to the following questions in the answer column. It may be necessary to consult your text or library references for one or two of the answers.

1. Give the kingdom in which each of the following groups of organisms is found:

    a. protozoans   d. bacteria

    b. algae   e. fungi

    c. cyanobacteria   f. microscopic invertebrates

2. Four kingdoms are represented by the organisms in the above question. Name the fifth kingdom.

3. What is the most significant characteristic seen in eukaryotes that is lacking in prokaryotes?

4. What characteristic in the microscopic invertebrates distinguishes them from protozoans?

5. Which protozoan phylum was not found in pond samples because phylum members are all parasitic?

6. Indicate whether the following are *present* or *absent* in the algae:

    a. cilia   b. flagella   c. chloroplasts

7. Indicate whether the following are *present* or *absent* in the protozoans:

    a. cilia   c. mitochondria

    b. chloroplasts   d. mitosis

8. Which photosynthetic pigment is common to all algae and cyanobacteria?

9. Name two photosynthetic pigments that are found in the cyanobacteria but not in the algae.

10. What photosynthetic pigment is found in bacteria but is lacking in all other photosynthetic organisms?

11. What type of movement is exhibited by the diatoms?

Answers

1.a. _____

  b. _____

  c. _____

  d. _____

  e. _____

  f. _____

2. _____

3. _____

4. _____

5. _____

6.a. _____

  b. _____

  c. _____

7.a. _____

  b. _____

  c. _____

  d. _____

8. _____

9.a. _____

  b. _____

10. _____

11. _____

C. Protozoan Characterization

Select the protozoan groups in the right-hand column that have the following characteristics:

1. move with flagella
2. move with cilia
3. move with pseudopodia
4. have nuclear membranes
5. lack nuclear membranes
6. all species are parasitic
7. produce resistant cysts

1. Sarcodina
2. Mastigophora
3. Ciliophora
4. Sporozoa
5. all of above
6. none of above

D. Characterization of Algae and Cyanobacteria

Select the groups in the right-hand column that have the following characteristics:

Pigments

1. chlorophyll a
2. chlorophyll b
3. chlorophyll c
4. fucoxanthin
5. c-phycocyanin
6. c-phycoerythrin

1. Euglenophycophyta
2. Chlorophycophyta
3. Chrysophycophyta
4. Phaeophycophyta
5. Pyrrophycophyta
6. Cyanobacteria
7. all of above
8. none of above

Food Storage

7. fats
8. oils
9. starches
10. laminarin
11. leucosin
12. paramylum
13. mannitol

Other Structures

14. pellicle, no cell wall
15. cell walls, box and lid
16. chloroplasts
17. phycobilisomes
18. thylakoids

Answers

Protozoa

1. _____
2. _____
3. _____
4. _____
5. _____
6. _____
7. _____

Algae

1. _____
2. _____
3. _____
4. _____
5. _____
6. _____
7. _____
8. _____
9. _____
10. _____
11. _____
12. _____
13. _____
14. _____
15. _____
16. _____
17. _____
18. _____

LABORATORY REPORT

8

Student: _____

Desk No.: _____ Section: _____

Fungi:
Yeasts and Molds

A. Yeast Study

Draw a few representative cells of *Saccharomyces cerevisiae* in the appropriate circles below. Blastospores (buds) and ascospores, if seen, should be shown and labeled.

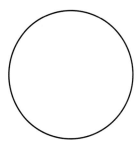

Prepared Slide

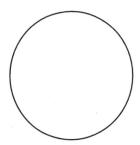

Living Cells

B. Mold Study

In the following table, list the genera of molds identified in this exercise. Under colony description, give the approximate diameter of the colony, its topside color and backside (bottom) color. For microscopic appearance, make a sketch of the organism as it appears on slide preparation.

GENUS	COLONY DESCRIPTION	MICROSCOPIC APPEARANCE (DRAWING)

C. Questions

Record the answers for the following questions in the answer column.

1. The science that is concerned with the study of fungi is called _____.

2. The kingdom to which the fungi belong is _____ .

3. Microscopic filaments of molds are called _____ .

4. A filamentlike structure formed by a yeast from a chain of blastospores is called a _____.

5. A mass of mold filaments, as observed by the naked eye, is called a _____.

6. Most molds have _____ hyphae (*septate* or *nonseptate*).

7. List three kinds of sexual spores that are the basis for classifying the molds.

8. What is the name of the rootlike structure that is seen in *Rhizopus?*

9. What type of hypha is seen in *Mucor* and *Rhizopus?*

10. What kind of asexual spores are seen in *Mucor* and *Rhizopus?*

11. What kind of asexual spores are seen in *Penicillium?*

12. What kind of asexual spores are seen in *Alternaria?*

13. Which subdivision of the Amastigomycota contains individuals that lack sexual spores?

14. What division of Myceteae consists of slime molds?

15. Fungi that exist both as yeasts and molds are said to be _____.

Answers

1. _____
2. _____
3. _____
4. _____
5. _____
6. _____
7.a. _____
  b. _____
  c. _____
8. _____
9. _____
10. _____
11. _____
12. _____
13. _____
14. _____
15. _____

LABORATORY REPORT

9

Student: _____

Desk No.: _____ Section: _____

Bacteria

A. Tabulation

After examining your TSA and blood agar plates, record your results in the following table and on a similar table that your instructor has drawn on the chalkboard. With respect to the plates, we are concerned with a quantitative evaluation of the degree of contamination and differentiation as to whether the organisms are bacteria or molds. Quantify your recording as follows:

0 no growth
+ 1 to 10 colonies
+ + 11 to 50 colonies

+ + + 51 to 100 colonies
+ + + + over 100 colonies

After shaking the tube of broth to disperse the organisms, look for cloudiness (turbidity). If the broth is clear, no bacterial growth occurred. Record no growth as 0. If tube is turbid, record + in last column.

STUDENT INITIALS	PLATE EXPOSURE METHOD		COLONY COUNTS		BROTH	
	TSA	Blood Agar	Bacteria	Mold	Source	Result

B. Questions

1. Using the number of colonies as an indicator, which habitat sampled by the class appears to be the most contaminated one? _____

   Why do you suppose this habitat contains such a high microbial count?_____

   _____

2. In a few words, describe some differences in the macroscopic appearance of bacteria and mold colonies:_____

   _____

   _____

   _____

   _____

3. How can you tell when a tube of broth contains bacterial growth? _____

   _____

   _____

4. a. Were any of the plates completely lacking in colonies? _____

   b. Do you think that the habitat sampled was really sterile?_____

   c. If your answer to *b* is *no,* then how can you account for the lack of growth on the plate?

   _____

   d. If your answer to *b* is *yes,* defend it: _____

   _____

   _____

   _____

Student: _____

Desk No.: _____ Section: _____

Motility Determination

A. Test Results

1. Which of the two organisms exhibited true motility on the slides?

   _____

2. Did the semisolid medium inoculations confirm the results obtained from the slides?

   _____

3. Sketch in the appearance of the two tube inoculations:

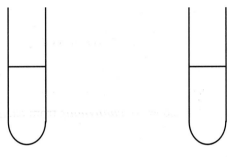

   Micrococcus luteus   *Proteus vulgaris*

B. Questions

1. How does Brownian movement differ from true motility?_____

   _____

   _____

2. How do you differentiate water current movement from true motility?_____

   _____

   _____

3. Make sketches that illustrate each of the following flagellar arrangements:

   Monotrichic   **Lophotrichic**

   Amphitrichic   **Peritrichic**

LABORATORY REPORT
20

Student: _____

Desk No.: _____ Section: _____

Culture Media Preparation

1. How do the following types of organisms differ in their carbon needs?

   Photoautotrophs:_____

   Photoheterotrophs:_____

2. Where do the above two types of organisms get their energy?_____

3. Where do **chemoheterotrophs** get their energy?_____

4. What is a growth factor?_____

   _____

5. Give two reasons why agar is such a good ingredient for converting liquid media to solid media.

   a. _____

   b. _____

6. Differentiate between the following two types of media:

   Synthetic medium:_____

   Nonsynthetic medium:_____

7. Differentiate between the following two types of media:

   Selective medium:_____

   Differential medium:_____

8. Briefly, list the steps that you would go through to make up a batch of nutrient agar slants:

   _____

   _____

   _____

   _____

Student: _____

Desk No.: _____ Section: _____

Pure Culture Techniques

A. Evaluation of Streak Plate

Show within the circle the distribution of the colonies on your streak plate. To identify the colonies, use red for *Serratia marcescens*, yellow for *Micrococcus luteus*, and purple for *Chromobacterium violaceum*. If time permits, your instructor may inspect your plate and enter a grade where indicated.

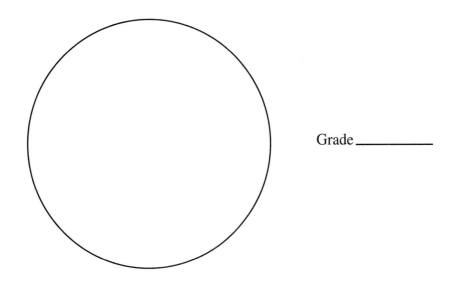

Grade _____

B. Evaluation of Pour Plates

Show the distribution of colonies on plates II and III, using only the quadrant section for plate II. If plate III has too many colonies, follow the same procedure. Use colors.

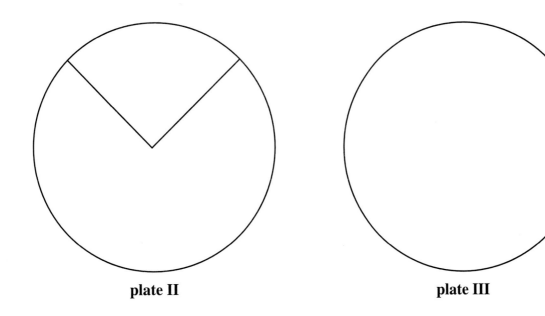

plate II   plate III

C. Subculture Evaluation

With colored pencils, sketch the appearance of the growth on the slant diagrams below. Also, draw a few cells of each organism as revealed by Gram staining in the adjacent circle.

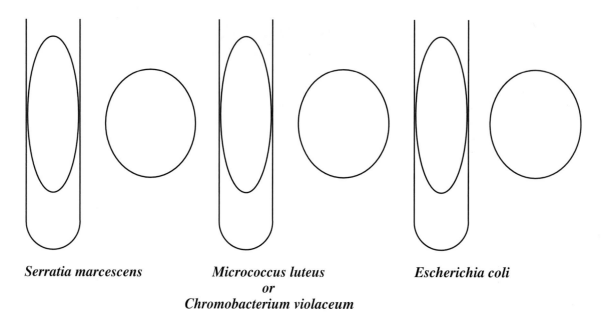

Serratia marcescens   *Micrococcus luteus*   *Escherichia coli*
or
Chromobacterium violaceum

D. Questions

1. Which method of separating organisms seems to achieve the best separation?

    _____

2. Which method requires the greatest skill?_____

3. Do you think you have pure cultures of each organism on the slants?_____

    Can you be absolutely sure by studying its microscopic appearance?_____

    _____

    Explain:_____

4. Give two reasons why the nutrient agar must be cooled to 50° C before inoculating and pouring.

    _____

    _____

    _____

5. Why should a Petri plate be discarded if media is splashed up the side to the top?

    _____

    _____

6. Give two reasons why it is important to invert plates during incubation:

    _____

    _____

7. Why is it important not to dig into the agar with the loop?_____

    _____

8. Why must the loop be flamed before entering a culture?_____

    Why must it be flamed after making an inoculation?_____

Student: _____

Desk No.: _____ Section: _____

Cultivation of Anaerobes

A. Tube Inoculations

After carefully comparing the appearance of the six cultures belonging to you and your laboratory partner, select the best tube for each organism and sketch its appearance in the tubes below. Indicate under each name the type of medium (FTM or TGYA).

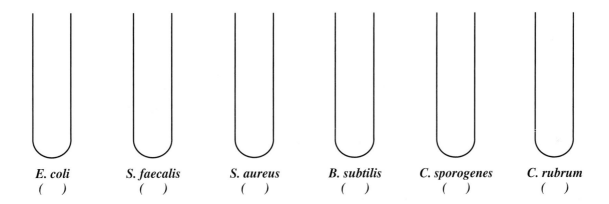

| *E. coli* | *S. faecalis* | *S. aureus* | *B. subtilis* | *C. sporogenes* | *C. rubrum* |
| ( ) | ( ) | ( ) | ( ) | ( ) | ( ) |

B. Plate Inoculations

After comparing the growths on the two plates of Brewer's anaerobic agar with the growths in the six tubes, classify each organism as to its oxygen requirements:

Escherichia coli: _____ *Bacillus subtilis:* _____

Streptococcus faecalis: _____ *Clostridium sporogenes:* _____

Staphylococcus aureus: _____ *Clostridium rubrum:* _____

C. Questions

1. What is the function of oxygen at the cellular level? _____
   _____

2. Why are facultative organisms able to grow with or without oxygen while aerobes grow only in its presence? _____
   _____

3. How do "indifferents" differ from "facultatives"? _____
   _____

4. What is the function of the following agents in the media used in this experiment?

   Sodium thioglycollate: _____

   Resazurin: _____

   Agar in FTM: _____

   Agar in TGYA shake: _____

5. How is oxygen removed from the air in a GasPak anaerobic jar?_____

_____

_____

_____

D. Spore Study

If a spore-stained slide is made of the three spore-formers, draw a few cells of each organism in the spaces provided below:

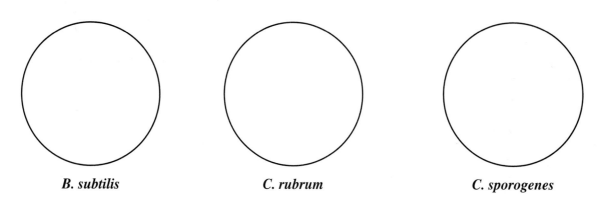

B. subtilis   *C. rubrum*   *C. sporogenes*

LABORATORY REPORT

23

Bacterial Population Counts

A. Quantitative Plating Method

1. Record your plate counts in this table:

DILUTION PLATED	ML PLATED	NUMBER OF COLONIES
1:10,000	1.0	
1:100,000	0.1	
1:1,000,000	1.0	
1:10,000,000	0.1	

2. How many cells per ml were there in the undiluted culture?_____

3. How would you inoculate a plate to get 1:100 dilution?_____

_____

4. How would you inoculate a plate to get 1:10 dilution?_____

_____

5. Give two reasons why it is necessary to shake the water blanks as recommended.

   a. _____

   b. _____

B. Turbidimetric Determinations

1. Record the percent transmittance and optical density values for your dilutions in the following table.

DILUTION	PERCENT TRANSMITTANCE	OPTICAL DENSITY
1:1		
1:2		
1:4		
1:8		
1:16		

2. Plot the optical densities versus the concentration of organisms. Complete the graph by drawing a line between plot points.

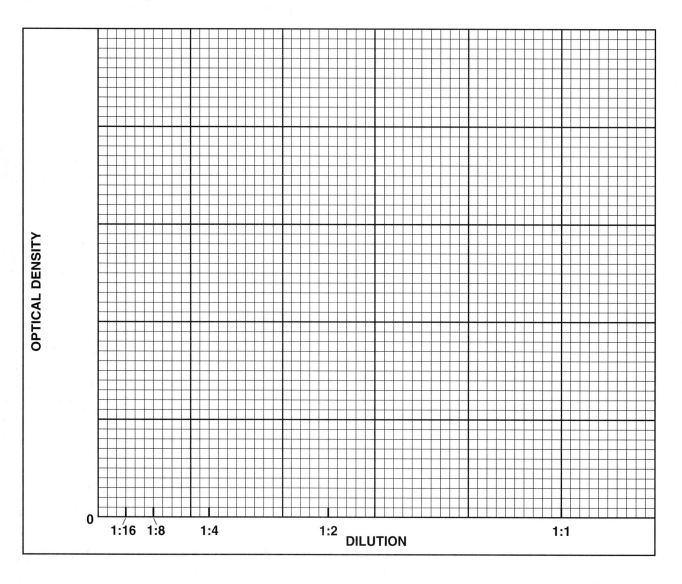

C. Questions

1. What is the maximum O.D. that is within the linear portion of the curve?_____

2. What is the corrected or true O.D. of the undiluted culture? (Hint: If the O.D. for the 1:2 dilution but not the 1:1 dilution is within the linear portion of the curve, then the O.D. of the 1:1 dilution should not be considered correct. The correct or true O.D. of the undiluted culture in this example could be estimated by multiplying the O.D. of the 1:2 dilution by 2.)_____

3. What is the correlation between corrected O.D. and cell number for your culture?_____

_____

_____

4. Why is it necessary to perform a plate count in conjunction with the turbidimetry procedure?

_____

5. If your medium were pale blue instead of amber-colored, as is the case of nutrient broth, would you set the wavelength control knob higher or lower than 686 nanometers?_____

Student: _____

Desk No.: _____ Section: _____

Ex. 25 Slide Culture:
Autotrophs

A. Microscopic Examination

While examining the two slides, move them around to different areas to note the various types of organisms that are present. Draw representative types.

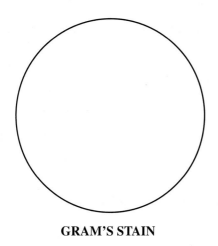

GRAM'S STAIN   **LIVING**

B. Questions

1. With respect to Gram's stain, which type (*gram-positive* or *gram-negative*) seems to predominate?

   _____

2. List as many different kinds of autotrophic protists as you can that can be cultured on this type of slide.

   _____

   _____

   _____

3. Some organisms that grow on this type of slide are chemosynthetic heterotrophs. What would be the source of their food?

   _____

   _____

Ex. 26 Slime Mold Culture

A. Observations

1. What happened when the flow of protoplasm on a plasmodium was interrupted by severance with a scalpel?

   _____

   _____

2. Describe your observations of the crushed spores on the hanging drop slide.

   _____

   _____

   _____

B. Questions

1. List two functions served by fructification (sporangia formation) in *Physarum.*

   a. _____

   b. _____

2. What is the principal function of the plasmodial stage of *Physarum?*

   _____

3. List two characteristics that the Myxobacterales and Gymnomycota have in common.

   a. _____

   b. _____

4. Postulate as to the evolutionary relationship between the myxobacteria and slime molds.

   _____

   _____

   _____

   _____

   _____

   _____

LABORATORY REPORT
27

Student: _____

Desk No.: _____ Section: _____

Photosynthetic Bacteria:
Isolation and Culture

A. Tabulation of Results

Record on the chalkboard, using a chart similar to the one below, the presence (+) or absence (−) of photosynthetic bacteria for each pond mud sample that was used.

STUDENT INITIALS	SOURCE OF POND MUD	CHROMATIUM MEDIUM	CHLOROBIUM MEDIUM

1. What percentage of samples contained *Chromatium?* _____

2. What percentage of samples contained *Chlorobium?* _____

3. What percentage of samples contained photosynthetic bacteria? _____

B. Microscopic Examination

After examining these organisms on hanging drop slides, describe their appearance.

_____

_____

_____

_____

C. Questions

1. Give the equations for photosynthesis in the following:

   Algae:_____

   Chromatium* and *Chlorobium:_____

2. On the following table, compare the photosynthetic bacteria with the algae and cyanobacteria. Place a check (✔) if item is present.

	ALGAE	CYANOBACTERIA	PHOTOSYNTHETIC BACTERIA
Chloroplasts			
Chromatophores			
Phycobilisomes			
Chlorophyll a			
Bacteriochlorophyll			

LABORATORY REPORT

28, 29

Isolation of Phage from Sewage and Flies

A. *Plaque Size Increase* (for both Exercises 28 and 29)

With a china marking pencil, circle and label three plaques on one of the plates and record their sizes in millimeters at 1-hour intervals.

TIME	PLAQUE SIZE (millimeters)		
	Plaque No. 1	Plaque No. 2	Plaque No. 3
When first seen			
1 hour later			
2 hours later			
3 hours later			

B. *Questions* (for Exercise 28 only)

1. Were any plaques seen on the negative control plate?_____

2. Do plates 1, 2, and 3 show a progressive increase in number of plaques with increased amount of sewage filtrate?_____

3. Did the phage completely "wipe out" all bacterial growth on any of the plates?_____ If so, which plates?_____

C. *Observations* (for Exercise 29 only)

Count all the plaques on each plate and record the counts in the following table. If the plaques are very numerous, use a Quebec colony counter and hand counting device. If this exercise was performed as a class project with individual students doing only one or two plates from a common fly-broth filtrate, record all counts on the chalkboard on a table similar to the one below.

Plate Number	1	2	3	4	5	6	7	8	9	10
E. coli (ml)	0.9	0.8	0.7	0.6	0.5	0.4	0.3	0.2	0.1	1.0
Filtrate (ml)	0.1	0.2	0.3	0.4	0.5	0.6	0.7	0.8	0.9	0
Number of plaques										

D. Questions (for Exercise 29 only)

1. Which plate was used as the negative control?_____

2. Were there any plaques on the negative control plate?_____

3. What would be the explanation for the presence of plaques on the negative control plate?

   _____

   _____

4. Were any plates completely "wiped out" by phage action?_____

   If so, which ones?_____

E. Terminology (for both Exercises 28 and 29)

1. Differentiate between the following:

   Lysis:_____

   Lysogeny:_____

2. Differentiate between the following:

   Virulent phage:_____

   Temperate phage:_____

LABORATORY REPORT
30

Student: _____

Desk No.: _____ Section: _____

Burst Size Determination:
A One-Step Growth Curve

A. *Plaque Counts*

Record the counts of plaques on each of the plates in the following table. Record the peak number of plaques as the *burst size*. The drop in plaque numbers after a peak results from adsorption of mature phage virions on other bacterial cells and cell debris.

15	25	30	35	40	45	50

Burst size: _____

B. *Dilution Interpretation*

Answer the following questions to clarify your understanding of the dilutions that occur in this experiment.

1. How many cells were present in each milliliter of the original bacterial culture?_____

2. How many bacterial cells (total) were dispensed into the ADS tube?_____

3. If the bacterial dilution per plate is 1:10,000,000, how many bacterial cells were distributed to each plate?_____

4. How many phage virions were present in 1 ml of the original phage suspension?

   _____

5. How many phage virions were present in the 0.1 ml of phage suspension that was added to the ADS tube?_____

6. What was the numerical ratio of phage virions to bacterial cells in the ADS tube?_____
   What is this ratio called?_____

7. How many bacterial cells were placed in the ADS-2 tube?_____

8. What effect does dilution have on adsorption?_____

LABORATORY REPORT
31–33

Student: _____

Desk No.: _____ Section: _____

Microbial Interrelationships
Ex. 31 Bacterial Commensalism

A. Results

Indicate the degree of turbidity (none, +, + +, + + +) in the following table. With colored pencils, draw the appearance of the gram-stained slides where indicated.

ORGANISMS	TURBIDITY	GRAM STAIN
Staphylococcus aureus		
Clostridium sporogenes		
S. aureus and *C. sporogenes*		

B. Questions

1. Does *C. sporogenes* grow well in nutrient broth?_____

2. How does *S. aureus* assist *C. sporogenes* in growth?_____

_____

Ex. 32 Bacterial Synergism

A. Results

Examine the six tubes of media, looking for acid and gas. In the presence of acid, bromthymol blue turns yellow. Record your results in the table below. Consult other students for their results and complete the table.

INDIVIDUAL ORGANISMS	LACTOSE		SUCROSE		COMBINATIONS	LACTOSE		SUCROSE	
	Acid	Gas	Acid	Gas		Acid	Gas	Acid	Gas
E. coli					*E. coli* and *P. vulgaris*				
P. vulgaris					*E. coli* and *S. aureus*				
S. aureus					*S. aureus* and *P. vulgaris*				

B. Questions

1. Did any of the three organisms produce gas in either lactose or sucrose broth when alone in the medium?_____

2. Which organisms act synergistically to produce gas in

   lactose?_____

   sucrose?_____

Ex. 33 Microbial Antagonism

A. Questions

1. Which organisms are antagonistic to *E. coli?*_____

   _____

2. Which organisms are antagonistic to *S. aureus?*_____

   _____

3. Name an antibiotic substance that is derived from each of the following types of organisms. Also, indicate whether the substance is effective against gram-positive or gram-negative organisms.

   A bacterium:_____

   An actinomyces:_____

   A fungus:_____

4. What role does microbial antagonism play in nature?_____

   _____

   _____

5. In what physiological way does penicillin affect penicillin-sensitive bacteria?_____

   _____

6. How do sulfonamides inhibit bacterial growth?_____

   _____

   _____

Student: _____

Desk No.: _____ Section: _____

Temperature:
Effects on Growth

A. Pigment Formation and Temperature

1. Draw the appearance of the growth of *Serratia marcescens* on the nutrient agar slants using colored pencils.

2. Which temperature seems to be closest to the optimum temperature for pigment formation?

   _____

3. What are the cellular substances that control pigment formation and are regulated by temperature?

   _____

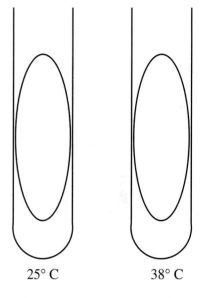

25° C   38° C

B. Growth Rate and Temperature

If a spectrophotometer is available, dispense the cultures into labeled cuvettes and determine the percent transmittance of each culture. Calculate the O.D. values from the percent transmittance, using the formula given in Exercise 23.

If no spectrophotometer is available, record only the visual reading as +, + +, + + +, and none.

Temp. °C	*SERRATIA MARCESCENS*			*ESCHERICHIA COLI*		
	Visual Reading	Spectrophotometer		Visual Reading	Spectrophotometer	
		%T	O.D.		%T	O.D.
5						
25						
38						
42						
55						

Growth curves of *Serratia marcescens* and *Escherichia coli* as related to temperature.

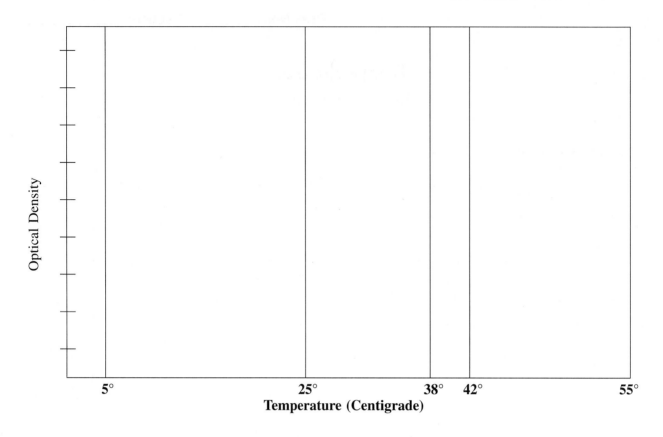

Temperature (Centigrade)

1. On the basis of the above graph, estimate the optimum growth temperature of the two organisms.

   *Serratia marcescens:*_____

   *Escherichia coli:*_____

2. To get more precise results for the above graph, what would you do?

   _____

   _____

3. Differentiate between the following:

   Thermophile:_____

   Mesophile:_____

   Psychrophile:_____

4. What is the optimum growth temperature range for most psychrophiles?_____

LABORATORY REPORT

35

Student: _____

Desk No.: _____ Section: _____

Temperature:
Lethal Effects

A. *Tabulation of Results*

Examine your five Petri plates, looking for evidence of growth. Record on the chalkboard, using a chart similar to the one below, the presence or absence of growth as (+) or (−). When all members of the class have recorded their results, complete this chart.

ORGANISM	60° C					70° C					80° C					90° C					100° C				
	C*	10	20	30	40	C*	10	20	30	40	C*	10	20	30	40	C*	10	20	30	40	C*	10	20	30	40
S. aureus																									
E. coli																									
B. megaterium																									

*C = control tube

1. If they can be determined from the above information, record the **thermal death point** for each of the organisms.

   *S. aureus:*_____ *E. coli:*_____ *B. megaterium:*_____

2. From the following table, determine the **thermal death time** for each organism at the tabulated temperatures.

ORGANISM	THERMAL DEATH TIME				
	60° C	70° C	80° C	90° C	100° C
S. aureus					
E. coli					
B. megaterium					

B. *Questions*

1. Give three reasons why endospores are much more resistant to heat than are vegetative cells.

   a. _____

   b. _____

   c. _____

2. Differentiate between the following:

   Thermoduric:_____

   _____

   Thermophilic:_____

   _____

3. List four diseases caused by spore-forming bacteria.

   a. _____   b. _____

   c. _____   d. _____

4. Since boiling water is unreliable in destroying endospores, how should one use heat in medical applications to ensure spore destruction? (three ways)

   a. _____

   b. _____

   c. _____

Student: _____

Desk No.: _____ Section: _____

pH and Microbial Growth

A. Tabulation of Results

If a spectrophotometer is available, dispense the cultures into labeled cuvettes and determine the percent transmittance of each culture. Calculate the O.D. values from the percent transmittance, using the formula given in Exercise 23. To complete the tables, get the results of the other three organisms from other members of the class, and delete the substitution organisms in the tables that were not used.

If no spectrophotometer is available, record only the visual reading as +, + +, + + +, and none.

pH	*Escherichia coli*			*Staphylococcus aureus*		
	Visual Reading	Spectrophotometer		Visual Reading	Spectrophotometer	
		%T	O.D.		%T	O.D.
3						
5						
7						
8						
9						
10						

pH	*Alcaligenes faecalis* or *Sporosarcina ureae*			*Saccharomyces cervisiae* or *Candida glabrata*		
	Visual Reading	Spectrophotometer		Visual Reading	Spectrophotometer	
		%T	O.D.		%T	O.D.
3						
5						
7						
8						
9						
10						

B. Growth Curves

Once you have computed all the O.D. values on the two tables, plot them on the following graph. Use different colored lines for each species.

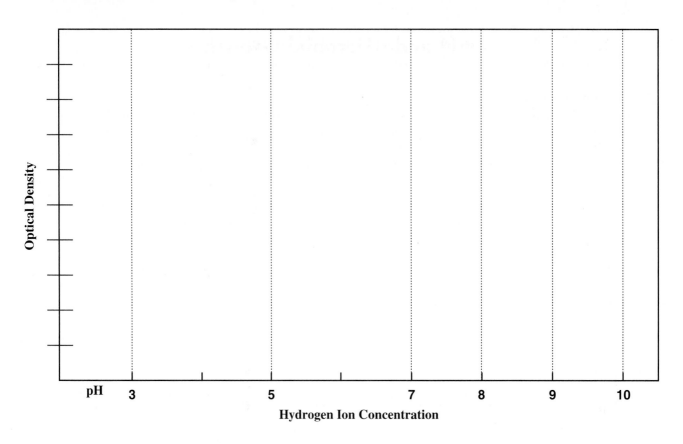

C. Questions

1. Which organism seems to grow best in acid media?_____

2. Which organism seems to grow best in alkaline media?_____

3. Which organism seems to tolerate the broadest pH range?_____

LABORATORY REPORT

37, 38

Student: _____

Desk No.: _____ Section: _____

Ex. 37 Osmotic Pressure and Bacterial Growth

A. Results

Record the amount of growth of each organism at the different salt concentrations, using, +, ++, +++, and none to indicate degree of growth.

ORGANISM	SODIUM CHLORIDE CONCENTRATION							
	0.5%		5%		10%		15%	
	48 hr	96 hr	48 hr	96 hr	48 hr	96 hr	48 hr	96 hr
Escherichia coli								
Staphylococcus aureus								
Halobacterium salinarium								

B. Questions

1. Evaluate the salt tolerance of the above organisms.

   Tolerates very little salt: _____

   Tolerates a broad range of salt concentration: _____

   Grows only in the presence of high salt concentration: _____

2. How would you classify *Halobacterium salinarium* as to salt needs? Check one.

   Obligate halophile _____ Facultative halophile _____

3. Differentiate between the following:

   Halophile: _____

   Osmophile: _____

4. Supply the following information concerning **mannitol salt agar** (Refer to the Difco Manual).

   Composition: _____

   _____

   For what organism is this medium selective? _____

   What ingredient makes it selective? _____

Ex. 38 Oligodynamic Action

A. Tabulation of Results

Measure the zone of inhibition from the edge of each piece of metal with a millimeter ruler. Record the measurements in the table. Spaces are provided for write-in of additional metals.

METAL	MILLIMETERS OF INHIBITION	
	E. coli	*S. aureus*
Copper		
Silver		
Aluminum		

B. Questions

1. Which metal seems to exhibit the greatest amount of oligodynamic action?

   _____

2. Which metal or metals seem to be ineffective? _____

3. Do these two organisms seem to differ in their susceptibility to oligodynamic action?

   _____

   Explain: _____

   _____

   _____

4. What specific chemical substances in bacterial cells are inactivated by heavy metals, affecting growth?

   _____

   _____

LABORATORY REPORT

39

Student: _____

Desk No.: _____ Section: _____

Ultraviolet Light:
Lethal Effects

A. Tabulation of Results

Your instructor will construct a table similar to the one below on the chalkboard for you to record your results. If substantial growth is present in the exposed area, record your results as +++. If three or fewer colonies survived, record +. Moderate survival should be indicated as ++. No growth should be recorded as −. Record all information in the table.

ORGANISM	EXPOSURE TIME							
S. aureus	10 sec	20 sec	40 sec	80 sec	2.5 min	5 min	10 min	•20 min
Survival								
B. megaterium	1 min	2 min	4 min	8 min	15 min	30 min	60 min	•6 min
Survival								

•Plates covered during exposure.

B. Questions

1. What length of time is required for the destruction of non-spore-forming bacteria such as *Staphylococcus aureus?*

   _____

2. Can you express, quantitatively, how much more resistant *B. megaterium* spores are to ultraviolet light than *S. aureus* vegetative cells (i.e., how many *times* more resistant are they)?

   _____

   _____

3. Why is it desirable to remove the cover from the Petri dish when making exposures?

   _____

4. In what specific way does ultraviolet light destroy microorganisms?

   _____

   _____

5. What adverse effect can result from overexposure of human tissues to ultraviolet light?

   _____

   _____

   _____

   _____

6. What wavelength of ultraviolet is most germicidal? _____

7. List several practical applications of ultraviolet light to microbial control.

_____

_____

_____

_____

_____

_____

_____

_____

LABORATORY REPORT
40

Evaluation of Disinfectants:
The Use-Dilution Method

A. Tabulation of Results

The instructor will draw a table on the chalkboard similar to the one below. Examine your tubes of nutrient broth and pins by shaking them and looking for growth (turbidity). If you are doubtful as to whether growth is present, compare the tubes with a tube of sterile nutrient broth. Record on the chalkboard a plus (+) sign if growth is present and a minus (−) sign if no growth is visible. After all students have recorded their results, complete the following chart.

DISINFECTANT		MINUTES											
		Staphylococcus aureus						Bacillus megaterium					
	Substitution	C*	1	5	10	30	60	C*	1	5	10	30	60
1:750 Zephiran													
5% phenol													
8% formaldehyde													

*C = control tube

B. Questions

1. What conclusions can be drawn from this experiment?

_____

_____

_____

_____

_____

_____

_____

_____

_____

_____

2. Distinguish between the following:

Disinfectant: _____

_____

Antiseptic: _____

_____

3. What factors other than time influence the action of a chemical agent on bacteria?

_____

_____

_____

_____

_____

_____

_____

_____

_____

4. Fill in the equation that explains how the **phenol coefficient** is determined:

P.C. = _____

5. What are some drawbacks that one encounters when attempting to apply the phenol coefficient to all disinfectants?

_____

_____

_____

_____

_____

_____

LABORATORY REPORT

41

Student: _____

Desk No.: _____ Section: _____

Evaluation of Alcohol:
Its Effectiveness as a Skin Degerming Agent

A. Tabulation of Results

Count the number of colonies that appear on each of the thumbprints and record them in the following table. If the number of colonies has increased in the second press, record a 0 in percent reduction. Calculate the percentages of reduction and record these data in the appropriate column. Use this formula:

$$\text{Percent Reduction} = \frac{(\text{Colony Count 1st press}) - (\text{Colony Count 2nd press})}{(\text{Colony Count 1st press})} \times 100$$

LEFT THUMB (Control)			RIGHT THUMB (Dipped)			RIGHT THUMB (Swabbed)		
Colony Count 1st Press	Colony Count 2nd Press	Percent Reduction	Colony Count 1st Press	Colony Count 2nd Press	Percent Reduction	Colony Count 1st Press	Colony Count 2nd Press	Percent Reduction
Av. % Reduction, Left (C)			Av. % Reduction, Right (D)			Av. % Reduction, Right (S)		

B. Questions

1. In general, what effect does alcohol have on the level of skin contaminants? _____

   _____

2. Is there any difference between the effects of dipping versus swabbing? _____

   Which method appears to be more effective? _____

3. There is definitely survival of some microorganisms even after alcohol treatment. Without staining or microscopic scrutiny, predict what types of microbes are growing on the medium where you made the

   right thumb impression after treatment. _____

   _____

   _____

   _____

   _____

LABORATORY REPORT

42

Student: _____

Desk No.: _____ Section: _____

Evaluation of Antiseptics:
The Filter Paper Disk Method

A. Tabulation of Results

With a millimeter scale, measure the zones of inhibition between the edge of the filter paper disk and the organisms. Record this information. Exchange your plates with other students' plates to complete the measurements for all chemical agents.

DISINFECTANT	MILLIMETERS OF INHIBITION	
	Staphylococcus aureus	*Pseudomonas aeruginosa*
5% phenol		
5% formaldehyde		
5% iodine		

B. Questions

1. What conclusions can be derived from these results? _____

_____

_____

_____

_____

_____

_____

_____

_____

2. What factors influence the size of the zone of inhibition? _____

_____

_____

_____

Evaluation of Antibiotics:
The Filter Paper Disk Method

Student: _____

Desk No.: _____ Section: _____

Antimicrobic Sensitivity Testing:
The Kirby-Bauer Method

A. Tabulation

List the antimicrobics that were used for each organism. Consult tables 43.2 and 43.3 to identify the various disks. After measuring and recording the zone diameters, consult Table VII in Appendix A for interpretation. Record the degrees of sensitivity (R, I, or S) in the sensitivity column. Exchange data with other class members to complete the entire chart.

	ANTIMICROBIC	ZONE DIA.	RATING (R, I, S)	ANTIMICROBIC	ZONE DIA.	RATING (R, I, S)
S. aureus						
P. aeruginosa						
Proteus vulgaris						
E. coli						

B. Questions

1. Which antimicrobics would be suitable for the control of the following organisms?

   S. aureus: _____

   E. coli: _____

   P. vulgaris: _____

   P. aeruginosa: _____

2. Differentiate between the following:

   Narrow spectrum antibiotic: _____

   _____

   Broad spectrum antibiotic: _____

   _____

3. Which antimicrobics used in this experiment would qualify as being excellent broad spectrum antimicrobics?

   _____

   _____

4. Differentiate between the following:

   Antibiotic: _____

   _____

   Antimicrobic: _____

   _____

5. How can drug resistance in microorganisms be circumvented? _____

   _____

   _____

   _____

Student: _____

Desk No.: _____ Section: _____

Effectiveness of Hand Scrubbing

A. Tabulation of Results

The instructor will draw a table on the chalkboard similar to the one below. Examine the six plates that your group inoculated from the basin of water. Select the two plates of a specific dilution that have approximately 30 to 300 colonies and count all of the colonies of each plate with a Quebec colony counter. Record the counts for each plate and their averages on the chalkboard. Once all the groups have recorded their counts, record the dilution factors for each group in the proper column. To calculate the organisms per milliliter multiply the average count by the dilution factor.

GROUP	0.1 ml COUNT		0.2 ml COUNT		0.4 ml COUNT		DILUTION FACTOR*	ORGANISMS PER MILLILITER
	Per Plate	Average	Per Plate	Average	Per Plate	Average		
A								
B								
C								
D								
E								

*Dilution factors: 0.1 ml = 10; 0.2 ml = 5; 0.4 ml = 2.5

B. Graph

After you have completed this tabulation, plot the number of organisms per milliliter that were present in each basin.

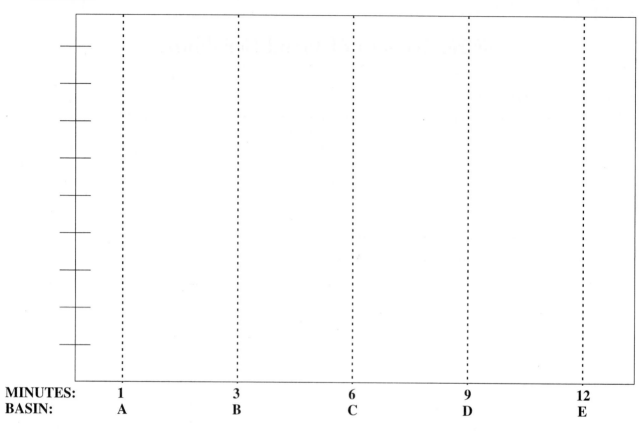

MINUTES:	1	3	6	9	12
BASIN:	A	B	C	D	E

C. Questions

1. What conclusions can be derived from this exercise?

_____

_____

_____

2. What might be an explanation of a higher count in Basin D than in B, ruling out contamination or faulty techniques?

_____

_____

3. Why is it so important that surgeons scrub their hands prior to surgery even though they wear rubber gloves?

_____

_____

LABORATORY REPORT

45

Preparation and Care of Stock Cultures

1. Why shouldn't cultures be stored at room temperature or in the incubator for any length of time?

   _____

   _____

   _____

2. Why should stock cultures be reinoculated to new media ("rotated") even if they are stored in the refrigerator? _____

   _____

3. For what types of inoculations do you use your

   reserve stock culture? _____

   _____

   working stock culture? _____

   _____

4. What is **lyophilization?** _____

   _____

   _____

   _____

   What advantage does this procedure have over the method we are using for maintaining stock cultures? _____

   _____

   _____

   _____

LABORATORY REPORT

48–50

Student: _____

Desk No.: _____ Section: _____

Physiological Characteristics of Bacteria

A. Media
List the media that are used for the following tests:

1. Butanediol production
2. Hydrogen sulfide production
3. Indole production
4. Starch hydrolysis
5. Urease production
6. Citrate utilization
7. Fat hydrolysis
8. Casein hydrolysis
9. Catalase production
10. Mixed acid fermentation
11. Glucose fermentation
12. Nitrate reduction

B. Reagents
Select the reagents that are used for the following tests:

1. Indole test
2. Voges-Proskauer test
3. Catalase test
4. Starch hydrolysis

Barritt's reagent—1
Gram's iodine—2
Hydrogen peroxide—3
Kovacs' reagent—4
None of these—5

C. Ingredients
Select the ingredients of the reagents for the following tests. Consult Appendix B. More than one ingredient may be present in a particular reagent.

1. Oxidase test
2. Voges-Proskauer test
3. Indole test
4. Nitrite test

α-naphthol—1
Dimethyl-α-naphthylamine—2
Dimethyl-ρ-phenylenediamine
    hydrochloride—3
ρ-dimethylamine benzaldehyde—4
Potassium hydroxide—5
Sulfanilic acid—6

D. Enzymes
What enzymes are involved in the following reactions?

1. Urea hydrolysis
2. Hydrogen gas production from formic acid
3. Casein hydrolysis
4. Indole production
5. Nitrate reduction
6. Starch hydrolysis
7. Fat hydrolysis
8. Gelatin hydrolysis (Ex. 47)
9. Hydrogen sulfide production

Answers

Media

1. _____
2. _____
3. _____
4. _____
5. _____
6. _____
7. _____
8. _____
9. _____
10. _____
11. _____
12. _____

Reagents	Ingredients
1. _____	1. _____
2. _____	2. _____
3. _____	3. _____
4. _____	4. _____

Enzymes

1. _____
2. _____
3. _____
4. _____
5. _____
6. _____
7. _____
8. _____
9. _____

347

E. Test Results

Indicate the appearance of the following positive test results.

1. Glucose fermentation, no gas
2. Citrate utilization
3. Urease production
4. Indole production
5. Acetoin production
6. Hydrogen sulfide production
7. Coagulation of milk
8. Peptonization in milk
9. Litmus reduction in milk
10. Nitrate reduction
11. Catalase production
12. Casein hydrolysis
13. Fat hydrolysis

Answers
1. _____
2. _____
3. _____
4. _____
5. _____
6. _____
7. _____
8. _____
9. _____
10. _____
11. _____
12. _____
13. _____

F. General Questions

1. Differentiate between the following:

   Respiration: _____

   Fermentation: _____

   Oxidation: _____

   Reduction: _____

   Catalase: _____

   Peroxidase: _____

2. List two or three difficulties one encounters in trying to differentiate bacteria on the basis of physiological characteristics. _____

   _____

   _____

3. Now that you have determined the morphological, cultural, and physiological characteristics of your unknown, what other kinds of tests might you perform on the organism to assist in identification?

   _____

   _____

LABORATORY REPORT

52

Student: _____

Desk No.: _____ Section: _____

Enterobacteriaceae Identification:
The API 20E System

A. Tabulation of Results

By referring to Charts I and II, Appendix D, determine the results of each test and record these results as positive (+) or negative (−) in the table below. Note that the results of the oxidase test must be recorded in the last column on the right side of the table.

ONPG	ADH	LDC	ODC	CIT	H2S	URE	TDA	IND	VP	GEL	GLU	MAN	INO	SOR	RHA	SAC	MEL	AMY	ARA	OXI
1	2	4	1	2	4	1	2	4	1	2	4	1	2	4	1	2	4	1	2	4

☐ ☐ ☐ ☐ ☐ ☐ ☐

NO2	N2 GAS	MOT	MAC	OF-O	OF-F
1	2	4	1	2	4

☐ ☐ Additional Digits

B. Construction of Seven-Digit Profile

Note in the above table that each test has a value of 1, 2, or 4. To compute the seven-digit profile for your unknown, total up the positive values for each group.

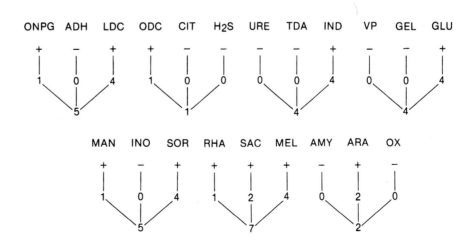

Example:
5 144 572 = *E. coli*

349

C. *Using the* API 20E Analytical Index *or the* API Characterization Chart

If the *API 20E Analytical Index* is available on the demonstration table, use it to identify your unknown, using the seven-digit profile number that has been computed. If no *Analytical Index* is available, use Characterization Chart III in Appendix D.

Name of Unknown: _____

D. Additional Tabulation Blank

If you need another form, use the one below:

| api® 20E | Reference Number _____ Patient _____ Date _____ |
| | Source/Site _____ Physician _____ Dept./Service _____ |

	ONPG 1	ADH 2	LDC 4	ODC 1	CIT 2	H₂S 4	URE 1	TDA 2	IND 4	VP 1	GEL 2	GLU 4	MAN 1	INO 2	SOR 4	RHA 1	SAC 2	MEL 4	AMY 1	ARA 2	OXI 4
5 h																					
24 h																					
48 h																					
Profile Number																					

	NO₂ 1	N₂ GAS 2	MOT 4	MAC 1	OF-O 2	OF-F 4	Additional Information
5 h							
24 h							Identification
48 h							
Additional Digits							00-42-012 E-3 (7/80)

E. Questions

1. What is the intended function of the API 20E system? _____

_____

_____

_____

2. In the "real world" who would use this system? _____

_____

_____

3. What might be an explanation for the failure of this system to work with some of the bacterial cultures we use? _____

_____

_____

Student: _____

Desk No.: _____ Section: _____

Enterobacteriaceae Identification:
The Enterotube II System

A. Tabulation of Results

Record the results of each test in the following table with a plus (+) or minus (−).

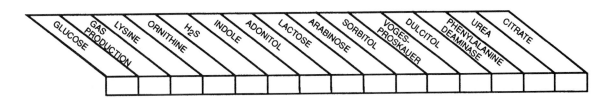

B. Identification by Chart Method

If no *Interpretation Guide* is available, apply the above results to Chart IV, Appendix D, to find the name of your unknown. Note that the spacing of the above table matches the size of the spaces on Chart IV. If this page is removed from the manual, folded, and placed on Chart IV, the results on the above table can be moved down the chart to make a quick comparison of your results with the expected results for each organism.

C. Using the Enterotube II Interpretation Guide

If the *Interpretation Guide* is available, determine the five-digit code number by circling the numbers (4, 2, or 1) under each test that is positive, and then totaling these numbers within each group to form a digit for that group. Note that there are two tally charts on the next page of this Laboratory Report for your use.

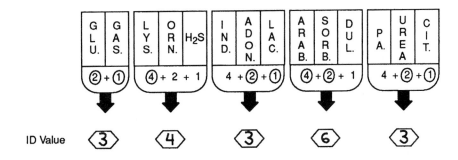

The "ID Value" 34363 can be found by thumbing the pages of the *Interpretation Guide*. The listing is as follows:

ID Value	Organism	Atypical Test Results
34363	*Klebsiella pneumoniae*	None

Conclusion: Organism was correctly identified as *Klebsiella pneumoniae*. In this case, the identification was made independent of the V-P test.

D. Tally Charts

ENTEROTUBE® II*

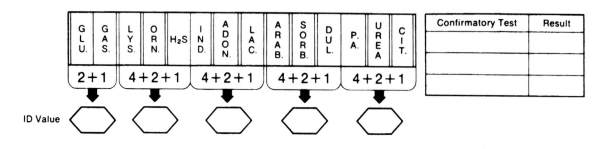

Culture Number, Case Number or Patient Name   Date   Organism Identified

*VP utilized as confirmatory test only.

ENTEROTUBE® II*

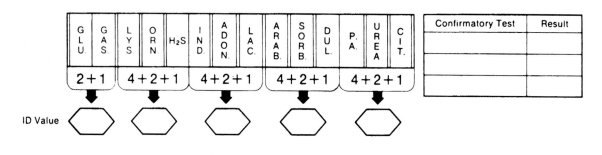

Culture Number, Case Number or Patient Name   Date   Organism Identified

*VP utilized as confirmatory test only.

E. Questions

1. What is the intended function of the Enterotube II System? _____
   _____
   _____
   _____

2. In the "real world" who would use this system? _____
   _____
   _____

3. What might be an explanation for the failure of this system to work with some of the bacterial cultures we use? _____
   _____
   _____

LABORATORY REPORT

54

Student: _____

Desk No.: _____ Section: _____

O/F Gram-Negative Rods Identification:
The Oxi/Ferm Tube II System

A. Tabulation of Results and Code Determination

Once you have marked the positive reactions on the side of the tube and circled the numbers that are assigned to each of the positive chambers, as indicated in the example below, add the numbers in each bracketed group to get the five-digit code.

The final step is to look up the code number in the *Oxi/Ferm Tube II Biocode Manual* to determine the genus and species. If confirmatory tests are necessary, the manual will tell you which ones to perform.

In the example below, the code number is 32303. If you look up this number in the *Biocode Manual* you will find on page 25 that the organism is *Pseudomonas aeruginosa*.

Use this procedure to identify your unknown by applying your results to the blank diagrams provided.

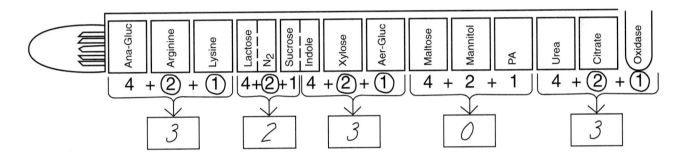

B. Results Pads

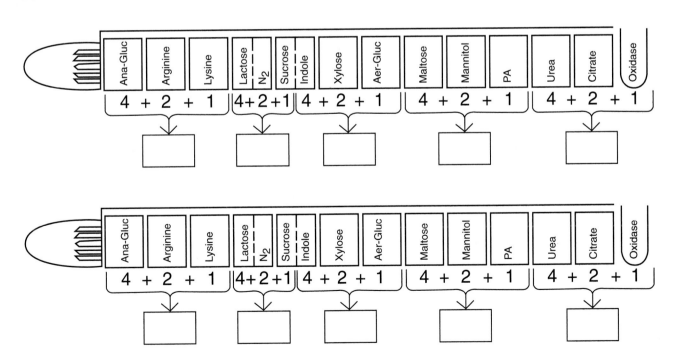

C. Questions

1. What is the intended function of the Oxi/Ferm Tube II System? _____
    _____
    _____
    _____

2. In the "real world" who would use this system? _____
    _____
    _____

3. What might be an explanation for the failure of this system to work with some of the bacterial cultures
    we use? _____
    _____
    _____

LABORATORY REPORT

55

Staphylococcus Identification:
The API Staph-Ident System

A. Tabulation of Results

By referring to Charts V and VI, Appendix D, determine the results of each test, and record these results as positive (+) or negative (−) in the Profile Determination Table below. Note that two more of these tables have been printed on the next page for tabulation of additional organisms.

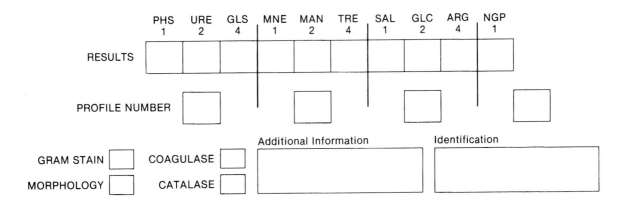

B. Construction of Four-Digit Profile

Note in the above table that each test has a value of 1, 2, or 4. To compute the four-digit profile for your unknown, total up the positive values for each group.

Example: 7 700 = *Staphylococcus aureus*

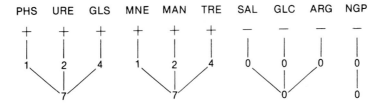

C. Final Determination

Refer to the Staph-Ident Profile Register (Chart VII, Appendix D) to find the organism that matches your profile number. Write the name of your unknown in the space below and list any additional tests that are needed for final confirmation. If the materials are available for these tests, perform them.

Name of Unknown: _____

Additional Tests: _____

	PHS 1	URE 2	GLS 4	MNE 1	MAN 2	TRE 4	SAL 1	GLC 2	ARG 4	NGP 1
RESULTS										

PROFILE NUMBER

GRAM STAIN   COAGULASE   Additional Information   Identification

MORPHOLOGY   CATALASE

	PHS 1	URE 2	GLS 4	MNE 1	MAN 2	TRE 4	SAL 1	GLC 2	ARG 4	NGP 1
RESULTS										

PROFILE NUMBER

GRAM STAIN   COAGULASE   Additional Information   Identification

MORPHOLOGY   CATALASE

D. Questions

1. What is the intended function of the API Staph-Ident System? _____
   _____
   _____
   _____

2. In the "real world" who would use this system? _____
   _____
   _____

3. What might be an explanation for the failure of this system to work with some of the bacterial cultures
   we use? _____
   _____
   _____

LABORATORY REPORT

56, 57

Ex. 56 Microbial Population Counts of Soil

A. Tabulation of Results

Select the best plate and count the organisms on a Quebec colony counter. Tabulate your results and the results of other students near you who cultured the other types of soil organisms. Be sure to count only representative types.

ORGANISMS	COUNT PER PLATE	DILUTION	ORGANISMS PER GRAM OF SOIL
Bacteria			
Actinomycetes			
Molds			

B. Conclusions

What generalizations can you make from this exercise?

_____

_____

_____

_____

C. Questions

1. Why would the number of bacteria present in the soil actually be higher than the number determined by your plate count? _____

_____

_____

2. What other types of microbes might be present in your soil sample? _____

_____

_____

Ex. 57 Ammonification in Soil

A. Tabulation of Results

Record the presence or absence of ammonia in the tubes of media:

+	slight ammonia (faint yellow)
++	moderate ammonia (deep yellow)
+++	much ammonia (brown precipitate)
−	no ammonification

INCUBATION TIME	AMOUNT OF AMMONIA		pH	
	Control	Peptone with Soil	Control	Peptone with Soil
4 days				
7 days				

B. Questions

1. In the natural situation, what compounds serve as the source of the ammonia released in ammonification? _____

   _____

2. In simple terms, what occurs during decay? _____

   _____

   _____

3. Differentiate between the following:

   Peptonization: _____

   _____

   Ammonification: _____

   _____

4. What happens to ammonia in soil when **denitrification** of ammonia takes place? _____

   _____

Student: _____

Desk No.: _____ Section: _____

Ex. 58 Nitrification in Soil

A. Tabulation of Results

Indicate the presence or absence of the various ions in the following table.

TIME	AMMONIUM MEDIUM		NITRITE MEDIUM	
	Ammonia	Nitrite	Nitrite	Nitrate
7 days				
14 days				

B. Microscopic Examination

After examining the organisms of a positive culture, draw a few representative cells.

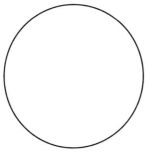

Nitrite Producers
(Nitrosomonas)

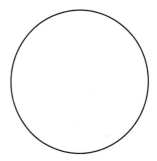

Nitrate Producers
(Nitrobacter)

C. Conclusions

What generalizations can you make from this exercise?

_____

_____

_____

D. Question

What is the significance of nitrification in terms of soil fertility? _____

_____

Ex. 59 Nitrogen-Fixing Bacteria

A. Questions

1. What enzyme is responsible for nitrogen fixation? _____

2. What metal is essential for this enzyme to function? _____

3. Why is nitrogen fixation so important? _____
   _____

B. Azotobacteraceae

Identify the characteristics of your isolate that led to your decision as to its identification.

1. Were the colonies pigmented? _____

   If pigmented, what color? _____

   Was fluorescence present? _____

   If fluorescent, what color? _____

2. Were you able to see cysts? _____

3. Is the organism motile? _____

4. Gram reaction? _____

5. Any other pertinent characteristic? _____
   _____

Azotobacteraceae

Name of Organism: _____
Draw a few cells of organisms in circle at right.

C. Rhizobiaceae

Draw a few representative cells of *Rhizobium* in the circle to the right.

1. What is the most important criterion for species identification in the family of nitrogen-fixers?
   _____
   _____

2. From the standpoint of amount of nitrogen fixation, is this group of nitrogen-fixers **more** or **less important** than the Azotobacteraceae?
   _____

Rhizobiaceae

LABORATORY REPORT

60

Isolation of an Antibiotic Producer from Soil

A. Results

Describe in detail how the experiment turned out.

_____

_____

_____

_____

_____

_____

_____

B. Questions

1. What four genera of microbes produce the vast majority of antibiotics? _____

_____

2. What is known about the importance of these antibiotics in the natural habitat of these microbes?

_____

_____

_____

Student: _____

Desk No.: _____ Section: _____

Bacteriological Examination of Water:
Qualitative Tests

A. Results of Presumptive Test (MPN Determination)

Record the number of positive tubes on the chalkboard and on the following table. When all students have recorded their results with the various water samples, complete this tabulation. Determine the MPN according to the instructions on page 210.

WATER SAMPLE (SOURCE)	NUMBER OF POSITIVE TUBES				MPN
	3 Tubes DSLB 10 ml	3 Tubes SSLB 1.0 ml	3 Tubes SSLB 0.1 ml	3 Tubes SSLB 0.01 ml	

B. Results of Confirmed Test

Record the results of the confirmed tests for each water sample that was positive on the presumptive test.

WATER SAMPLE (SOURCE)	POSITIVE	NEGATIVE

C. Results of Completed Test

Record the results of completed tests for each water sample that was positive on the confirmed test.

WATER SAMPLE (SOURCE)	LACTOSE FERMENTATION RESULTS	MORPHOLOGY	EVALUATION

D. Questions

1. Does a positive presumptive test mean that the water is absolutely unsafe to drink? _____

   Explain: _____

2. What might cause a false positive presumptive test? _____

   _____

3. List three characteristics required of a good sewage indicator: a. _____

   b. _____ c. _____

4. What enteric **bacterial** diseases are transmitted in polluted water? _____

   _____

5. Name one or more **protozoan** diseases transmitted by polluted water. _____

   _____

6. Why don't health departments routinely test for pathogens instead of using a sewage indicator?

   _____

   _____

7. Give the functions of the various media used in these tests:

   Lactose broth: _____

   Levine EMB agar: _____

   _____

   Nutrient agar slant: _____

   _____

8. What media, other than the ones used here, can be used for confirmatory tests? _____

   _____

   _____

Ex. 62 Membrane Filter Method

A. Tabulation

A table similar to the one below will be provided for you, either on the chalkboard or as a photocopy. Record your coliform count on it. Once all data are available, complete this table.

SAMPLE	SOURCE	COLIFORM COUNT	AMOUNT OF WATER FILTERED	MPN*
A				
B				
C				
D				
E				
F				
G				
H				

$$*MPN = \frac{Coliform\ Count \times 100}{Amount\ of\ Water\ Filtered}$$

B. Questions

1. Give two limitations of the membrane filter technique:

   _____

   _____

2. Even if the membrane filter removed all bacteria from water being tested, is the water that passes through sterile? _____ Explain: _____

   _____

3. List some other applications of membrane filter technology in microbiology. _____

   _____

   _____

   _____

Ex. 63 Standard Plate Count:
A Quantitative Test

A. Tabulation of Results

After you have made your plate counts, record your results on the chalkboard and on the following table. After all students have recorded their results, complete this table.

SAMPLE	SOURCE	PLATE COUNT (AVERAGE)	DILUTION	ORGANISMS PER ML
A				
B				
C				
D				
E				
F				
G				
H				
I				

What generalizations, if any, can be drawn from these results?

_____

_____

_____

_____

_____

_____

_____

B. Questions

1. What kinds of organisms in water will not produce colonies on TGEA?

   _____

   _____

2. Would a high plate count indicate that the water is unsafe to drink? _____

   Explain: _____

   _____

Student: _____

Desk No.: _____ Section: _____

Ex. 64 Standard Plate Count of Milk

A. Tabulation of Results

After you have made your plate count, record your results on the following table. Get the results of the other milk sample from some other member of the class.

TYPE OF MILK	PLATE COUNT	DILUTION	ORGANISMS PER ML
High-quality			
Poor-quality			

B. Questions

1. Do plate count figures represent numbers of organisms or numbers of clumps of bacteria?

   _____

2. What are some factors that will produce errors in the SPC technique?

   _____

   _____

   _____

3. What might be the explanation of a very high count in raw milk that has been properly refrigerated from the time of collection?

   _____

   _____

   _____

4. What is the most common source of bacteria in milk? _____

5. Why is milk a more suitable vector of disease than water? _____

   _____

6. What infectious diseases of cows can be transmitted to humans via milk? _____

   _____

   _____

Ex. 65 Direct Microscopic Count of Organisms in Milk: The Breed Count

A. Microscope Factor

Show your computations in determining the MF of your microscope.

1. Field diameter: _____ millimeters

2. Field area (πr^2): _____ square millimeters

   _____ square centimeters

3. Number of fields in one square centimeter: _____

4. Microscope factor (100 × no. of fields): _____

B. Cell Counts

Record the counts of each field, total them, and determine their average counts. Get the results of the other type of milk from a class member.

	GOOD MILK			POOR MILK		
	Bacterial Cells	Bacterial Clumps	Leukocytes	Bacterial Cells	Bacterial Clumps	Leukocytes
1						
2						
3						
4						
5						
Total						
Av.						

Determine the number of cells, clumps, and leukocytes per ml for each type of milk.

Good Milk

Bacterial Cells: _____

Bacterial Clumps: _____

Leukocytes: _____

Poor Milk

Bacterial Cells: _____

Bacterial Clumps: _____

Leukocytes: _____

C. Questions

1. Should milk from a healthy cow be completely free of bacteria? _____

2. Are leukocytes in milk always an indication of infection (mastitis)? _____

3. Why are direct counts higher than standard plate counts? _____

_____

Reductase Test

1. How would you grade the two samples of milk that you tested? Give the MBRT for each one.

   Sample A: _____

   Sample B: _____

2. Is milk with a short reduction time necessarily unsafe to drink? _____

   Explain: _____

   _____

   _____

   _____

3. What other dye can be substituted for methylene blue in this test? _____

   _____

4. What advantage do you see in this method over the direct count method? _____

   _____

   _____

   _____

   _____

5. What kinds of organisms may be plentiful in a milk sample, yet give a negative reductase test?

   _____

   _____

   _____

   _____

Student: _____

Desk No.: _____ Section: _____

Bacterial Counts of Foods

A. Tabulation of Results

Record your count and the bacterial counts of various other foods made by other students.

TYPE OF FOOD	PLATE COUNT	DILUTION	ORGANISMS PER ML

B. Questions

1. Why is there such great variability in organisms per ml between different kinds of food? _____

   _____

   _____

   _____

2. What dangers and undesirable results may occur from ground meats of high bacterial counts?

   _____

   _____

3. What bacterial pathogens might be present in frozen foods? _____

   _____

   _____

4. What harm can result from repeated thawing and freezing of foods? _____

   _____

   _____

5. What precautions are taken to prevent the spoilage of foods? _____
   _____
   _____

6. Which methods in question 5 are most effective? _____
   _____

   Least effective? _____

LABORATORY REPORT

68, 69

Student: _____

Desk No.: _____ Section: _____

Ex. 68 Microbial Spoilage of Canned Food

A. Results

Record your observations of the effects of each organism on the cans of vegetables. Share results with other students.

ORGANISM	PEAS		CORN	
	Gas Production + or −	Odor	Gas Production + or −	Odor
E. coli				
B. coagulans				
B. stearothermophilus				
C. sporogenes				
C. thermosaccharolyticum				

B. Microscopy

After making gram-stained and spore-stained slides of all organisms from the canned food extracts, sketch in representatives of each species:

E. coli	B. coagulans	B. stearothermophilus	C. sporogenes	C. thermosac-charolyticum

C. Questions

1. Which organisms, if any, caused "flat sour spoilage"? _____

   _____

2. Which organisms, if any, caused "T.A. spoilage"? _____

   _____

3. Which organisms, if any, caused "stinker spoilage"? _____

   _____

4. Does "flat sour" cause a health problem? _____

5. Describe how typical spoilage resulting in botulism occurs. _____

   _____

   _____

Ex. 69 Microbial Spoilage of Refrigerated Meat

A. Results

1. After performing the colony count on your best plates, how many psychrophilic-psychrotrophic bacteria do you find were present in each gram of the meat sample? _____

2. Describe the appearance of the organism that you stained with Gram's stain. _____
_____

B. Questions

1. Differentiate between the following:

   Psychrophile: _____
   _____

   Psychrotroph: _____
   _____

2. Why was it necessary to incubate the plates for 2 weeks? _____
   _____

3. List some genera of bacteria that might be psychrophilic or psychrotrophic in the meat sample:

   Gram-negative types: _____
   _____

   Gram-positive types: _____
   _____

4. List three or four pathogenic psychrotrophs that might be found in refrigerated meat: _____
   _____
   _____

5. What are the probable sources of psychrophilic microbes found growing in meat? _____
   _____
   _____

6. What types of measures can be taken to prevent spoilage of meats by psychrophilic-psychrotrophic bacteria? _____
   _____
   _____
   _____

Student: _____

Desk No.: _____ Section: _____

Ex. 70 Microbiology of Fermented Beverages

A. Results

Record here your observations of the fermented product:

Aroma: _____

pH: _____

H$_2$S production: _____

B. Questions

1. Why must the fermentor be sealed? _____

   _____

   Why with a balloon? _____

2. What compound in the grape juice is being fermented? _____

3. Why would production of hydrogen sulfide by the yeast be of importance? _____

   _____

   _____

4. Why are we concerned about the pH of the fruit juice and the wine? _____

   _____

   _____

5. What happens to wine if *Acetobacter* takes over? _____

   _____

6. What process can be used to prevent the action of *Acetobacter* in the production of wine and beer?

   _____

Ex. 71 Microbiology of Fermented Milk Products

A. Results

1. Record here your observations of the fermented product:

   Color: _____

   Aroma: _____

   Texture: _____

   Taste: _____

2. In the space provided below sketch in the microscopic appearance of the organisms as seen on the slide stained with methylene blue.

B. Questions

1. What type of fermentation is involved in the production of yogurt? _____

_____

2. What is the nutritional value of yogurt? _____

_____

Ex. 72 Mutant Isolation by Gradient Plate Method

A. Results

1. How many colonies of *E. coli* did you count in the high concentration area of the plate?

   _____

2. When the streptomycin-resistant colonies were smeared with a loop toward the high concentration side and reincubated, did they continue to grow in the new area?

   _____

   What does this result indicate? _____

   _____

Ex. 73 Mutant Isolation by Replica Plating

A. Tabulation of Results

Count the colonies that occur on both plates and record the information on the chalkboard on a table similar to the one below. After all students have recorded their counts, complete this table.

STUDENT INITIALS	NUMBER OF COLONIES		STUDENT INITIALS	NUMBER OF COLONIES	
	Nutrient Agar	Streptomycin Agar		Nutrient Agar	Streptomycin Agar
TOTALS			TOTALS		
AVERAGE PER PLATE			AVERAGE PER PLATE		

B. Questions

1. After determining the average number of streptomycin-resistant colonies per plate, calculate the mutability rate of the organisms, assuming that there were 100,000,000 organisms per milliliter in the original culture. Show your mathematical computations.

    Mutability rate: _____

2. What does replica plating of mutants attempt to prove that is not established by the gradient plate method?

    _____

    _____

    _____

3. How can the frequency of mutation in bacteria be increased?

    _____

    _____

    _____

    _____

4. In what way does the presence of an antibiotic increase the numbers of resistant forms of bacteria?

    _____

    _____

    _____

    _____

    _____

LABORATORY REPORT

74

Student: _____

Desk No.: _____ Section: _____

Bacterial Mutagenicity and Carcinogenesis:
The Ames Test

A. Tabulation of Results

Record the results of your tests in the following table and on a similar table on the chalkboard. Also record the results of substances tested by other students. A positive result will exhibit a zone of colonies similar to the zone shown on the plate in illustration 5, figure 74.1.

TEST SUBSTANCE	RESULT (+ or −)	TEST SUBSTANCE	RESULT (+ or −)	TEST SUBSTANCE	RESULT (+ or −)

B. Questions

1. Did you observe a zone of inhibition between the growing colonies and the impregnated disk on your positive plates? _____

   What is the cause of such a zone? _____

   _____

2. Differentiate between the following:

   Prototroph: _____

   Auxotroph: _____

   (If you performed Exercise 75 and turned in Laboratory Report 75, you may skip this question.)

3. Define **back mutation:** _____

   _____

   _____

4. List two characteristics of the Ames test that made this test so much superior to previous mutagenesis tests: _____

   _____

   _____

5. Does the fact that a chemical substance is carcinogenic in animals necessarily mean that it is also carcinogenic in humans? _____

LABORATORY REPORT

75

Bacterial Conjugation

A. Control Inoculations

Look for colonies on the two TSA and two minimal agar control plates and answer these questions:

1. Did the auxotrophic strains (thr+met– and thr–met+) grow on the trypticase soy agar?

   _____

2. Did the auxotrophic strains grow on the minimal agar? _____

B. Mating Inoculations

Select the plate that has between 30 and 300 colonies and count all colonies, recording the results on the table below.

1:100	1:1000	1:10,000

C. Calculations

1. Calculate the number of recombinants per ml that formed in the M × F tube by multiplying the count on the minimal agar plate by the dilution.

   Recombinants per ml: _____

2. Determine the percentage of recombination, using this formula:

   $$\text{Percentage of Recombinants} = \frac{\text{Number of Recombinants per ml}}{\text{Number of Organisms per ml*}} \times 100$$

   *Remember, the original cultures had 10^8 org./ml

D. Questions

1. Differentiate between the following:

   Prototroph: _____

   _____

   Auxotroph: _____

   _____

2. What is the genetic composition of the organisms that grew on the minimal agar as a result of conjugation?

   _____

LABORATORY REPORT
76

Student: _____

Desk No.: _____ Section: _____

A Synthetic Epidemic

A. Tabulation of Results

Record in the table below the information that has been tabulated on the chalkboard. The SHAKER is the person designated by the instructor to shake hands with another class member. The SHAKEE is the individual chosen by the shaker. For **Procedure A** a blue color is positive; yellow or brown is negative. For **Procedure B** red colonies (*S. marcescens*) is positive; no red colonies is negative.

SHAKER Round 1	RESULT + or –	SHAKEE Round 1	RESULT + or –	SHAKER Round 2	RESULT + or –	SHAKEE Round 2	RESULT + or –
1.				1.			
2.				2.			
3.				3.			
4.				4.			
5.				5.			
6.				6.			
7.				7.			
8.				8.			
9.				9.			
10.				10.			
11.				11.			
12.				12.			
13.				13.			
14.				14.			
15.				15.			
16.				16.			
17.				17.			
18.				18.			
19.				19.			
20.				20.			
21.				21.			
22.				22.			
23.				23.			
24.				24.			

B. Questions

1. Who in the group was "patient zero," the starter of the epidemic? _____

2. How many carriers resulted after

   Round 1? _____   Round 2? _____

3. If this were a real infectious agent, such as a cold virus or influenza, list some other factors in transmission besides the ones we tested:

   _____

   _____

4. How would it have been possible to stop this infection cycle? _____

   _____

   _____

Student: _____

Desk No.: _____ Section: _____

The Staphylococci:
Isolation and Identification

A. Tabulation

At the beginning of the third laboratory period, the instructor will construct a chart similar to this one on the chalkboard. After examining your mannitol salt agar and staphylococcus medium 110 plates, record the presence (+) or absence (−) of staphylococcus growth in the appropriate columns. After performing coagulase tests on the various isolates record the results also as (+) or (−) in the appropriate columns.

STUDENT INITIALS	NOSE			FOMITES			
	Staph Colonies		Coagulase	Item	Staph Colonies		Coagulase
	MSA	SM110			MSA	SM110	

B. Microscopy

Provide drawings here of the various isolates as seen under oil immersion (gram staining).

UNKNOWN-CONTROL	NOSE	FOMITE

C. Percentages

From the data in the table on the previous page, determine the incidence (percentage) of individuals and fomites that harbor coagulase-positive and coagulase-negative staphylococci in this experiment.

SOURCE	TOTAL TESTED	TOTAL POSITIVE	PERCENTAGE POSITIVE	TOTAL NEGATIVE	PERCENTAGE NEGATIVE
Humans (Nose)					
Fomites					

D. Record of Test Results

Record here the results of each test performed in this experiment. Under GRAM STAIN indicate cellular arrangement as well as Gram reaction.

ISOLATE	GRAM STAIN	ALPHA TOXIN	MANNITOL (ACID)	COAGULASE
Unknown-Control No.____				
Nose Isolate No. 1				
Nose Isolate No. 2				
Fomite Isolate				

E. Final Determination

Record here the name of your unknown-control. If API Staph-Ident miniaturized multitest strips are available, confirm your conclusions by testing each isolate. See Exercise 55.

Name of unknown-control: _____

Staph-Ident results: _____

F. Questions

1. What are nosocomial infections? _____

   _____

2. What bacterial organism causes most nosocomial infections? _____

78

Student: _____

Desk No.: _____ Section: _____

The Streptococci: _—gram(+) cocci_
Isolation and Identification

A. Tabulation of Pharynx Isolates

The instructor will construct a chart similar to this one on the chalkboard. After examining the blood agar plates that were inoculated with pharynx organisms, record the types and size range of colonies that are present on your plates. Record these data first on this table, then on the chalkboard. After all students have recorded their results on the board, complete the tabulation of their results here, also. The names of the organisms will not be recorded until all tests are completed.

STUDENT INITIALS	TYPE OF HEMOLYSIS (ALPHA, ALPHA-PRIME, BETA)	SIZE RANGE OF COLONIES (MM)	NAMES OF ORGANISMS

B. Microscopy

Provide drawings here of the various pharyngeal isolates as seen under oil immersion (gram staining).

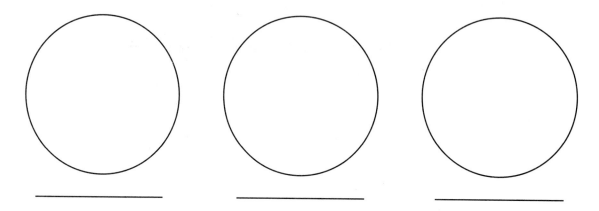

_____ _____ _____

C. Percentages

From the data in the table on the previous page, calculate the percentages for each type of streptococci that were isolated from classmates.

S. pyogenes: _____   **Group C streptococci:** _____

S. agalactiae: _____   **Group D enterococci:** _____

S. pneumoniae: _____   **Group D nonenterococci:** _____

Viridans streptococci: _____

D. Record of Test Results

Record here all information pertaining to the identification of pharyngeal isolates and unknowns.

Essential for Exam

SOURCE OF UNKNOWN	Hemolysis	Bacitracin Susceptibility	CAMP Reaction	Hippurate Hydrolysis	SXT Sensitivity	Bile Esculin Hydrolysis	Tolerance to 6.5% NaCl	Optochin Susceptibility	Bile Solubility
Strep pnemoniae								(+)	+
agalactiae			(−)	(+)			(−)		
pyogenes		(+)				(−)			
Enterococcus faecalis				(−)		(+)	(+)	(−)	

E. Final Determination

Record here the identities of your various isolates and unknowns:

Pharyngeal isolates: _____

_____

Unknowns: _____

_____

_____

LABORATORY REPORT

79

Gram-Negative Intestinal Pathogens

A. Unknown Identification

1. What was the genus of your unknown?

    _____   _____
    Genus   No.

2. What problems, if any, did you encounter?

    _____

    _____

    _____

    _____

    _____

    _____

3. Now that you know the genus of your unknown, what steps would you follow to determine the species?

    _____

    _____

    _____

B. General Questions

1. Why are bile salts and sodium desoxycholate used in certain selective media in this exercise?

    _____

    _____

2. How can one identify coliforms on MacConkey agar? _____

    _____

    _____

3. How does one differentiate *Salmonella* from *Shigella* colonies on XLD agar? _____

    _____

    _____

4. What characteristics do the salmonella and shigella have in common?

    _____

    _____

    _____

    _____

    _____

5. How do the salmonella and shigella differ?

_____

_____

_____

_____

6. What restrictions might be placed on a person who is a typhoid carrier?

_____

_____

_____

_____

_____

LABORATORY REPORT

80

Urinary Tract Pathogens

A. Quantitative Evaluation

After counting the colonies on the TSA plate, record the count as follows:

Number of colonies: _____

Dilution: _____

No. of organisms/ml of urine: _____

Gram-Stained Slide. If organisms are seen on a gram-stained slide of an uncentrifuged sample, sketch in color in the circle below.

Conclusion: Do the plate count and gram-stained slide of the uncentrifuged sample provide presumptive evidence of a urinary infection?

_____

B. Microscopic Study (Centrifuged Sample)

Illustrate in the circle below the microscopic appearance of a centrifuged sample.

Conclusion: Describe here the morphological appearance of the predominant organism seen:

_____

_____

_____

_____

C. Culture Analyses

After studying the organisms on the three plates and thioglycollate medium, what organism do you believe is causing the infection?

Organism: _____

What further testing should be performed for confirmation?

_____

_____

_____

Student: _____

Desk No.: _____ Section: _____

Ex. 81 Slide Agglutination Test:
Serological Typing

1. Record the unknown number that proved to be a salmonella. _____

2. Why was phenolized saline used instead of plain physiological saline? _____

   _____

   _____

3. If your results were negative for both cultures, what might be the explanation? _____

   _____

   _____

Ex. 82 Slide Agglutination (Latex) Test:
For *S. aureus* Identification

1. If your test turned out to be positive for *S. aureus*, record the degree of positivity here: _____

2. What other test for *S. aureus* is highly correlated with this test? _____

3. What two kinds of antibodies are attached to the latex particles in the Difco latex reagent?

   a. _____ b. _____

4. What role, if any, do the staphylococci cells play in this reaction? _____

   _____

   _____

LABORATORY REPORT

83, 84

Ex. 83 Tube Agglutination Test:
The Heterophile Antibody Test

1. What was the titer of the serum that you tested? _____
2. For what disease is this diagnostic test used? _____
3. Below what titer would this test be considered to be negative? _____
4. What is the name of the virus that causes this disease? _____
5. What is unusual about a heterophile antibody? _____
_____

6. *If you are not going to perform the Widal test (Ex. 84),* answer all the questions below except for question 1. It may be necessary for you to read the introduction to Ex. 84 for some of the answers.

Ex. 84 Tube Agglutination Test:
The Widal Test

1. What was the titer of the serum that you tested? _____
2. What is the exact meaning of the "titer" of a serum? _____
_____
_____

3. Differentiate between the following:

Serum: _____
_____

Antiserum: _____
_____

Antitoxin: _____
_____

4. How would you prepare antiserum for an organism such as *E. coli?* _____
_____
_____
_____
_____
_____

5. Indicate the type of antigen (*soluble protein, red blood cells,* or *bacteria*) that is used for each of the following serological tests:

   Agglutination: _____

   Precipitation: _____

   Hemolysis: _____

6. For which one of the above tests is complement necessary? _____

_____

Student: _____

Desk No.: _____ Section: _____

Phage Typing

1. To which phage types was this strain of *S. aureus* susceptible?

   _____

2. To what lytic group does this strain of staphylococcus belong?

   _____

3. In what way can bacteriophage alter the genetic structure of a bacterium?

   _____

   _____

   _____

   _____

   _____

Student: _____

Desk No.: _____ Section: _____

Ex. 86 White Blood Cell Study:
The Differential WBC Count

A. Tabulation of Results

As you move the slide in the pattern indicated in figure 86.4, record all the different types of cells in the following table. Refer to figures 86.1 and 86.2 for cell identification. Use this method of tabulation: ⊬⊬⊦ ⊬⊬⊦ 11. Identify and tabulate 100 leukocytes. Divide the total of each kind of cell by 100 to determine percentages.

NEUTROPHILS	LYMPHOCYTES	MONOCYTES	EOSINOPHILS	BASOPHILS
Total				
Percent				

B. Questions

1. Were your percentages for each type within the normal ranges? _____
   _____

2. What errors might one be likely to make when doing this count for the first time? _____
   _____
   _____
   _____

3. Differentiate between the following:

   Cellular immunity: _____

   Humoral immunity: _____

4. Do cellular and humoral immunity work independently? _____ Explain: _____
   _____
   _____
   _____
   _____
   _____
   _____

Ex. 87 Total WBC Count

A. Calculations

Using the formula provided on page 283, calculate the number of leukocytes per cubic millimeter.

Total Count: _____

B. Questions

1. What is the normal WBC count range? _____

2. List several causes of a high WBC count: _____

_____

_____

_____

_____

Ex. 88 Blood Grouping

1. On the basis of this test, what is your blood type? _____

2. What antibodies are present in each type of blood?

   Type A: _____ Type B: _____ Type AB: _____ Type O: _____

3. Why does a person of type A blood go into a transfusion reaction when given type B blood? _____

_____

_____

_____

4. Why can Rh-positive blood be given only once to a person who is Rh-negative? _____

_____

_____

_____

LABORATORY REPORT

89

The Snyder Caries Susceptibility Test

1. What degree of caries susceptibility was indicated for you as a result of this test? _____
   _____

2. Is this substantiated by the amount of dental work you have had or should have had on your teeth?
   _____

3. What factors could affect the reliability of this test? _____
   _____
   _____
   _____

4. To increase the validity of this test how many times should it be performed and at what time of the
   day? _____
   _____
   _____

Descriptive Chart

STUDENT: _____

LAB SECTION: _____

Habitat: _____ Culture No.: _____

Source: _____

Organism: _____

MORPHOLOGICAL CHARACTERISTICS

Cell Shape:

Arrangement:

Size:

Spores:

Gram's Stain:

Motility:

Capsules:

Special Stains:

CULTURAL CHARACTERISTICS

Colonies:

   Nutrient Agar:

   Blood Agar:

Agar Slant:

Nutrient Broth:

Gelatin Stab:

Oxygen Requirements:

Optimum Temp.:

PHYSIOLOGICAL CHARACTERISTICS

	TESTS	RESULTS
Fermentation	Glucose	
	Lactose	
	Sucrose	
	Mannitol	
Hydrolysis	Gelatin Liquefaction	
	Starch	
	Casein	
	Fat	
IMViC	Indole	
	Methyl Red	
	V-P (acetylmethylcarbinol)	
	Citrate Utilization	
	Nitrate Reduction	
	H$_2$S Production	
	Urease	
	Catalase	
	Oxidase	
	DNase	
	Phenylalanase	

	REACTION	TIME
Litmus Milk	Acid	_____
	Alkaline	_____
	Coagulation	_____
	Reduction	_____
	Peptonization	_____
	No Change	_____

Descriptive Chart

STUDENT: _____

LAB SECTION: _____

Habitat: _____ Culture No.: _____

Source: _____

Organism: _____

MORPHOLOGICAL CHARACTERISTICS

Cell Shape:

Arrangement:

Size:

Spores:

Gram's Stain:

Motility:

Capsules:

Special Stains:

CULTURAL CHARACTERISTICS

Colonies:

   Nutrient Agar:

   Blood Agar:

Agar Slant:

Nutrient Broth:

Gelatin Stab:

Oxygen Requirements:

Optimum Temp.:

PHYSIOLOGICAL CHARACTERISTICS

	TESTS	RESULTS
Fermentation	Glucose	
	Lactose	
	Sucrose	
	Mannitol	
Hydrolysis	Gelatin Liquefaction	
	Starch	
	Casein	
	Fat	
IMViC	Indole	
	Methyl Red	
	V-P (acetylmethylcarbinol)	
	Citrate Utilization	
	Nitrate Reduction	
	H_2S Production	
	Urease	
	Catalase	
	Oxidase	
	DNase	
	Phenylalanase	

	REACTION	TIME
Litmus Milk	Acid	_____
	Alkaline	_____
	Coagulation	_____
	Reduction	_____
	Peptonization	_____
	No Change	_____

407

Descriptive Chart

STUDENT: _____

LAB SECTION: _____

Habitat: _____ Culture No.: _____

Source: _____

Organism: _____

MORPHOLOGICAL CHARACTERISTICS

Cell Shape:

Arrangement:

Size:

Spores:

Gram's Stain:

Motility:

Capsules:

Special Stains:

CULTURAL CHARACTERISTICS

Colonies:

   Nutrient Agar:

   Blood Agar:

Agar Slant:

Nutrient Broth:

Gelatin Stab:

Oxygen Requirements:

Optimum Temp.:

PHYSIOLOGICAL CHARACTERISTICS

	TESTS	RESULTS
Fermentation	Glucose	
	Lactose	
	Sucrose	
	Mannitol	
Hydrolysis	Gelatin Liquefaction	
	Starch	
	Casein	
	Fat	
IMViC	Indole	
	Methyl Red	
	V-P (acetylmethylcarbinol)	
	Citrate Utilization	
	Nitrate Reduction	
	H_2S Production	
	Urease	
	Catalase	
	Oxidase	
	DNase	
	Phenylalanase	

	REACTION	TIME
Litmus Milk	Acid	_____
	Alkaline	_____
	Coagulation	_____
	Reduction	_____
	Peptonization	_____
	No Change	

Descriptive Chart

STUDENT: _____

LAB SECTION: _____

Habitat: _____ Culture No.: _____

Source: _____

Organism: _____

MORPHOLOGICAL CHARACTERISTICS

Cell Shape:

Arrangement:

Size:

Spores:

Gram's Stain:

Motility:

Capsules:

Special Stains:

CULTURAL CHARACTERISTICS

Colonies:

  Nutrient Agar:

  Blood Agar:

Agar Slant:

Nutrient Broth:

Gelatin Stab:

Oxygen Requirements:

Optimum Temp.:

PHYSIOLOGICAL CHARACTERISTICS

	TESTS	RESULTS
Fermentation	Glucose	
	Lactose	
	Sucrose	
	Mannitol	
Hydrolysis	Gelatin Liquefaction	
	Starch	
	Casein	
	Fat	
IMViC	Indole	
	Methyl Red	
	V-P (acetylmethylcarbinol)	
	Citrate Utilization	
	Nitrate Reduction	
	H_2S Production	
	Urease	
	Catalase	
	Oxidase	
	DNase	
	Phenylalanase	

	REACTION	TIME
Litmus Milk	Acid	_____
	Alkaline	_____
	Coagulation	_____
	Reduction	_____
	Peptonization	_____
	No Change	

Appendix
Tables

Table I International Atomic Weights

Element	Symbol	Atomic Number	Atomic Weight
Aluminum	Al	13	26.97
Antimony	Sb	51	121.76
Arsenic	As	33	74.91
Barium	Ba	56	137.36
Beryllium	Be	4	9.013
Bismuth	Bi	83	209.00
Boron	B	5	10.82
Bromine	Br	35	79.916
Cadmium	Cd	48	112.41
Calcium	Ca	20	40.08
Carbon	C	6	12.010
Chlorine	Cl	17	35.457
Chromium	Cr	24	52.01
Cobalt	Co	27	58.94
Copper	Cu	29	63.54
Fluorine	F	9	19.00
Gold	Au	79	197.2
Hydrogen	H	1	1.0080
Iodine	I	53	126.92
Iron	Fe	26	55.85
Lead	Pb	82	207.21
Magnesium	Mg	12	24.32
Manganese	Mn	25	54.93
Mercury	Hg	80	200.61
Nickel	Ni	28	58.69
Nitrogen	N	7	14.008
Oxygen	O	8	16.0000
Palladium	Pd	46	106.7
Phosphorus	P	15	30.98
Platinum	Pt	78	195.23
Potassium	K	19	39.096
Radium	Ra	88	226.05
Selenium	Se	34	78.96
Silicon	Si	14	28.06
Silver	Ag	47	107.880
Sodium	Na	11	22.997
Strontium	Sr	38	87.63
Sulfur	S	16	32.066
Tin	Sn	50	118.70
Titanium	Ti	22	47.90
Tungsten	W	74	183.92
Uranium	U	92	238.07
Vanadium	V	23	50.95
Zinc	Zn	30	65.38
Zirconium	Zr	40	91.22

Table II Four-Place Logarithms

N	0	1	2	3	4	5	6	7	8	9
10	0000	0043	0086	0128	0170	0212	0253	0294	0334	0374
11	0414	0453	0492	0531	0569	0607	0645	0682	0719	0755
12	0792	0828	0864	0899	0934	0969	1004	1038	1072	1106
13	1139	1173	1206	1239	1271	1303	1335	1367	1399	1430
14	1461	1492	1523	1553	1584	1614	1644	1673	1703	1732
15	1761	1790	1818	1847	1875	1903	1931	1959	1987	2014
16	2041	2068	2095	2122	2148	2175	2201	2227	2253	2279
17	2304	2330	2355	2380	2405	2430	2455	2480	2504	2529
18	2553	2577	2601	2625	2648	2672	2695	2718	2742	2765
19	2788	2810	2833	2856	2878	2900	2923	2945	2967	2989
20	3010	3032	3054	3075	3096	3118	3139	3160	3181	3201
21	3222	3243	3263	3284	3304	3324	3345	3365	3385	3404
22	3424	3444	3464	3483	3502	3522	3541	3560	3579	3598
23	3617	3636	3655	3674	3692	3711	3729	3747	3766	3784
24	3802	3820	3838	3856	3874	3892	3909	3927	3945	3962
25	3979	3997	4014	4031	4048	4065	4082	4099	4116	4133
26	4150	4166	4183	4200	4216	4232	4249	4265	4281	4298
27	4314	4330	4346	4362	4378	4393	4409	4425	4440	4456
28	4472	4487	4502	4518	4533	4548	4564	4579	4594	4609
29	4624	4639	4654	4669	4683	4698	4713	4728	4742	4757
30	4771	4786	4800	4814	4829	4843	4857	4871	4886	4900
31	4914	4928	4942	4955	4969	4983	4997	5011	5024	5038
32	5051	5065	5079	5092	5105	5119	5132	5145	5159	5172
33	5185	5198	5211	5224	5237	5250	5263	5276	5289	5302
34	5315	5328	5340	5353	5366	5378	5391	5403	5416	5428
35	5441	5453	5465	5478	5490	5502	5514	5527	5539	5551
36	5563	5575	5587	5599	5611	5623	5635	5647	5658	5670
37	5682	5694	5705	5717	5729	5740	5752	5763	5775	5786
38	5798	5809	5821	5832	5843	5855	5866	5877	5888	5899
39	5911	5922	5933	5944	5955	5966	5977	5988	5999	6010
40	6021	6031	6042	6053	6064	6075	6085	6096	6107	6117
41	6128	6138	6149	6160	6170	6180	6191	6201	6212	6222
42	6232	6243	6253	6263	6274	6284	6294	6304	6314	6325
43	6335	6345	6355	6365	6375	6385	6395	6405	6415	6425
44	6435	6444	6454	6464	6474	6484	6493	6503	6513	6522
45	6532	6542	6551	6561	6571	6580	6590	6599	6609	6618
46	6628	6637	6646	6656	6665	6675	6684	6693	6702	6712
47	6721	6730	6739	6749	6758	6767	6776	6785	6794	6803
48	6812	6821	6830	6839	6848	6857	6866	6875	6884	6893
49	6902	6911	6920	6928	6937	6946	6955	6964	6972	6981
50	6990	6998	7007	7016	7024	7033	7042	7050	7059	7067
51	7076	7084	7093	7101	7110	7118	7126	7135	7143	7152
52	7160	7168	7177	7185	7193	7202	7210	7218	7226	7235
53	7243	7251	7259	7267	7275	7284	7292	7300	7308	7316
54	7324	7332	7340	7348	7356	7364	7372	7380	7388	7396
N	0	1	2	3	4	5	6	7	8	9

Table II *(continued)*

N	0	1	2	3	4	5	6	7	8	9
55	7404	7412	7419	7427	7435	7443	7451	7459	7466	7474
56	7482	7490	7497	7505	7513	7520	7528	7536	7543	7551
57	7559	7566	7574	7582	7589	7597	7604	7612	7619	7627
58	7634	7642	7649	7657	7664	7672	7679	7686	7694	7701
59	7709	7716	7723	7731	7738	7745	7752	7760	7767	7774
60	7782	7789	7796	7803	7810	7818	7825	7832	7839	7846
61	7853	7860	7868	7875	7882	7889	7896	7903	7910	7917
62	7924	7931	7938	7945	7952	7959	7966	7973	7980	7987
63	7993	8000	8007	8014	8021	8028	8035	8041	8048	8055
64	8062	8069	8075	8082	8089	8096	8102	8109	8116	8122
65	8129	8136	8142	8149	8156	8162	8169	8176	8182	8189
66	8195	8202	8209	8215	8222	8228	8235	8241	8248	8254
67	8261	8267	8274	8280	8287	8293	8299	8306	8312	8319
68	8325	8331	8338	8344	8351	8357	8363	8370	8376	8382
69	8388	8395	8401	8407	8414	8420	8426	8432	8439	8445
70	8451	8457	8463	8470	8476	8482	8488	8494	8500	8506
71	8513	8519	8525	8531	8537	8543	8549	8555	8561	8567
72	8573	8579	8585	8591	8597	8603	8609	8615	8621	8627
73	8633	8639	8645	8651	8657	8663	8669	8675	8681	8686
74	8692	8698	8704	8710	8716	8722	8727	8733	8739	8745
75	8751	8756	8762	8768	8774	8779	8785	8791	8797	8802
76	8808	8814	8820	8825	8831	8837	8842	8848	8854	8859
77	8865	8871	8876	8882	8887	8893	8899	8904	8910	8915
78	8921	8927	8932	8938	8943	8949	8954	8960	8965	8971
79	8976	8982	8987	8993	8998	9004	9009	9015	9020	9025
80	9031	9036	9042	9047	9053	9058	9063	9069	9074	9079
81	9085	9090	9096	9101	9106	9112	9117	9122	9128	9133
82	9138	9143	9149	9154	9159	9165	9170	9175	9180	9186
83	9191	9196	9201	9206	9212	9217	9222	9227	9232	9238
84	9243	9248	9253	9258	9263	9269	9274	9279	9284	9289
85	9294	9299	9304	9309	9315	9320	9325	9330	9335	9340
86	9345	9350	9355	9360	9365	9370	9375	9380	9385	9390
87	9395	9400	9405	9410	9415	9420	9425	9430	9435	9440
88	9445	9450	9455	9460	9465	9469	9474	9479	9484	9489
89	9494	9499	9504	9509	9513	9518	9523	9528	9533	9538
90	9542	9547	9552	9557	9562	9566	9571	9576	9581	9586
91	9590	9595	9600	9605	9609	9614	9619	9624	9628	9633
92	9638	9643	9647	9652	9657	9661	9666	9671	9675	9680
93	9685	9689	9694	9699	9703	9708	9713	9717	9722	9727
94	9731	9736	9741	9745	9750	9754	9759	9763	9768	9773
95	9777	9782	9786	9791	9795	9800	9805	9809	9814	9818
96	9823	9827	9832	9836	9841	9845	9850	9854	9859	9863
97	9868	9872	9877	9881	9886	9890	9894	9899	9903	9908
98	9912	9917	9921	9926	9930	9934	9939	9943	9948	9952
99	9956	9961	9965	9969	9974	9978	9983	9987	9991	9996
100	0000	0004	0009	0013	0017	0022	0026	0030	0035	0039
N	0	1	2	3	4	5	6	7	8	9

Table III Temperature Conversion Table Centigrade to Fahrenheit

°C	0	1	2	3	4	5	6	7	8	9
−50	**−58.0**	**−59.8**	**−61.6**	**−63.4**	**−65.2**	**−67.0**	**−68.8**	**−70.6**	**−72.4**	**−74.2**
−40	−40.0	−41.8	−43.6	−45.4	−47.2	−49.0	−50.8	−52.6	−54.4	−56.2
−30	−22.0	−23.8	−25.6	−27.4	−29.2	−31.0	−32.8	−34.6	−36.4	−38.2
−20	− 4.0	− 5.8	− 7.6	− 9.4	−11.2.	−13.0	−14.8	−16.6	−18.4	−20.2
−10	+14.0	+12.2	+10.4	+ 8.6	+ 6.8	+ 5.0	+ 3.2	+ 1.4	− 0.4	− 2.2
− 0	+32.0	+30.2	+28.4	+26.6	+24.8	+23.0	+21.2	+19.4	+17.6	+15.8
0	**32.0**	**33.8**	**35.6**	**37.4**	**39.2**	**41.0**	**42.8**	**44.6**	**46.4**	**48.2**
10	50.0	51.8	53.6	55.4	57.2	59.0	60.8	62.6	64.4	66.2
20	68.0	69.8	71.6	73.4	75.2	77.0	78.8	80.6	82.4	84.2
30	86.0	87.8	89.6	91.4	93.2	95.0	96.8	98.6	100.4	102.2
40	104.0	105.8	107.6	109.4	111.2	113.0	114.8	116.6	118.4	120.2
50	122.0	123.8	125.6	127.4	129.2	131.0	132.8	134.6	136.4	138.2
60	**140.0**	**141.8**	**143.6**	**145.4**	**147.2**	**149.0**	**150.8**	**152.6**	**154.4**	**156.2**
70	158.0	159.8	161.6	163.4	165.2	167.0	168.8	170.6	172.4	174.2
80	176.0	177.8	179.6	181.4	183.2	185.0	186.8	188.6	190.4	192.2
90	194.0	195.8	197.6	199.4	201.2	203.0	204.8	206.6	208.4	210.2
100	212.0	213.8	215.6	217.4	219.2	221.0	222.8	224.6	226.4	228.2
110	**230.0**	**231.8**	**233.6**	**235.4**	**237.2**	**239.0**	**240.8**	**242.6**	**244.4**	**246.2**
120	248.0	249.8	251.6	253.4	255.2	257.0	258.8	260.6	262.4	264.2
130	266.0	267.8	269.6	271.4	273.2	275.0	276.8	278.6	280.4	282.2
140	284.0	285.8	287.6	289.4	291.2	293.0	294.8	296.6	298.4	300.2
150	302.0	303.8	305.6	307.4	309.2	311.0	312.8	314.6	316.4	318.2
160	**320.0**	**321.8**	**323.6**	**325.4**	**327.2**	**329.0**	**330.8**	**332.6**	**334.4**	**336.2**
170	338.0	339.8	341.6	343.4	345.2	347.0	348.8	350.6	352.4	354.2
180	356.0	357.8	359.6	361.4	363.2	365.0	366.8	368.6	370.4	372.2
190	374.0	375.8	377.6	379.4	381.2	383.0	384.8	386.6	388.4	390.2
200	392.0	393.8	395.6	397.4	399.2	401.0	402.8	404.6	406.4	408.2
210	**410.0**	**411.8**	**413.6**	**415.4**	**417.2**	**419.0**	**420.8**	**422.6**	**424.4**	**426.2**
220	428.0	429.8	431.6	433.4	435.2	437.0	438.8	440.6	442.4	444.2
230	446.0	447.8	449.6	451.4	453.2	455.0	456.8	458.6	460.4	462.2
240	464.0	465.8	467.6	469.4	471.2	473.0	474.8	476.6	478.4	480.2
250	482.0	483.8	485.6	487.4	489.2	491.0	492.8	494.6	496.4	498.2

$$°F = °C \times 9/5 + 32 \qquad °C = °F - 32 \times 5/9$$

Table IV Autoclave Steam Pressures and Corresponding Temperatures

Steam Pressure lb/sq in	Temperature °C	Temperature °F	Steam Pressure lb/sq in	Temperature °C	Temperature °F	Steam Pressure lb/sq in	Temperature °C	Temperature °F
0	100.0	212.0						
1	101.9	215.4	11	116.4	241.5	21	126.9	260.4
2	103.6	218.5	12	117.6	243.7	22	127.8	262.0
3	105.3	221.5	13	118.8	245.8	23	128.7	263.7
4	106.9	224.4	14	119.9	247.8	24	129.6	265.3
5	108.4	227.1	15	121.0	249.8	25	130.4	266.7
6	109.8	229.6	16	122.0	251.6	26	131.3	268.3
7	111.3	232.3	17	123.0	253.4	27	132.1	269.8
8	112.6	234.7	18	124.1	255.4	28	132.9	271.2
9	113.9	237.0	19	125.0	257.0	29	133.7	272.7
10	115.2	239.4	20	126.0	258.8	30	134.5	274.1

Figures are for steam pressure only and the presence of any air in the autoclave invalidates temperature readings from the above table.

Table V Autoclave Temperatures as Related to the Presence of Air

Gauge Pressure, lb	Pure steam, complete air discharge		Two-thirds air discharge, 20-in. vacuum		One-half air discharge, 15-in. vacuum		One-third air discharge, 10-in. vacuum		No air discharge	
	°C	°F	°C	°F	°C	°F	°C	°F	°C	°F
5	109	228	100	212	94	202	90	193	72	162
10	115	240	109	228	105	220	100	212	90	193
15	121	250	115	240	112	234	109	228	100	212
20	126	259	121	250	118	245	115	240	109	228
25	130	267	126	259	124	254	121	250	115	240
30	135	275	130	267	128	263	126	259	121	250

Table VI MPN Determination from Multiple Tube Test

NUMBER OF TUBES GIVING POSITIVE REACTION OUT OF			MPN Index per 100 ml	95 PERCENT CONFIDENCE LIMITS	
3 of 10 ml each	3 of 1 ml each	3 of 0.1 ml each		Lower	Upper
0	0	1	3	<0.5	9
0	1	0	3	<0.5	13
1	0	0	4	<0.5	20
1	0	1	7	1	21
1	1	0	7	1	23
1	1	1	11	3	36
1	2	0	11	3	36
2	0	0	9	1	36
2	0	1	14	3	37
2	1	0	15	3	44
2	1	1	20	7	89
2	2	0	21	4	47
2	2	1	28	10	150
3	0	0	23	4	120
3	0	1	39	7	130
3	0	2	64	15	380
3	1	0	43	7	210
3	1	1	75	14	230
3	1	2	120	30	380
3	2	0	93	15	380
3	2	1	150	30	440
3	2	2	210	35	470
3	3	0	240	36	1,300
3	3	1	460	71	2,400
3	3	2	1,100	150	4,800

From *Standard Methods for the Examination of Water and Wastewater,* Twelfth edition. (New York: The American Public Health Association, Inc.), p. 608.

Table VII Significance of zones of inhibition in Kirby-Bauer Method of antimicrobic sensitivity testing (1995)

Antimicrobial Agent	Disk Potency	R Resistant mm	I Intermediate mm	S Sensitive mm
Amikacin	30 mcg	<14	15–16	>17
Amoxicillin/Clavulinic Acid	30 mcg			
Staphylococci		<19	14–17	>20
Other gram-positive organisms		<13	14–17	>18
Ampicillin	75 mcg			
Gram-negative enterics		<13	14–16	>17
Staphylococci		<28		>29
Enterococci		<16		>17
Streptococci (not *S. pneumoniae*)		<21	22–29	>30
Haemophilus spp.		<18	19–21	>22
Listeria monocytogenes		<19		>20
Azlocillin (*Pseudomonas aeruginosa*)	75 mcg	<17		>18
Carbenicillin (*P. aeruginosa*)	100 mcg	<13	14–16	>17
Other gram-negative organisms		<19	20–22	>23
Cefaclor	30 mcg	<14	15–17	>18
Cephalothin	30 mcg	<14	15–17	>18
Chloramphenicol	30 mcg	<12	13–17	>18
S. pneumoniae		<20		>21
Clarithromycin	15 mcg	<13	14–17	>18
S. pneumoniae		<16	17–20	>21
Clindamycin	2 mcg	<14	15–20	>21
S. pneumoniae		<20		>21
Erythromycin	15 mcg	<13	14–22	>23
S. pneumoniae		<15	16–20	>21
Gentamicin	10 mcg	<12	13–14	>15
Impenem	10 mcg	<13	14–15	>16
Haemophilus spp.				>16
Kanamycin	30 mcg	<13	14–17	>18
Lomefloxacin	10 mcg	<18	19–21	>22
Loracarbef	30 mcg	<14	15–17	>18
Mezlocillin (*P. aeruginosa*)	75 mcg	<15		>16
Other gram-negative organisms		<17	18–20	>19
Minocycline	30 mcg	<14	15–18	>19
Moxalactam	30 mcg	<14	15–22	>23
Nafcillin	1 mcg	<10	11–12	>13
Nalidixic Acid	30 mcg	<13	14–18	>19
Netilmicin	30 mcg	<12	13–14	>15
Norfloxacin	10 mcg	<12	13–16	>17
Ofloxacin	5 mcg	<12	13–15	>16
Penicillin G (Staphylococci)	10 units	<28		>29
Enterococci		<14		>15
Streptococci (not *S. pneumoniae*)		<19	20–27	>28
Neisseria gonorrhoeae		<26	27–46	>47
L. monocytogenes		<19		>20
Piperacillin/Tazobactam	100/10 mcg			
Staphylococci		<17		>18
P. aeruginosa		<17		>18
Other gram-negative organisms		<14	15–19	>20
Rifampin	5 mcg	<16	17–19	>20
Haemophilus spp.		<16	17–19	>20
S. pneumoniae		<16	17–18	>19
Streptomycin	10 mcg	<11	12–14	>15
Sulfisoxazole	300 mcg	<12	13–16	>17
Tetracycline	30 mcg	<14	15–18	>19
S. pneumoniae		<17	18–21	>22
Tobramycin	10 mcg	<12	13–14	>15
Trimethoprim/Sulfamethoxazole	1.25/23.75	<10	11–15	>16
Vancomycin	30 mcg	<14	15–16	>17

Table VIII Indicators of Hydrogen Ion Concentration

Many of the following indicators are used in the media of certain exercises in this manual. This table indicates the pH range of each indicator and the color changes that occur. To determine the exact pH within a particular range one should use a set of standard colorimetric tubes that are available from the prep room. Consult your lab instructor.

Indicator	Full Acid Color	Full Alkaline Color	pH Range
Cresol Red	red	yellow	0.2 – 1.8
Metacresol Purple (acid range)	red	yellow	1.2 – 2.8
Thymol Blue	red	yellow	1.2 – 2.8
Bromphenol Blue	yellow	blue	3.0 – 4.6
Bromcresol Green	yellow	blue	3.8 – 5.4
Chlorcresol Green	yellow	blue	4.0 – 5.6
Methyl Red	red	yellow	4.4 – 6.4
Chlorphenol Red	yellow	red	4.8 – 6.4
Bromcresol Purple	yellow	purple	5.2 – 6.8
Bromthymol Blue	yellow	blue	6.0 – 7.6
Neutral Red	red	amber	6.8 – 8.0
Phenol Red	yellow	red	6.8 – 8.4
Cresol Red	yellow	red	7.2 – 8.8
Metacresol Purple (alkaline range)	yellow	purple	7.4 – 9.0
Thymol Blue (alkaline range)	yellow	blue	8.0 – 9.6
Cresolphthalein	colorless	red	8.2 – 9.8
Phenolphthalein	colorless	red	8.3 – 10.0

Indicators

All the indicators used in this manual can be made by (1) dissolving a measured amount of the indicator in 95% ethanol, (2) adding a measured amount of water, and (3) filtering with filter paper. The following chart provides the correct amounts of indicator, alcohol, and water for various indicator solutions.

Indicator Solution	Indicator (gm)	95% Ethanol (ml)	Distilled H$_2$O (ml)
Bromcresol green	0.4	500	500
Bromcresol purple	0.4	500	500
Bromthymol blue	0.4	500	500
Cresol red	0.2	500	500
Methyl red	0.2	500	500
Phenolphthalein	1.0	50	50
Phenol red	0.2	500	500
Thymol blue	0.4	500	500

Stains and Reagents

Acid-Alcohol (for Ziehl-Neelsen stain)

3 ml concentrated hydrochloric acid in 100 ml of 95% ethyl alcohol.

Acid-Alcohol (for fluorochrome staining)

HCl . . . . .2.5 ml
Ethyl alcohol, 70% . . . . .500.0 ml
NaCl . . . . .2.5 gm

Alcohol, 70% (from 95%)

Alcohol, 95% . . . . .368.0 ml
Distilled water .132.0 ml

Auramine-Rhodamine Stain
(for mycobacteria)

Auramine 0 . . . . .3.0 gm
Rhodamine B . . . . .1.5 gm
Glycerol . . . . .150.0 ml
Phenol crystals (liquefied at 50° C) . . . .20.0 ml
Distilled water .100.0 ml

Clarify by filtration with glass wool or Whatman #2 filter paper. Do not heat. Store at room temperature.

Barritt's Reagent (Voges-Proskauer test)

Solution A: 6 gm alpha-naphthol in 100 ml 95% ethyl alcohol.

Solution B: 16 gm potassium hydroxide in 100 ml water.

Note that no creatine is used in these reagents as is used in O'Meara's reagent for the V-P test.

Carbolfuchsin Stain (Ziehl's)

Solution A: Dissolve 0.3 gm of basic fuchsin (90% dye content) in 10 ml 95% ethyl alcohol.

Solution B: Dissolve 5 gm of phenol in 95 ml of water.

Mix solutions A and B.

Crystal Violet Stain (Hucker modification)

Solution A: Dissolve 2.0 gm of crystal violet (85% dye content) in 20 ml of 95% ethyl alcohol.

Solution B: Dissolve 0.8 gm ammonium oxalate in 80.0 ml distilled water.

Mix solutions A and B.

Diphenylamine Reagent (nitrate test)

Dissolve 0.7 gm diphenylamine in a mixture of 60 ml of concentrated sulfuric acid and 28.8 ml of distilled water.

Cool and add slowly 11.3 ml of concentrated hydrochloric acid. After the solution has stood for 12 hours some of the base separates, showing that the reagent is saturated.

Ferric Chloride Reagent (Ex. 78)

$FeCl_3 \cdot 6H_2O$ ....12 gm
2% Aqueous HCl....................100 ml

Make up the 2% aq. HCl by adding 5.4 ml of concentrated HCl (37%) to 94.6 ml H_2O. Inoculate with two or three colonies of beta hemolytic streptococci, incubate at 35° C for 20 or more hours. Centrifuge the medium to pack the cells, and pipette 0.8 ml of the clear supernate into a Kahn tube. Add 0.2 ml of the ferric chloride reagent to the Kahn tube and mix well. If a heavy precipitate remains longer than 10 minutes, the test is positive.

Gram's Iodine (Lugol's)

Dissolve 2.0 gm of potassium iodide in 300 ml of distilled water and then add 1.0 gm iodine crystals.

Iodine, 5% Aqueous Solution (Ex. 42)

Dissolve 4 gm of potassium iodide in 300 ml of distilled water and then add 2.0 gm iodine crystals.

Kovacs' Reagent (indole test)

n-amyl alcohol.....................75.0 ml
Hydrochloric acid (conc.) ....25.0 ml
ρ-dimethylamine-benzaldehyde ....5.0 gm

Lactophenol Cotton Blue Stain

Phenol crystals.......................20 gm
Lactic acid ....20 ml
Glycerol ....40 ml
Cotton blue ....0.05 gm

Dissolve the phenol crystals in the other ingredients by heating the mixture gently under a hot water tap.

Leifson's Flagellar Stain

Prepare three separate solutions as follows:

Solution A: Dissolve 0.9 gm of pararosaniline acetate and 0.3 gm of pararosaniline hydro-

chloride in 100 ml of 95% ethyl alcohol. Let stand overnight at room temperature to complete dissolution.

Solution B: Dissolve 3 gm of tannic acid in 100 ml of distilled water.

Solution C: Dissolve 1.5 gm of sodium chloride in 100 ml of distilled water.

Mix equal volumes of solutions A, B, and C and let stand for 2 hours. Store in a stoppered bottle in a refrigerator (up to 2 months). Disregard precipitate that forms in bottom of bottle. Do not filter. Will store indefinitely, if frozen. Frozen stain solution must be thoroughly mixed after thawing since the water separates from the alcohol. After mixing, the precipitate should be allowed to settle to the bottom.

Note: Pararosaniline compounds should be certified for flagellar staining.

Malachite Green Solution (spore stain)

Dissolve 5.0 gm malachite green oxalate in 100 ml distilled water.

McFarland Nephelometer Barium Sulfate Standards (Ex. 55)

Prepare 1% aqueous barium chloride and 1% aqueous sulfuric acid solutions.

Add the amounts indicated in table 1 to clean, dry ampoules. Ampoules should have the same diameter as the test tube to be used in subsequent density determinations.

Seal the ampoules and label them.

Table 1 Amounts for Standards

Tube	Barium Chloride 1% (ml)	Sulfuric Acid 1% (ml)	Corresponding Approx. Density of Bacteria (million/ml)
1	0.1	9.9	300
2	0.2	9.8	600
3	0.3	9.7	900
4	0.4	9.6	1200
5	0.5	9.5	1500
6	0.6	9.4	1800
7	0.7	9.3	2100
8	0.8	9.2	2400
9	0.9	9.1	2700
10	1.0	9.0	3000

Methylene Blue (Loeffler's)

Solution A: Dissolve 0.3 gm of methylene blue (90% dye content) in 30.0 ml ethyl alcohol (95%).

Solution B: Dissolve 0.01 gm potassium hydroxide in 100.0 ml distilled water. Mix solutions A and B.

Naphthol, alpha

5% alpha-naphthol in 95% ethyl alcohol
Caution: Avoid all contact with human tissues. Alpha-naphthol is considered to be carcinogenic.

Nessler's Reagent (ammonia test)

Dissolve about 50 gm of potassium iodide in 35 ml of cold ammonia-free distilled water. Add a saturated solution of mercuric chloride until a slight precipitate persists. Add 400 ml of a 50% solution of potassium hydroxide. Dilute to 1 liter, allow to settle, and decant the supernatant for use.

Nigrosine Solution (Dorner's)

Nigrosine, water soluble . . . . .10 gm
Distilled water. . . . . .100 ml

Boil for 30 minutes. Add as a preservative 0.5 ml formaldehyde (40%). Filter twice through double filter paper and store under aseptic conditions.

Nitrate Test Reagent
(see Diphenylamine)

Nitrite Test Reagents

Solution A: Dissolve 8 gm sulfanilic acid in 1000 ml 5N acetic acid (1 part glacial acetic acid to 2.5 parts water).

Solution B: Dissolve 5 gm dimethyl-alpha-naphthylamine in 1000 ml 5N acetic acid. Do not mix solutions.

Caution: Although at this time it is not known for sure, there is a possibility that dimethyl-α-naphthylamine in solution B may be carcinogenic. For reasons of safety, avoid all contact with tissues.

Oxidase Test Reagent

Mix 1.0 gm of dimethyl-ρ-phenylenediamine hydrochloride in 100 ml of distilled water.

Preferably, the reagent should be made up fresh, daily. It should not be stored longer than one week in the refrigerator. Tetramethyl-ρ-phenylenediamine dihydrochloride (1%) is even more sensitive, but is considerably more expensive and more difficult to obtain.

Phenolized Saline

Dissolve 8.5 gm sodium chloride and 5.0 gm phenol in 1 liter distilled water.

Physiological Saline

Dissolve 8.5 gm sodium chloride in 1 liter distilled water.

Potassium permanganate (for fluorochrome staining)

$KMnO_4$. . . . . .2.5 gm
Distilled water . . . . .500.0 ml

Safranin (for gram staining)

Safranin O (2.5% sol'n in 95% ethyl alcohol) . . . . .10.0 ml
Distilled water . . . . .100.0 ml

Trommsdorf's Reagent (nitrite test)

Add slowly, with constant stirring, 100 ml of a 20% aqueous zinc chloride solution to a mixture of 4.0 gm of starch in water. Continue heating until the starch is dissolved as much as possible, and the solution is nearly clear. Dilute with water and add 2 gm of potassium iodide. Dilute to 1 liter, filter, and store in amber bottle.

Vaspar

Melt together 1 pound of Vaseline and 1 pound of paraffin. Store in small bottles for student use.

Voges-Proskauer Test Reagent (see Barritt's)

White Blood Cell (WBC) Diluting Fluid

Hydrochloric acid . . . . .5 ml
Distilled water . . . . .495 ml

Add 2 small crystals of thymol as a preservative.

Conventional Media   The following media are used in the experiments of this manual. All of these media are available in dehydrated form from either Difco Laboratories, Detroit, Michigan, or Baltimore Biological Laboratory (BBL), a division of Becton, Dickinson & Co., Cockeysville, Maryland. Compositions, methods of preparation, and usage will be found in their manuals, which are supplied upon request at no cost. The source of each medium is designated as (B) for BBL and (D) for Difco.

Bile esculin (D)
Brewer's anaerobic agar (D)
Desoxycholate citrate agar (B,D)
Desoxycholate lactose agar (B,D)
DNase test agar (B,D)
Endo agar (B,D)
Eugonagar (B,D)
Fluid thioglycollate medium (B,D)
Heart infusion agar (D)
Hektoen Enteric Agar (B,D)
Kligler iron agar (B,D)
Lead acetate agar (D)
Levine EMB agar (B,D)
Lipase reagent (D)
Litmus milk (B,D)
Lowenstein-Jensen medium (B,D)
MacConkey Agar (B,D)
Mannitol salt agar (B,D)
MR-VP medium (D)
Mueller-Hinton medium (B,D)
Nitrate broth (D)
Nutrient agar (B,D)

Nutrient broth (B,D)
Nutrient gelatin (B,D)
Phenol red sucrose broth (B,D)
Phenylalanine agar (D)
Phenylethyl alcohol medium (B)
Russell double sugar agar (B,D)
Sabouraud's glucose (dextrose) agar (D)
Semisolid medium (B)
Simmons citrate agar (B,D)
Snyder test agar (D)
Sodium hippurate (D)
Spirit blue agar (D)
SS agar (B,D)
m-Staphylococcus broth (D)
Staphylococcus medium 110 (D)
Starch agar (D)
Trypticase soy agar (B)
Trypticase soy broth (B)
Tryptone glucose extract agar (B,D)
Urea (urease test) broth (B,D)
Veal infusion agar (B,D)
Xylose Lysine Desoxycholate Agar (B,D)

Special Media   The following media are not included in the manuals that are supplied by Difco and BBL; therefore, methods of preparation are presented here.

Ammonium Medium (for *Nitrosomonas*)

$(NH_4)_2SO_4$ . . . . .2.0 gm
$MgSO_4 \cdot 7H_2O$ . . . . .0.5 gm
$FeSO_4 \cdot 7H_2O$ . . . . .0.03 gm
NaCl . . . . .0.3 gm
$MgCO_3$ . . . . .10.0 gm
K_2HPO_4 . . . . .1.0 gm
Water . . . . .1000.0 ml

For the isolation of *Nitrosomonas* from soil, sterilization is not necessary if the inoculations are made as soon as the medium is made up. The pH should be adjusted to 7.3. If sterilization is desirable for storage or other reasons, adjust the pH aseptically after sterilization with sterile 1N HCl.

Bile Esculin Slants (Ex. 78)

Heart infusion agar . . . . .40.0 gm
Esculin . . . . .1.0 gm
Ferric chloride . . . . .0.5 gm
Distilled water . . . . .1000.0 ml
Dispense into sterile 15 × 125 mm screw-capped tubes, sterilize in autoclave at 121° C for 15 minutes, and slant during cooling.

Blood Agar

Trypticase soy agar powder . . . . .40 gm
Distilled water . . . . .1000 ml
    Final pH of 7.3

Defibrinated sheep or rabbit blood . . . . . . .50 ml

Liquefy and sterilize 1000 ml of trypticase soy agar in a large Erlenmeyer flask. While the TSA is being sterilized, warm up 50 ml of defibrinated blood to 50° C. After cooling the TSA to 50° C, aseptically transfer the blood to the flask and mix by gently rotating the flask (cold blood many cause lumpiness).

Pour 10–12 ml of the mixture into sterile Petri plates. If bubbles form on the surface of the medium, flame the surface gently with a Bunsen burner before the medium solidifies. It is best to have an assistant to lift off the Petri plate lids while pouring the medium into the plates. A full flask of blood agar is somewhat cumbersome to handle with one hand.

Bromthymol Blue Carbohydrate Broths

Make up stock indicator solution:
Bromthymol blue......................8 gm
95% ethyl alcohol ....250 ml
Distilled water.......................250 ml

Indicator is dissolved first in alcohol and then water is added.

Make up broth:
Sugar base (lactose, sucrose, glucose, etc.) ..5 gm
Tryptone ....10 gm
Yeast extract.........................5 gm
Indicator solution ....2 ml
Distilled water.....................1000 ml

Final pH 7.0

Chlorobium Agar Shake Deeps (Ex. 27)

The incompatibility of the ingredients of this medium at autoclave temperatures requires that the medium be made up in five different portions that can be sterilized separately and mixed at a lower temperature. Ideally, the medium is made up the day it is to be used and kept liquid in a water bath until the inoculations are made.

Solution A
(make in 2000 ml flask)

NH_4Cl...............................1.0 gm
KH_2PO_4.............................1.0 gm
$MgCl_2$...............................0.5 gm
Agar ....15.0 gm
Water.............................700.0 ml

Solution B

$NaHCO_3$.............................2.0 gm
Water.............................100.0 ml

Solution C

$Na_2S \cdot 9H_2O$......................1.0 gm
Water.............................100.0 ml

Solution D

$FeCl_3 \cdot 6H_2O$....................5.0 mg
Water.............................100.0 ml

Solution E
(1N H_3PO_4)

After sterilizing all five solutions in the autoclave at 121° C for 15 minutes, cool them to 50° C and pour solutions B, C, and D into solution A flask. Use solution E to adjust the pH of the medium to 7.3.

Dispense aseptically to sterile 16 mm dia. soft glass test tubes (10 ml per tube). Keep the tubes of media liquefied in a 50° C water bath until inoculations are completed.

Chlorobium Enrichment Medium (Ex. 27)

Due to the incompatibility of the ingredients of this medium at high temperatures, it is best to prepare it without sterilization just before it is to be used. Its unstable nature also precludes storage of the medium for any length of time in the refrigerator.

NH_4Cl...............................1.0 gm
KH_2PO_4.............................1.0 gm
$MgCl_2$...............................0.5 gm
$NaHCO_3$.............................2.0 gm
$Na_2S \cdot 9H_2O$......................1.0 gm
$FeCl_3 \cdot 6H_2O$..................0.0005 gm
Water ....1000 ml

Chromatium Agar Shake Deeps (Ex. 27)

This medium is similar in composition to that used for chlorobium; thus, the problems are similar. The medium must be made up in three different portions that can be sterilized separately and mixed at a lower temperature. It is best to make up the medium the same day it is to be used. It must be kept liquid in a water bath until the inoculations are made.

Solution A
(make in 2000 ml flask)

NH_4Cl...............................0.1 gm
KH_2PO_4.............................0.1 gm
$MgCl_2$..............................0.05 gm
Agar ....20.0 gm
Water.............................780.0 ml

Solution B

NaHCO$_3$.............................2.0 gm
Water..............................100.0 ml

Solution C

Na$_2$S · 9H$_2$O........................1.0 gm
Water..............................100.0 ml

After sterilizing each of the three solutions in the autoclave at 121° C for 15 minutes, cool them to 50° C and pour solutions B and C into flask A.

Dispense aseptically to sterile 16 mm dia. soft glass test tubes (10 ml per tube). Keep the tubes of media liquefied in a 50° C water bath until inoculations are completed.

Chromatium Enrichment Medium
(Ex. 27)

This medium is similar in composition to the enrichment medium used for chlorobium; thus, the problems are similar. It is best to prepare this medium without sterilization and use it immediately. It does not store well, even in the refrigerator, because of its unstable nature.

NH$_4$Cl..............................0.1 gm
KH$_2$PO$_4$............................0.1 gm
MgCl$_2$..............................0.05 gm
NaHCO$_3$.............................2.0 gm
Na$_2$S · 9H$_2$O........................1.0 gm
Water..............................1000 ml

Deca-Strength Phage Broth (Ex. 28)

Peptone...........................100 gm
Yeast extract......................50 gm
NaCl...............................25 gm
K$_2$HPO$_4$...........................80 gm
Distilled water....................1000 ml

Final pH 7.6

Emmons' Culture Medium for Fungi

C. W. Emmons developed the following recipe as an improvement over Sabouraud's glucose agar for the cultivation of fungi. Its principal advantage is that a neutral pH does not inhibit certain molds that have difficulty growing on Sabouraud's agar (pH 5.6). Instead of relying on a low pH to inhibit bacteria, it contains chloramphenicol, which does not adversely affect the fungi.

Glucose............................20 gm
Neopeptone.........................10 gm
Agar...............................20 gm
Chloramphenicol....................40 mg
Distilled water....................1000 ml

After the glucose, peptone, and agar are dissolved, heat to boiling, add the chloramphenicol which has been suspended in 10 ml of 95% alcohol and remove quickly from the heat. Autoclave for only 10 minutes.

Glucose–Minimal Salts Agar
(Ex. 74, Ames test)

This medium is made from glucose, agar, and Vogel-Bonner medium E (50×).

Vogel-Bonner Medium E (50×)

Distilled water (45° C)...............670 ml
MgSO$_4$ · 7H$_2$O.......................10 gm
Citric acid monohydrate..............100 gm
K$_2$HPO$_4$ (anhydrous)................500 gm
Sodium ammonium phosphate
    (NaHNH$_4$PO$_4$ · 4H$_2$O)............175 gm

Add salts in the order indicated to warm water (45° C) in a 2-liter beaker or flask placed on a magnetic stirring hot plate. Allow each salt to dissolve completely before adding the next. Adjust the volume to 1 liter. Distribute into two 1-liter glass bottles. Autoclave, loosely capped, for 20 minutes at 121° C.

Plates of Glucose-Minimal Salts Agar

Agar...............................15 gm
Distilled water....................930 ml
50× V-B salts......................20 ml
40% glucose........................50 ml

Add 15 gm of agar to 930 ml of distilled water in a 2-liter flask. Autoclave for 20 minutes using slow exhaust. When the solution has cooled slightly, add 20 ml of sterile 50× V-B salts and 50 ml of sterile 40% glucose. For mixing, a large magnetic stir bar can be added to the flask before autoclaving. After all the ingredients have been added, the solution should be stirred thoroughly. Pour 30 ml into each Petri plate. **Important:** The 50× V-B salts and 40% glucose should be autoclaved separately.

Glucose Peptone Acid Agar

Glucose............................10 gm
Peptone............................5 gm
Monopotassium phosphate............1 gm
Magnesium sulfate (MgSO$_4$ · 7H$_2$O)....0.5 gm
Agar...............................15 gm
Water..............................1000 ml

While still liquid after sterilization, add sufficient sulfuric acid to bring the pH down to 4.0.

Glycerol Yeast Extract Agar

Glycerol . . . . .5 ml
Yeast extract . . . . .2 gm
Dipotassium phosphate . . . . .1 gm
Agar . . . . .15 gm
Water . . . . .1000 ml

m Endo MF Broth (Ex. 62)

This medium is extremely hygroscopic in the dehydrated form and oxidizes quickly to cause deterioration of the medium after the bottle has been opened. Once a bottle has been opened it should be dated and discarded after one year. If the medium becomes hardened within that time it should be discarded. Storage of the bottle inside a larger bottle that contains silica gel will extend shelf life.

Failure of Exercise 62 can often be attributed to faulty preparation of the medium. It is best to make up the medium the day it is to be used. It should not be stored over 96 hours prior to use. The Millipore Corporation recommends the following method for preparing this medium. (These steps are not exactly as stated in the Millipore Application Manual AM302.)

1. Into a 250 ml screw-cap Erlenmeyer flask place the following:

   Distilled water . . . . .50 ml
   95% ethyl alcohol . . . . .2 ml
   Dehydrated medium (*m* Endo MF
      broth) . . . . .4.8 gms

   Shake the above mixture by swirling the flask until the medium is dissolved and then add another 50 ml of distilled water.
2. Cap the flask loosely and immerse it into a pan of boiling water. As soon as the medium begins to simmer, remove the flask from the water bath. Do not boil the medium any further.
3. Cool the medium to 45° C, and adjust the pH to between 7.1 and 7.3.
4. If the medium must be stored for a few days, place it in the refrigerator at 2°–10° C, with screw-cap tightened securely.

Milk Salt Agar (15% NaCl)

Prepare three separate beakers of the following ingredients:

1. Beaker containing 200 grams of sodium chloride.

2. Large beaker (2000 ml size) containing 50 grams of skim milk powder in 500 ml of distilled water.
3. Glycerol-peptone agar medium:

   $MgSO_4 \cdot 7H_2O$ .5.0 gm
   $MgNO_3 \cdot 6H_2O$ .1.0 gm
   $FeCl_3 \cdot 7H_2O$ .0.025 gm
   Difco proteose-peptone #3 . . . . .5.0 gm
   Glycerol . . . . .10.0 gm
   Agar . . . . .30.0 gm
   Distilled water . . . . .500.0 ml

Sterilize the above three beakers separately. The milk solution should be sterilized at 113°–115° C (8 lb pressure) in autoclave for 20 minutes. The salt and glycerol-peptone agar can be sterilized at conventional pressure and temperature. After the milk solution has cooled to 55° C, add the sterile salt, which should also be cooled down to a moderate temperature. If the salt is too hot, coagulation may occur. Combine the milk-salt and glycerol-peptone agar solutions by gently swirling with a glass rod. Dispense aseptically into Petri plates.

Minimal Agar (Ex. 75)

Glucose . . . . .2 gm
$(NH_4)_2SO_4$ . . . . .1 gm
K_2HPO_4 . . . . .7 gm
$MgSO_4$ . . . . .0.5 gm
Agar . . . . .15 gm
Distilled water . . . . .1000 ml

Nitrite Medium (for *Nitrobacter*)

$NaNO_2$ . . . . .1.0 gm
$MgSO_4 \cdot 7H_2O$ .0.5 gm
$FeSO_4 \cdot 7H_2O$ .0.03 gm
NaCl . . . . .0.3 gm
Na_2CO_3 . . . . .1.0 gm
K_2HPO_4 . . . . .1.0 gm
Water . . . . .1000.0 ml

For the isolation of *Nitrobacter* from soil, sterilization is not necessary if the inoculations are made as soon as the medium is made up. The pH should be adjusted to 7.3. If sterilization is desirable for storage or other reasons, adjust the pH aseptically after sterilization with sterile 1N HCl.

Nitrogen-Free Glucose Agar (Ex. 59)

Add 15 grams of agar to the basal salts portion of the above recipe, bring to a boil, and sterilize in the autoclave at 121° C for 15 minutes. The glucose is dissolved in 100 ml of water and sterilized separately in similar manner. Mix the two solutions aseptically, and dispense into sterile Petri plates.

Nitrogen-Free Medium
(Ex. 59, *Azotobacter*)

```
K₂HPO₄ ............................1.0 gm
MgSO₄ · 7H₂O ....................0.2 gm
FeSO₄ · 7H₂O ....................0.05 gm
CaCl₂ · 2H₂O .....................0.1 gm
Na₂MoO₄ · 2H₂O ...............0.001 gm
*Glucose .........................10.0 gm
Distilled water...................1000 ml
```
*Sterilize separately.
Adjust pH to 7.2.

If this medium is to be used immediately to isolate *Azotobacter* from soil, sterilization is not necessary. When it must be stored for any length of time, it should be sterilized.

If it is to be sterilized, the glucose should be dissolved separately in 100 ml of water and sterilized at 121° C for 15 minutes. The remainder of the medium is sterilized in a similar manner.

After sterilization, the two solutions are mixed aseptically and dispensed into sterile 8-oz prescription bottles (50 ml per bottle).

Phage Growth Medium (Ex. 29)

```
KH₂PO₄ ...........................1.5 gm
Na₂HPO₄ ..........................3.0 gm
NH₄Cl.............................1.0 gm
MgSO₄ · 7H₂O ....................0.2 gm
Glycerol ........................10.0 gm
Acid-hydrolyzed casein............5.0 gm
dl-Tryptophan ...................0.01 gm
Gelatin ..........................0.02 gm
Tween-80 .........................0.2 gm
Distilled water................1000.0 ml
```

Sterilize in autoclave at 121° C for 20 minutes.

Phage Lysing Medium (Ex. 29)

Add sufficient sodium cyanide (NaCN) to the above growth medium to bring the concentration up to 0.02M. For 1 liter of lysing medium this will amount to about 1 gram (actually 0.98 gm) of NaCN. When an equal amount of this lysing medium is added to the growth medium during the last 6 hours of incubation, the concentration of NaCN in the combined medium is 0.01M.

Russell Double Sugar Agar (Ex. 79)

```
Beef extract.......................1 gm
Proteose Peptone No. 3 (Difco)........12 gm
Lactose...........................10 gm
Dextrose...........................1 gm
Sodium chloride....................5 gm
Agar..............................15 gm
Phenol red (Difco).............0.025 gm
Distilled water.................1000 ml
```

Final pH 7.5 at 25° C

Dissolve ingredients in water, and bring to boiling. Cool to 50°–60° C, and dispense about 8 ml per tube (16 mm dia tubes). Slant tubes to cool. Butt depth should be about ½".

Skim Milk Agar

```
Skim milk powder..................100 gm
Agar..............................15 gm
Distilled water.................1000 ml
```

Dissolve the 15 gm of agar into 700 ml of distilled water by boiling. Pour into a large flask and sterilize at 121° C, 15 lb pressure.

In a separate container, dissolve the 100 gm of skim milk powder into 300 ml of water heated to 50° C. Sterilize this milk solution at 113°–115° C (8 lb pressure) for 20 minutes.

After the two solutions have been sterilized, cool to 55° C and combine in one flask, swirling gently to avoid bubbles. Dispense into sterile Petri plates.

Sodium Chloride (6.5%) Tolerance Broth (Ex. 78)

```
Heart infusion broth ..............25 gm
NaCl..............................60 gm
Indicator (1.6 gm bromcresol purple in 100 ml
    95% ethanol) ..................1 ml
Dextrose...........................1 gm
Distilled water.................1000 ml
```

Add all reagents together up to 1000 ml (final volume). Dispense in 15×125 mm screw-capped tubes and sterilize in an autoclave 15 minutes at 121° C.

A positive reaction is recorded when the indicator changes from purple to yellow or when growth is obvious even though the indicator does not change.

Sodium Hippurate Broth (Ex. 78)

Heart infusion broth . . . . .25 gm
Sodium hippurate. . . . . .10 gm
Distilled water. .1000 ml

Sterilize in autoclave at 121° C for 15 minutes after dispensing in 15 × 125 mm screw-capped tubes. Tighten caps to prevent evaporation.

Soft Nutrient Agar (for bacteriophage)

Dehydrated nutrient broth . . . . .8 gm
Agar . . . . .7 gm
Distilled water. .1000 ml

Sterilize in autoclave at 121° C for 20 minutes.

Spirit Blue Agar (Ex. 49)

This medium is used to detect lipase production by bacteria. Lipolytic bacteria cause the medium to change from pale lavender to deep blue.

Spirit blue agar (Difco) . . . . .35 gm
Lipase reagent (Difco). . . . . .35 ml
Distilled water. .1000 ml

Dissolve the spirit blue agar in 1000 ml of water by boiling. Sterilize in autoclave for 15 minutes at 15 psi (121° C). Cool to 55° C and slowly add the 35 ml of lipase reagent, agitating to obtain even distribution. Dispense into sterile Petri plates.

Streptomycin Agar (Ex. 73)

To 1000 ml sterile liquid nutrient agar (50° C), aseptically add 100 mg of streptomycin sulfate. Pour directly into sterile Petri plates.

Top Agar (Ex. 74, Ames test)

Tubes containing 2 ml of top agar are made up just prior to using from bottles of top agar base and his/bio stock solution.

His/Bio Stock Solution

D-Biotin (F.W. 247.3) . . . . .30.9 mg
L-Histidine · HCl (F.W. 191.7). . . . . .24.0 mg
Distilled water. . . . . .250 ml

Dissolve by heating the water to the boiling point. This can be done in a microwave oven. Sterilize by filtration through 0.22 μm membrane filter, or autoclave for 20 minutes at 121° C. Store in a glass bottle at 4° C.

Top Agar Base

Agar . . . . .6 gm
Sodium chloride (NaCl) . . . . .5 gm
Distilled water. . . . . .1000 ml

The agar may be dissolved in a steam bath or microwave oven, or by autoclaving briefly. Mix thoroughly and transfer 100-ml aliquots to 250-ml glass bottles with screw caps. Autoclave for 20 minutes with loosened caps. Slow exhaust. Cool the agar and tighten caps.

Just before using, add 10 ml of the his/bio stock solution to bottle of 100 ml of liquefied top agar base (45° C). After thoroughly mixing, distribute, aseptically, 2 ml of this mixture to sterile tubes (13 mm × 100 mm). Hold tubes at 45° C until used.

Tryptone Agar

Tryptone . . . . .10 gm
Agar . . . . .15 gm
Distilled water. .1000 ml

Tryptone Broth

Tryptone . . . . .10 gm
Distilled water. .1000 ml

Tryptone Yeast Extract Agar

Tryptone . . . . .10 gm
Yeast extract . . . . .5 gm
Dipotassium phosphate . . . . .3 gm
Sucrose . . . . .50 gm
Agar . . . . .15 gm
Water . . . . .1000 ml

pH 7.4

D

Chart I Interpretation of Test Results of API 20E System

Tube		Positive	Negative	Comments
		Interpretation of Reactions		
		Positive	**Negative**	**Comments**
ONPG		Yellow	Colorless	(1) Any shade of yellow is a positive reaction. (2) VP tube, before the addition of reagents, can be used as a negative control.
ADH	Incubation 18–24 h	Red or Orange	Yellow	Orange reactions occurring at 36–48 hours should be interpreted as negative.
	36–48 h	Red	Yellow or Orange	
LDC	18–24 h	Red or Orange	Yellow	Any shade of orange within 18–24 hours is a positive reaction. At 36–48 hours, orange decarboxylase reactions should be interpreted as negative.
	36–48 h	Red	Yellow or Orange	
ODC	18–24 h	Red or Orange	Yellow	Orange reactions occurring at 36–48 hours should be interpreted as negative.
	36–48 h	Red	Yellow or Orange	
CIT		Turquoise or Dark Blue	Light Green or Yellow	(1) Both the tube and cupule should be filled. (2) Reaction is read in the aerobic (cupule) area.
H2S		Black Deposit	No Black Deposit	(1) H_2S production may range from a heavy black deposit to a very thin black line around the tube bottom. Carefully examine the bottom of the tube before considering the reaction negative. (2) A "browning" of the medium is a negative reaction unless a black deposit is present. "Browning" occurs with TDA-positive organisms.
URE	18–24 h	Red or Orange	Yellow	A method of lower sensitivity has been chosen. *Klebsiella, Proteus,* and *Yersinia* routinely give positive reactions.
	36–48 h	Red	Yellow or Orange	
TDA	Add 1 drop 10% ferric chloride			(1) Immediate reaction. (2) Indole-positive organisms may produce a golden orange color due to indole production. This is a negative reaction.
		Brown-Red	Yellow	
IND	Add 1 drop Kovacs' reagent			(1) The reaction should be read within 2 minutes after the addition of the Kovacs' reagent and the results recorded. (2) After several minutes, the HCl present in Kovacs' reagent may react with the plastic of the cupule resulting in a change from a negative (yellow) color to a brownish-red. This is a negative reaction.
		Red Ring	Yellow	
VP	Add 1 drop of 40% potassium hydroxide, then 1 drop of 6% alpha–naphthol.			(1) Wait 10 minutes before considering the reaction negative. (2) A pale pink color (after 10 min) should be interpreted as negative. A pale pink color appears immediately after the addition of reagents but turns dark pink or red after 10 min should be interpreted as positive.
		Red	Colorless	
				Motility may be observed by hanging drop or wet mount preparation.
GEL		Diffusion of the pigment	No diffusion	(1) The solid gelatin particles may spread throughout the tube after inoculation. Unless diffusion occurs, the reaction is negative. (2) Any degree of diffusion is a positive reaction.
GLU		Yellow or Gray	Blue or Blue-Green	**Comments for all Carbohydrates** — **Fermentation** (Enterobacteriaceae, *Aeromonas, Vibrio*) (1) Fermentation of the carbohydrates begins in the most anaerobic portion (bottom) of the tube. Therefore, these reactions should be read from the bottom of the tube to the top. (2) A yellow color at the bottom of the tube only indicates a weak or delayed positive reaction. **Oxidation** (Other Gram-negatives) (1) Oxidative utilization of the carbohydrates begins in the most aerobic portion (top) of the tube. Therefore, these reactions should be read from the top to the bottom of the tube. (2) A yellow color in the upper portion of the tube and a blue in the bottom of the tube indicates oxidative utilization of the sugar. This reaction should be considered positive **only** for non-Enterobacteriaceae gram-negative rods. This is a negative reaction for fermentative organisms such as Enterobacteriaceae.
MAN INO SOR RHA SAC MEL AMY ARA		Yellow	Blue or Blue-Green	
GLU Nitrate Reduction	After reading GLU reaction, add 2 drops 0.8% sulfanilic acid and 2 drops 0.5% N, N-dimethylalpha-naphthylamine			(1) Before addition of reagents, observe GLU tube (positive or negative) for bubbles. Bubbles are indicative of reduction of nitrate to the nitrogenous (N_2) state. (2) A positive reaction may take 2–3 minutes for the red color to appear. (3) Confirm a negative test by adding zinc dust or 20-mesh granular zinc. A pink-orange color after 10 minutes confirms a negative reaction. A yellow color indicates reduction of nitrates to nitrogenous (N_2) state.
	NO_2	Red	Yellow	
	N_2 gas	Bubbles; Yellow after reagents and zinc	Orange after reagents and zinc	
MAN INO SOR Catalase	After reading carbohydrate reaction, add 1 drop 1.5% H_2O_2			(1) Bubbles may take 1–2 minutes to appear. (2) Best results will be obtained if the test is run in tubes which have no gas from fermentation.
		Bubbles	No bubbles	

Courtesy of Analytab Products, Plainview, N.Y.

Chart II Symbol Interpretation of API 20E System

Tube	Chemical/Physical Principles	Components		Ref.
		Reactive Ingredients	Quantity	
ONPG	Hydrolysis of ONPG by beta-galactosidase releases yellow orthonitrophenol from the colorless ONPG; ITPG (isopropylthiogalactopyranoside) is used as inducer.	ONPG ITPG	0.2 mg 8.0 μg	12 13 14
ADH	Arginine dihydrolase transforms arginine into ornithine, ammonia, and carbon dioxide. This causes a pH rise in the acid-buffered system and a change in the indicator from yellow to red.	Arginine	2.0 mg	15
LDC	Lysine decarboxylase transforms lysine into a basic primary amine, cadaverine. This amine causes a pH rise in the acid-buffered system and a change in the indicator from yellow to red.	Lysine	2.0 mg	15
ODC	Ornithine decarboxylase transforms ornithine into a basic primary amine, putrescine. This amine causes a pH rise in the acid-buffered system and a change in the indicator from yellow to red.	Ornithine	2.0 mg	15
CIT	Citrate is the sole carbon source. Citrate utilization results in a pH rise and a change in the indicator from green to blue.	Sodium Citrate	0.8 mg	21
H$_2$S	Hydrogen sulfide is produced from thiosulfate. The hydrogen sulfide reacts with iron salts to produce a black precipitate.	Sodium Thiosulfate	80.0 μg	6
URE	Urease releases ammonia from urea; ammonia causes the pH to rise and changes the indicator from yellow to red.	Urea	0.8 mg	7
TDA	Tryptophane deaminase forms indolepyruvic acid from tryptophane. Indolepyruvic acid produces a brownish-red color in the presence of ferric chloride.	Tryptophane	0.4 mg	22
IND	Metabolism of tryptophane results in the formation of indole. Kovacs' reagent forms a colored complex (pink to red) with indole.	Tryptophane	0.2 mg	10
VP	Acetoin, an intermediary glucose metabolite, is produced from sodium pyruvate and indicated by the formation of a colored complex. Conventional VP tests may take up to 4 days, but by using sodium pyruvate, API has shortened the required test time. Creatine intensifies the color when tests are positive.	Sodium Pyruvate Creatine	2.0 mg 0.9 mg	3
GEL	Liquefaction of gelatin by proteolytic enzymes releases a black pigment which diffuses throughout the tube.	Kohn Charcoal Gelatin	0.6 mg	9
GLU MAN INO SOR RHA SAC MEL AMY ARA	Utilization of the carbohydrate results in acid formation and a consequent pH drop. The indicator changes from blue to yellow.	Glucose Mannitol Inositol Sorbitol Rhamnose Sucrose Melibiose Amygdalin (l +) Arabinose	2.0 mg 2.0 mg 2.0 mg 2.0 mg 2.0 mg 2.0 mg 2.0 mg 2.0 mg 2.0 mg	5 6 12
GLU Nitrate Reduction	Nitrites form a red complex with sulfanilic acid and N, N–dimethylalpha-naphthylamine. In case of negative reaction, addition of zinc confirms the presence of unreduced nitrates by reducing them to nitrites (pink-orange color). If there is no color change after the addition of zinc, this is indicative of the complete reduction of nitrates through nitrites to nitrogen gas or to an anaerogenic amine.	Potassium Nitrate	80.0 μg	6
MAN INO SOR Catalase	Catalase releases oxygen gas from hydrogen peroxide.			24

Courtesy of Analytab Products, Plainview, N.Y.

Chart III Characterization of Gram-Negative Rods—The API 20E System

	ORGANISM	ONPG	ADH	LDC	ODC	CIT	H₂S	URE	TDA	IND	VP	GEL	GLU	MAN	INO	SOR	RHA	SAC	MEL	AMY	ARA	OXI
Escherichieae	E. coli	98.2	1.0	90.2	67.3	0	1.0	0	0	85.0	0	0	100	98.4	0.1	95.5	84.5	41.1	88.4	0.1	95.0	0
	Shigella dysenteriae	27.8	0	0	0	0	0	0	0	33.0	0	0	100	0.1	0	0	22.2	0	61.1	0	16.7	0
	Sh. flexneri	5.3	0	0	0		0	0	0	15.0	0	0	100	94.7	0	78.9	0	0	21.1	0	36.8	0
	Sh. boydii	5.0	0	0	0	0	0	0	0	20.0	0	0	100	60.0	0	53.3	1.0	0	33.3	0	66.7	0
	Sh. sonnei	96.7	0	0	80.0	0	0	0	0	0	0	0	100	99.0	0	39.9	80.0	0	50.0	0	96.7	0
	Edwardsiella tarda	0	0	99.0	99.0	0	55.0	0	0	100	0	0	100	0	0	0	0	0	50.0	0	1.1	0
Salmonelleae	Salmonella enteritidis	1.9	1.0	89.2	95.4	15.4	76.9	0	0	3.1	0	0	100	98.7	4.6	95.2	95.4	4.6	96.9	0	94.5	0
	Sal. typhi	0	0	90.0	0	0	0.1	0	0	0	0	0	100	99.0	0	99.0	1.8	0	100	0	27.0	0
	Sal. paratyphi A	0	0	0	100	0	0.2	0	0	0	0	0	100	99.0	0	99.0	99.0	0	40.0	0	80.0	0
	Arizona-S. arizonae	94.7	1.0	95.0	98.5	15.0	85.0	0	0	0	0	0	100	99.0	0	87.0	96.1	0	89.5	0	95.0	0
	Citrobacter freundii	97.0	10.0	0	60.0	10.0	81.0	0	0	6.0	0	0	100	98.0	1.0	96.0	87.0	59.0	77.0	30.0	98.0	0
	C. diversus-Levinea	97.0	10.0	0	90.0	10.0	0	0	0	91.0	0	0	100	97.0	14.5	88.0	99.0	51.0	47.0	34.0	99.0	0
	C. amalonaticus	97.0	10.0	0	95.0	10.0	0	0	0	99.0	0	0	100	97.0	0.1	93.0	99.0	29.4	53.0	80.0	93.8	0
Klebsielleae	Klebsiella pneumoniae	99.0	0	80.0	0	13.9	0	10.0	0	0	72.0	0	100	98.0	30.0	95.0	91.0	99.0	99.0	98.0	99.0	0
	K. oxytoca	98.0	0	83.0	0	13.0	0	10.0	0	100	60.0	1.0	100	99.0	29.0	92.0	98.0	99.0	99.0	98.0	99.0	0
	K. ozaenae	85.0	0	38.0	0	1.0	0	0	0	0	0	0	100	69.0	1.0	76.0	69.0	15.0	92.0	99.0	84.0	0
	K. rhinoscleromatis	0	0	0	0	0	0	0	0	0	0	0	100	99.0	1.0	86.0	53.0	33.0	66.0	99.0	95.0	0
	Enterobacter aerogenes	99.0	0	98.0	98.0	8.9	0	0	0	0	56.0	0	100	99.0	28.0	90.0	90.0	85.0	97.0	96.0	98.0	0
	Ent. cloacae	97.0	51.9	0	65.0	9.0	0	0	0	0	80.0	0	100	99.0	1.0	92.0	90.0	98.0	92.0	65.0	95.0	0
	Ent. agglomerans	90.0	0	0	0	5.4	0	0	0	50.0	20.0	0	100	99.0	1.0	80.0	60.0	60.0	70.0	70.0	95.0	0
	Ent. gergoviae	99.0	0	61.0	99.0	8.2	0	75.0	0	0	75.0	0	100	99.0	1.0	8.3	99.0	99.0	99.0	99.0	99.0	0
	Ent. sakazakii	97.0	51.6	0	59.0	8.6	0	0	0	4.0	85.0	0	100	99.0	4.0	8.5	90.0	95.0	90.0	76.0	95.0	0
	Serratia liquefaciens	85.0	0	85.0	95.0	8.9	0	1.0	0	0	50.0	60.0	100	99.0	1.0	99.0	30.0	85.0	80.7	80.0	92.9	0
	Ser. marcescens	83.0	0	88.0	94.0	8.0	0	1.0	0	0	58.0	72.0	100	96.0	1.0	97.0	2.0	98.0	37.0	72.0	18.0	0
	Ser. rubidaea	96.0	0	60.5	0.1	8.2	0	0	0	0	70.0	75.6	100	99.0	10.0	75.0	13.4	99.0	82.6	96.0	85.8	0
	Ser. odorifera 1	99.0	0	95.0	99.0	9.5	0	0	0	90.0	63.0	62.0	100	99.0	10.0	99.0	85.0	100	99.0	90.0	99.0	0
	Ser. odorifera 2	99.0	0	92.0	100	9.1	0	0	0	90.0	80.0	78.0	100	99.0	10.0	99.0	95.0	0	99.0	85.4	99.0	0
	Hafnia alvei	60.0	0	99.0	99.0	1.0	0	0	0	0	25.0	0	99.0	99.0	0	35.0	75.0	0	50.0	30.0	95.0	0

Chart III (continued)

ORGANISM	ONPG	ADH	LDC	ODC	CIT	H₂S	URE	TDA	IND	VP	GEL	GLU	MAN	INO	SOR	RHA	SAC	MEL	AMY	ARA	OXI
Proteus vulgaris	0.5	0	0	0	4.1	75.3	91.0	95.0	75.3	0	75.3	100	0	0.1	0	0	83.0	1.0	20.0	4.0	0
Prot. mirabilis	1.0	0	0	90.0	5.8	66.0	97.0	90.0	1.0	0	93.0	100	0	0	0	1.0	9.6	10.0	1.0	27.0	0
Providencia alcalifaciens	0	0	0	0	9.8	0	0	95.0	94.0	0	0	100	0	0	0	0	0	0	0	25.0	0
Prov. stuartii	1.0	0	0	0	8.5	0	0	95.0	86.0	0	0	100	0	8.0	0	0.8	3.7	34.0	0	30.0	0
Prov. stuartii URE +	1.0	0	0	0	6.9	0	99.0	99.0	95.0	0	0	100	15.0	5.0	0	0.5	65.0	20.0	0	20.0	0
Prov. rettgeri	1.0	0	0	0	7.1	0	80.0	95.0	90.0	0	0	100	85.0	1.0	30.0	40.0	5.0	0	40.0	10.0	0
Morganella morganii	1.0	0	0	87.0	0.2	0	78.0	92.0	92.0	0	0	98.0	0	0	0	0	0	0	0	1.0	0
Yersinia enterocolitica	81.0	0	0	36.0	0	0	59.0	0	54.0	0.4	0	100	99.0	1.0	95.0	9.0	78.0	40.4	31.0	76.6	0
Y. pseudotuberculosis	80.0	0	0	0	0	0	88.0	0	0	0	0	100	94.0	0	76.0	58.0	0	5.0	0	52.0	0
Y. pestis	93.0	0	0	0	0	0	0	0	0	1.0	0	93.0	87.0	0	56.0	0	0	0.6	25.0	87.0	0
API Group 1	99.0	0	58.8	99.0	9.2	0	0	0	99.0	0	0	100	99.0	0	75.4	82.4	82.4	94.1	97.0	94.1	0
API Group 2	99.0	2.0	7.3	0	0	0	0	0	0	0	0	100	99.0	0	2.3	30.8	5.6	90.0	38.5	92.3	0
Pseudomonas maltophilia	62.0	0	5.0	0	7.6	0	0	0	0	0	50.0	0.5	0	0	0	0	0	0	0	22.0	4.8
Ps. cepacia	61.0	0	5.0	5.0	7.5	0	0	0	0	1.0	46.0	33.0	1.0	0	1.0	0	7.0	0	1.0	1.0	90.7
Ps. paucimobilis	40.0	0	0	0	1.0	0	0	0	0	0	0	0.5	0	0	0	0	0.5	0	0	0.5	50.0
A. calco var. anitratus	0	0	0	0	2.8	0	0	0	0	1.0	0.1	85.0	0	0	0	0	0.1	77.0	0	60.0	0
A. calco var. lwoffii	0	0	0	0	0	0	0	0	0	0.1	0	0	0	0	0	0	0	0	0	0	0
CDC Group VE-1	90.0	1.0	0	0	7.7	0	0	0	0	1.0	1.3	33.0	0	1.0	0	1.0	0.1	1.0	1.0	16.0	0
CDC Group VE-2	0	0	0	0	7.9	0	0	0	0	1.0	0.1	4.5	0	1.0	0	0	0	1.0	0	5.0	0

Row groupings (left margin): Proteeae (Proteus vulgaris through Morganella morganii); Yersiniae (Yersinia enterocolitica through Y. pestis); Other Gram-negatives (API Group 1 through CDC Group VE-2).

Courtesy of Analytab Products, Plainview, N.Y.

Chart IV Characterization of Enterobacteriaceae—The Enterotube II System

Groups		Organism	Glucose (Reactions)	Gas Production	Lysine	Ornithine	H2S	Indole	Adonitol	Lactose	Arabinose	Sorbitol	Voges-Proskauer	Dulcitol	Phenylalanine Deaminase	Urea	Citrate
ESCHERICHIEAE		Escherichia	+ 100.0	+J 92.0	d 80.6	d 57.8	−K 4.0	+ 96.3	− 5.2	+J 91.6	+ 91.3	± 80.3	d 0.0	d 49.3	− 0.1	− 0.1	− 0.2
		Shigella	+ 100.0	−A 2.1	− 0.0	∓B 20.0	− 0.0	∓ 37.8	− 0.0	−B 0.3	± 67.8	∓ 29.1	− 0.0	d 5.4	− 0.0	− 0.0	− 0.0
EDWARDSIELLEAE		Edwardsiella	+ 100.0	+ 99.4	+ 100.0	+ 99.0	+ 99.6	+ 99.0	− 0.0	− 0.0	∓ 10.7	− 0.2	− 0.0	− 0.0	− 0.0	− 0.0	− 0.0
SALMONELLEAE		Salmonella	+ 100.0	+C 91.9	+H 94.6	+I 92.7	+E 91.6	− 1.1	− 0.0	− 0.8	± 89.2	+ 94.1	− 0.0	dD 86.5	− 0.0	− 0.0	dF 80.1
		Arizona	+ 100.0	+ 99.7	+ 99.4	+ 100.0	+ 98.7	− 2.0	− 0.0	d 69.8	+ 99.1	+ 97.1	− 0.0	− 0.0	− 0.0	− 0.0	+ 96.8
	CITROBACTER	freundii	+ 100.0	+ 91.4	− 0.0	d 17.2	± 81.6	− 6.7	− 0.0	d 39.3	+ 100.0	+ 98.2	− 0.0	d 59.8	− 0.0	dw 89.4	+ 90.4
		amalonaticus	+ 100.0	+ 97.0	− 0.0	+ 97.0	− 0.0	+ 99.0	− 0.0	± 70.0	+ 99.0	+ 97.0	− 0.0	∓ 11.0	− 0.0	± 81.0	+ 94.0
		diversus	+ 100.0	+ 97.3	− 0.0	+ 99.8	− 0.0	+ 100.0	+ 100.0	d 40.3	+ 98.0	+ 98.2	− 0.0	± 52.2	− 0.0	dw 85.8	+ 99.7
PROTEEAE	PROTEUS	vulgaris	+ 100.0	±G 86.0	− 0.0	− 0.0	+ 95.0	+ 91.4	− 0.0	− 0.0	− 0.0	− 0.0	− 0.0	− 0.0	+ 100.0	+ 95.0	d 10.5
		mirabilis	+ 100.0	+G 96.0	− 0.0	+ 99.0	+ 94.5	− 3.2	− 0.0	− 2.0	− 0.0	− 0.0	.∓ 16.0	− 0.0	+ 99.6	± 89.3	± 58.7
	MORGANELLA	morganii	+ 100.0	±G 86.0	− 0.0	+ 97.0	− 0.0	+ 99.5	− 0.0	− 0.0	− 0.0	− 0.0	− 0.0	− 0.0	+ 95.0	+ 97.1	−L 0.0
	PROVIDENCIA	alcalifaciens	+ 100.0	dG 85.2	− 0.0	− 1.2	− 0.0	+ 99.4	+ 94.3	− 0.3	− 0.7	− 0.6	− 0.0	− 0.0	+ 97.4	− 0.0	+ 97.9
		stuartii	+ 100.0	− 0.0	− 0.0	− 0.0	− 0.0	+ 98.6	∓ 12.4	− 3.6	− 4.0	− 3.4	− 0.0	− 0.0	+ 94.5	∓ 20.0	+ 93.7
		rettgeri	+ 100.0	∓G 12.2	− 0.0	− 0.0	− 0.0	+ 95.9	+ 99.0	d 10.0	− 0.0	− 1.0	− 0.0	− 0.0	+ 98.0	+ 100.0	+ 96.0
KLEBSIELLEAE	ENTEROBACTER	cloacae	+ 100.0	+ 99.3	− 0.0	+ 93.7	− 0.0	− 0.0	∓ 28.0	+ 94.0	+ 99.4	+ 100.0	+ 100.0	d 15.2	− 0.0	± 74.6	+ 98.9
		sakazakii	+ 100.0	+ 97.0	− 0.0	+ 97.0	− 0.0	∓ 16.0	− 0.0	+ 100.0	+ 100.0	− 0.0	+ 97.0	6.0	− 0.0	− 0.0	+ 94.0
		gergoviae	+ 100.0	+ 93.0	± 64.0	+ 100.0	− 0.0	− 0.0	− 0.0	∓ 42.0	+ 100.0	− 0.0	+ 100.0	− 0.0	− 0.0	+ 100.0	+ 96.0
		aerogenes	+ 100.0	+ 95.9	+ 97.5	+ 95.9	− 0.0	− 0.8	+ 97.5	+ 92.5	+ 100.0	+ 98.3	+ 100.0	4.1	− 0.0	− 0.0	+ 92.6
		agglomerans	+ 100.0	∓ 24.1	− 0.0	− 0.0	− 0.0	∓ 19.7	− 7.5	d 52.9	+ 97.5	d 26.3	± 64.8	d 12.9	∓ 27.6	d 34.1	d 84.2
	HAFNIA	alvei	+ 100.0	+ 98.9	+ 99.6	+ 98.6	− 0.0	− 0.0	− 0.0	d 2.8	+ 99.3	− 0.0	± 65.0	± 2.4	− 0.0	− 3.0	d 5.6
	SERRATIA	marcescens	+ 100.0	±G 52.6	+ 99.6	+ 99.6	− 0.0	−w 0.1	∓ 56.0	− 1.3	− 0.0	+ 99.1	+ 98.7	− 0.0	− 0.0	dw 39.7	+ 97.6
		liquefaciens	+ 100.0	d 72.5	± 64.2	+ 100.0	− 0.0	−w 1.8	− 8.3	d 15.6	+ 97.3	+ 97.3	∓ 49.5	− 0.0	− 0.9	dw 3.7	+ 93.6
		rubidaea	+ 100.0	dG 35.0	+ 61.0	− 0.0	− 0.0	−w 2.0	± 88.0	+ 100.0	+ 100.0	− 8.0	+ 92.0	− 0.0	− 0.0	dw 4.0	± 88.0
	KLEBSIELLA	pneumoniae	+ 100.0	+ 96.0	+ 97.2	− 0.0	− 0.0	− 0.0	± 89.0	+ 98.7	+ 99.9	+ 99.4	+ 93.7	∓ 33.0	− 0.0	+ 95.4	+ 96.8
		oxytoca	+ 100.0	+ 96.0	+ 97.2	− 0.0	− 0.0	+ 100.0	± 89.0	∓ 98.7	+ 100.0	+ 98.0	+ 93.7	∓ 33.0	− 0.0	∓ 95.4	∓ 96.8
		ozaenae	+ 100.0	d 55.0	∓ 35.8	− 1.0	− 0.0	− 0.0	+ 91.8	d 26.2	+ 100.0	± 78.0	− 0.0	− 0.0	− 0.0	d 14.8	d 28.1
		rhinoscleromatis	+ 100.0	− 0.0	− 0.0	− 0.0	− 0.0	− 0.0	+ 98.0	d 6.0	+ 100.0	+ 98.0	− 0.0	− 0.0	− 0.0	− 0.0	− 0.0
YERSINIAE	YERSINIA	enterocolitica	+ 100.0	− 0.0	− 0.0	+ 90.7	− 0.0	∓ 26.7	− 0.0	− 0.0	+ 98.7	+ 98.7	− 0.1	− 0.0	− 0.0	+ 90.7	− 0.0
		pseudotuberculosis	+ 100.0	− 0.0	− 0.0	− 0.0	− 0.0	− 0.0	− 0.0	− 0.0	± 55.0	− 0.0	− 0.0	− 0.0	− 0.0	+ 100.0	− 0.0

Courtesy of Roche Diagnostics, Nutley, N.J.

E. *S. enteritidis* bioserotype Paratyphi A and some rare biotypes may be H$_2$S negative.

F. *S. typhi*, *S. enteritidis* bioserotype Paratyphi A and some rare biotypes are citrate-negative and *S. cholerae-suis* is usually delayed positive.

G. The amount of gas produced by *Serratia*, *Proteus* and *Providencia alcalifaciens* is slight; therefore, gas production may not be evident in the ENTEROTUBE II.

H. *S. enteritidis* bioserotype Paratyphi A is negative for lysine decarboxylase.

I. *S. typhi* and *S. gallinarum* are ornithine decarboxylase-negative.

J. The Alkalescens-Dispar (A-D) group is included as a biotype of *E. coli*. Members of the A-D group are generally anaerogenic, non-motile and do not ferment lactose.

K. An occasional strain may produce hydrogen sulfide.

L. An occasional strain may appear to utilize citrate.

Chart V Reaction Interpretations for API Staph-Ident

MICROCUPULE		INTERPRETATION OF REACTIONS		
NO.	SUBSTRATE	POSITIVE	NEGATIVE	COMMENTS AND REFERENCES
1	PHS	Yellow	Clear or straw-colored	A positive result should be recorded only if significant color development has occurred.(3)
2	URE	Purple to Red-Orange	Yellow or Yellow-Orange	Phenol red has been added to the urea formulation to allow detection of alkaline end products resulting from urea utilization.(1)
3	GLS	Yellow	Clear or straw-colored	A positive result should be recorded only if significant color development has occurred.
4 5 6 7	MNE MAN TRE SAL	Yellow or Yellow-Orange	Red or Orange	Cresol red has been added to each carbohydrate to allow detection of acid production if the respective carbohydrates are utilized. (1,7)
8	GLC	Yellow	Clear or straw-colored	A positive result should be recorded only if significant color development has occurred.
9	ARG	Purple to Red-Orange	Yellow or Yellow-Orange	Phenol red has been added to the arginine formulation to allow detection of alkaline end products resulting from arginine utilization.(1)
10	NGP	Add 1–2 drops of STAPH-IDENT REAGENT Plum-Purple (Mauve)	 Yellow or colorless	Color development will begin within 30 seconds of reagent addition. (1,5)

Courtesy of Analytab Products, Plainview, N.Y.

Abbreviation	Test
PHS	Phosphatase
URE	Urea utilization
GLS	β-Glucosidase
MNE	Mannose utilization
MAN	Mannitol utilization
TRE	Trehalose utilization
SAL	Salicin utilization
GLC	β-Glucuronidase
ARG	Arginine utilization
NGP	β-Galactosidase

Chart VI Biochemistry of API Staph-Ident Tests

MICROCUPULE		CHEMICAL/PHYSICAL PRINCIPLES	REACTIVE INGREDIENTS	QUANTITY
NO.	SUBSTRATE			
1	PHS	Hydrolysis of p-nitrophenyl-phosphate, disodium salt, by alkaline phosphatase releases yellow paranitrophenol from the colorless substrate.	p-nitrophenyl-phosphate, disodium salt	0.2%
2	URE	Urease releases ammonia from urea; ammonia causes the pH to rise and changes the indicator from yellow to red.	Urea	1.6%
3	GLS	Hydrolysis of p-nitrophenyl-β-D-glucopyranoside by β-glucosidase releases yellow para-nitrophenol from the colorless substrate.	p-nitrophenyl-β-D-glucopyranoside	0.2%
4	MNE	Utilization of carbohydrate results in acid formation and a consequent pH drop. The indicator changes from red to yellow.	Mannose	1.0%
5	MAN		Mannitol	1.0%
6	TRE		Trehalose	1.0%
7	SAL		Salicin	1.0%
8	GLC	Hydrolysis of p-nitrophenyl-β-D-glucuronide by β-glucuronidase releases yellow para-nitrophenol from the colorless substrate.	p-nitrophenyl-β-D-glucuronide	0.2%
9	ARG	Utilization of arginine produces alkaline end products which change the indicator from yellow to red.	Arginine	1.6%
10	NGP	Hydrolysis of 2–naphthol-β-D-galactopyranoside by β-galactosidase releases free β-naphthol which complexes with STAPH-IDENT REAGENT to produce a plum-purple (mauve) color.	2–naphthol-β-D-galactopyranoside	0.3%

Courtesy of Analytab Products, Plainview, N.Y.

Chart VII API Staph-Ident Profile Register*

Profile	Identification		Profile	Identification	
0 040	STAPH CAPITIS		4 700	STAPH AUREUS	COAG +
0 060	STAPH HAEMOLYTICUS			STAPH SCIURI	COAG −
0 100	STAPH CAPITIS		4 710	STAPH SCIURI	
0 140	STAPH CAPITIS		5 040	STAPH EPIDERMIDIS	
0 200	STAPH COHNII		5 200	STAPH SCIURI	
0 240	STAPH CAPITIS		5 210	STAPH SCIURI	
0 300	STAPH CAPITIS		5 300	STAPH AUREUS	COAG +
0 340	STAPH CAPITIS			STAPH SCIURI	COAG −
0 440	STAPH HAEMOLYTICUS		5 310	STAPH SCIURI	
0 460	STAPH HAEMOLYTICUS		5 600	STAPH SCIURI	
0 600	STAPH COHNII		5 610	STAPH SCIURI	
0 620	STAPH HAEMOLYTICUS		5 700	STAPH AUREUS	COAG +
0 640	STAPH HAEMOLYTICUS			STAPH SCIURI	COAG −
0 660	STAPH HAEMOLYTICUS		5 710	STAPH SCIURI	
1 000	STAPH EPIDERMIDIS		5 740	STAPH AUREUS	
1 040	STAPH EPIDERMIDIS		6 001	STAPH XYLOSUS	XYL + ARA +
1 300	STAPH AUREUS			STAPH SAPROPHYTICUS	XYL − ARA −
1 540	STAPH HYICUS (An)		6 011	STAPH XYLOSUS	
1 560	STAPH HYICUS (An)		6 021	STAPH XYLOSUS	
2 000	STAPH SAPROPHYTICUS	NOVO R	6 101	STAPH XYLOSUS	
	STAPH HOMINIS	NOVO S	6 121	STAPH XYLOSUS	
2 001	STAPH SAPROPHYTICUS		6 221	STAPH XYLOSUS	
2 040	STAPH SAPROPHYTICUS	NOVO R	6 300	STAPH AUREUS	
	STAPH HOMINIS	NOVO S	6 301	STAPH XYLOSUS	
2 041	STAPH SIMULANS		6 311	STAPH XYLOSUS	
2 061	STAPH SIMULANS		6 321	STAPH XYLOSUS	
2 141	STAPH SIMULANS		6 340	STAPH AUREUS	COAG +
2 161	STAPH SIMULANS			STAPH WARNERI	COAG −
2 201	STAPH SAPROPHYTICUS		6 400	STAPH WARNERI	
2 241	STAPH SIMULANS		6 401	STAPH XYLOSUS	XYL + ARA +
2 261	STAPH SIMULANS			STAPH SAPROPHYTICUS	XYL − ARA −
2 341	STAPH SIMULANS		6 421	STAPH XYLOSUS	
2 361	STAPH SIMULANS		6 460	STAPH WARNERI	
2 400	STAPH HOMINIS	NOVO S	6 501	STAPH XYLOSUS	
	STAPH SAPROPHYTICUS	NOVO R	6 521	STAPH XYLOSUS	
2 401	STAPH SAPROPHYTICUS		6 600	STAPH WARNERI	
2 421	STAPH SIMULANS		6 601	STAPH SAPROPHYTICUS	XYL − ARA −
2 441	STAPH SIMULANS			STAPH XYLOSUS	XYL + ARA +
2 461	STAPH SIMULANS		6 611	STAPH XYLOSUS	
2 541	STAPH SIMULANS		6 621	STAPH XYLOSUS	
2 561	STAPH SIMULANS		6 700	STAPH AUREUS	
2 601	STAPH SAPROPHYTICUS		6 701	STAPH XYLOSUS	
2 611	STAPH SAPROPHYTICUS		6 721	STAPH XYLOSUS	
2 661	STAPH SIMULANS		6 731	STAPH XYLOSUS	
2 721	STAPH COHNII (SSP1)		7 000	STAPH EPIDERMIDIS	
2 741	STAPH SIMULANS		7 021	STAPH XYLOSUS	
2 761	STAPH SIMULANS		7 040	STAPH EPIDERMIDIS	
3 000	STAPH EPIDERMIDIS		7 141	STAPH INTERMEDIUS (An)	
3 040	STAPH EPIDERMIDIS		7 300	STAPH AUREUS	
3 140	STAPH EPIDERMIDIS		7 321	STAPH XYLOSUS	
3 540	STAPH HYICUS (An)		7 340	STAPH AUREUS	
3 541	STAPH INTERMEDIUS (An)		7 401	STAPH XYLOSUS	
3 560	STAPH HYICUS (An)		7 421	STAPH XYLOSUS	
3 601	STAPH SIMULANS	NOVO S	7 501	STAPH INTERMEDIUS (An)	COAG +
	STAPH SAPROPHYTICUS	NOVO R		STAPH XYLOSUS	COAG −
4 060	STAPH HAEMOLYTICUS		7 521	STAPH XYLOSUS	
4 210	STAPH SCIURI		7 541	STAPH INTERMEDIUS (An)	
4 310	STAPH SCIURI		7 560	STAPH HYICUS (An)	
4 420	STAPH HAEMOLYTICUS		7 601	STAPH XYLOSUS	
4 440	STAPH HAEMOLYTICUS		7 621	STAPH XYLOSUS	
4 460	STAPH HAEMOLYTICUS		7 631	STAPH XYLOSUS	
4 610	STAPH SCIURI		7 700	STAPH AUREUS	
4 620	STAPH HAEMOLYTICUS		7 701	STAPH XYLOSUS	
4 660	STAPH HAEMOLYTICUS		7 721	STAPH XYLOSUS	
			7 740	STAPH AUREUS	

*Date of Publication: March, 1984
Courtesy of Analytab Products, Plainview, N.Y.

441

E Appendix
The Streptococci:
Classification, Habitat, Pathology, and
Biochemical Characteristics

To fully understand the characteristics of the various species of medically important streptococci, this appendix has been included as an adjunct to Exercise 78. The table of streptococcal characteristics on this page is the same one that is shown on page 257 of Exercise 78. It is also the basis for much of the discussion that follows.

The first system that was used for grouping the streptococci was based on the type of hemolysis and was proposed by J. H. Brown in 1919. In 1933, R. C. Lancefield proposed that these bacteria be separated into groups A, B, C, etc., on the basis of precipitation-type serological testing. Both hemolysis and serological typing still play predominant roles today in our classification system. Note below that the Lancefield groups are categorized with respect to the type of hemolysis that is produced on blood agar.

Beta Hemolytic Groups

Using a streak-stab technique, a blood agar plate is incubated aerobically at 37° C for 24 hours. Isolates

Table 1 Physiological Tests for Streptococcal Differentiation

GROUP	Type of Hemolysis	Bacitracin Susceptibility	CAMP Reaction or Hippurate Hydrolysis	SXT Sensitivity	Bile Esculin Hydrolysis	Tolerance to 6.5% NaCl	Optochin Susceptibility	Bile Solubility	
Group A S. pyogenes	beta	+	−	R	−	−	−	−	
Group B S. agalactiae	beta	−*	+	R	−	±	−	−	
Group C S. equi S. equisimilis S. zooepidemicus	beta	−*	−	S	−	−	−	−	
Group D (enterococci) S. faecalis S. faecium etc.	alpha beta none	−	−	R	+	+	−	−	
Group D (nonenterococci) S. bovis etc.	alpha none	−	−	R/S	+	−	−	−	
Viridans S. mitis S. salivarius S. mutans etc.	alpha none	−*	−*	S	−	−	−	−	
Pneumococci S. pneumoniae	alpha	±	−		−	−	+	+	

*Exceptions occur occasionally   **See comments on pp. 445 and 446 concerning correct genus.
Note: R = resistant; S = sensitive; blank = not significant

that have colonies surrounded by clear zones completely free of red blood cells are characterized as being *beta hemolytic.* Three serological groups of streptococci fall in this category: groups A, B, and C; a few species in group D are also beta hemolytic.

Group A Streptococci

This group is represented by only one species: *Streptococcus pyogenes.* Approximately 25% of all upper respiratory infections (URIs) are caused by this species; another 10% of URIs are caused by other streptococci; most of the remainder (65%) are caused by viruses. Since no unique clinical symptoms can be used to differentiate viral from streptococcal URIs, and since successful treatment relies on proper identification, it becomes mandatory that throat cultures be taken in an attempt to prove the presence or absence of streptococci. It should be added that if streptococcal URIs are improperly treated, serious sequelae such as pneumonia, acute endocarditis, rheumatic fever, and glomerularnephritis can result.

S. pyogenes is the only beta hemolytic streptococcus that is primarily of *human origin.* Although the pharynx is the most likely place to find this species, it may be isolated from the skin and rectum. Asymptomatic pharyngeal and anal carriers are not uncommon. Outbreaks of postoperative streptococcal infections have been traced to both pharyngeal and anal carriers among hospital personnel.

These coccoidal bacteria (0.6–1.0 μm diameter) occur as pairs and as short to moderate-length chains in clinical specimens; in broth cultures, the chains are often longer.

When grown on blood agar, the colonies are small (0.5 mm dia.), transparent to opaque, and domed; they have a smooth or semimatt surface and an entire edge; complete hemolysis (beta-type) occurs around each colony, usually two to four times the diameter of the colony.

S. pyogenes produces two hemolysins: streptolysin S and streptolysin O. The beta-type hemolysis on blood agar is due to the complete destruction of red blood cells by the streptolysin S.

There is no group of physiological tests that can be used with *absolute* certainty to differentiate *S. pyogenes* from other streptococci; however, if an isolate is beta hemolytic and sensitive to bacitracin, one can be 95% certain that the isolate is *S. pyogenes.* The characteristics of this organism are the first ones tabulated in table I on the previous page.

Group B Streptococci

The only recognized species of this group is *S. agalactiae.* Although this organism is frequently found in milk and associated with *mastitis in cattle,* the list of human infections caused by it is as long as the one for *S. pyogenes:* abscesses, acute endocarditis, impetigo, meningitis, neonatal sepsis, and pneumonia are just a few. Like *S. pyogenes,* this pathogen may also be found in the pharynx, skin, and rectum; however, it is more likely to be found in the genital and intestinal tracts of healthy adults and infants. It is not unusual to find the organism in vaginal cultures of third-trimester pregnant women.

Cells are spherical to ovoid (0.6–1.2 μm dia) and occur in chains of seldom less than four cells; long chains are frequently present. Characteristically, the chains appear to be composed of paired cocci.

Colonies of *S. agalactiae* on blood agar often produce double zone hemolysis. After 24 hours incubation colonies exhibit zones of beta hemolysis. After cooling, a second ring of hemolysis forms which is separated from the first by a ring of red blood cells.

Reference to table I emphasizes the significant characteristics of *S. agalactiae.* Note that this organism gives a positive CAMP reaction, hydrolyzes hippurate, and is not (usually) sensitive to bacitracin. It is also resistant to SXT. Presumptive identification of this species relies heavily on a positive CAMP test or hippurate hydrolysis, even if beta hemolysis is not clearly demonstrated.

Group C Streptococci

Three species fall in this group: *S. equisimilis, S. equi,* and *S. zooepidemicus.* Although all of these species may cause human infections, the diseases are not usually as grave as those caused by groups A and B. Some group C species have been isolated from impetiginous lesions, abscesses, sputum, and the pharynx. There is no evidence that they are associated with acute glomerularnephritis, rheumatic fever, or even pharyngitis.

Presumptive differentiation of this group from *S. pyogenes* and *S. agalactiae* is based primarily on (1) resistance to bacitracin, (2) inability to hydrolyze hippurate or bile esculin, and (3) a negative CAMP test. There are other groups that have some of these same characteristics, but they will not be studied here. Tables 12.16 and 12.17 on page 1049 of *Bergey's Manual,* vol. 2, provide information about these other groups.

Alpha Hemolytic Groups

Streptococcal isolates that have colonies with zones of incomplete lysis around them are said to be **alpha hemolytic.** These zones are often greenish; sometimes they are confused with beta hemolysis. *The only way to be certain that such zones are not beta* *hemolytic is to examine the zones under 60× microscopic magnification.* Figure 78.4, page 255, illustrates the differences between alpha and beta hemolysis. If some red blood cells are seen in the zone, the isolate is classified as being alpha hemolytic.

The grouping of streptococci on the basis of alpha hemolysis is not as clear-cut as it is for beta hemolytic groups. Note in table I that the bottom four groups that have alpha hemolytic types may also have beta hemolytic or nonhemolytic strains. Thus, we see that hemolysis in these four groups can be a misleading characteristic in identification.

Alpha hemolytic isolates from the pharynx are usually *S. pneumoniae,* viridans streptococci, or group D. Our primary concern here in this experiment is to identify isolates of *S. pneumoniae.* To accomplish this goal, it will be necessary to differentiate any alpha hemolytic isolate from group D and viridans streptococci.

Streptococcus pneumoniae
(Pneumococcus)

This organism is the most frequent cause of bacterial pneumonia, a disease that has a high mortality rate among the aged and debilitated. It is also frequently implicated in conjunctivitis, otitis media, pericarditis, subacute endocarditis, meningitis, septicemia, empyema, and peritonitis. Thirty to 70% of normal individuals carry this organism in the pharynx.

Spherical or ovoid, these cells (0.5–1.25 μm dia) occur typically as pairs, sometimes singly, often in short chains. Distal ends of the cells are pointed or lancet-shaped and are heavily encapsulated with polysaccharide on primary isolation.

Colonies on blood agar are small, mucoidal, opalescent, and flattened with entire edges surrounded by a zone of greenish discoloration (alpha hemolysis). In contrast, the viridans streptococcal colonies are smaller, gray to whitish gray, and opaque with entire edges.

Presumptive identification of *S. pneumoniae* can be made with the optochin and bile solubility tests. On the optochin test, the pneumococci exhibit sensitivity to ethylhydrocupreine (optochin). With the bile solubility test, pneumococci are dissolved in bile (2% sodium desoxycholate). Table I reveals that except for bacitracin susceptibility (±), *S. pneumoniae* is negative on all other tests used for differentiation of streptococci.

Viridans Group

Streptococci that fall in this group are primarily alpha hemolytic; some are nonhemolytic. Approximately 10 species are included in this group. All of them are highly adapted parasites of the upper respiratory tract. Although usually regarded as having low pathogenicity, they are opportunistic and sometimes cause serious infections. Two species (*S. mutans* and *S. sanguis*) are thought to be the primary cause of dental caries, since they have the ability to form dental plaque. Viridans streptococci are implicated more often than any other bacteria in subacute bacterial endocarditis.

When it comes to differentiation of bacteria of this group from the pneumococci and enterococci, we will use the optochin, bile solubility, and salt-tolerance tests. See table I.

Group D Streptococci (Enterococci)

Members of this group are, currently, considered by most taxonomists to belong to the genus *Enterococcus.* During the preparation of volume 2 of *Bergey's Manual* Schleifer and Kilper-Balz presented conclusive evidence that *S. faecalis, S. faecium,* and *S. bovis* were so distantly related to the other groups of streptococci that they should be transferred to another genus. Since the term *Enterococcus* had been previously suggested by others, Schleifer and Kilper-Balz recommended that this be the name of a new genus to include all of the Group D streptococci, nonenterococci included. The fact that these papers came too late for *Bergey's Manual* to include this new genus caused the genus *Streptococcus* to be retained. To avoid confusion in our use of *Bergey's Manual,* we have retained the same terminology used in *Bergey's Manual.*

The enterococci of serological group D may be alpha hemolytic, beta hemolytic, or nonhemolytic. The principal species of this enterococcal group are *S. faecalis, S. faecium, S. durans,* and *S. avium.*

Subacute endocarditis, pyelonephritis, urinary tract infections, meningitis, and biliary infections are caused by these organisms. All five of these species have been isolated from the intestinal tract. Approximately 20% of subacute bacterial endocarditis and 10% of urinary tract infections are caused by members of this group. Differentiation of this group from other streptococci in systemic infections is mandatory because *S. faecalis, S. faecium,* and *S. durans* are resistant to penicillin and require combined antibiotic therapy.

Since *S. faecalis* can be isolated from many food products (not connected with fecal contamination), it can be a transient in the pharynx and show up as an isolate in throat cultures. Morphologically, the cells are ovoid (0.5–1.0 μm dia) occurring as pairs in short chains. Hemolytic reactions of *S. faecalis* on blood agar will vary with the type of blood used in the medium. Some strains produce beta hemolysis on agar with horse, human, and rabbit blood; on sheep blood agar the colonies will always exhibit alpha hemolysis. Other streptococci are consistently either beta, alpha, or nonhemolytic.

Cells of *S. faecium* are morphologically similar to *S. faecalis* except that motile strains are often encountered. A strong alpha-type hemolysis is usually seen around colonies of *S. faecium* on blood agar.

Although presumptive differentiation of group D enterococcal streptococci from groups A, B, and C is not too difficult with physiological tests, it is more laborious to differentiate the individual species within group D. As indicated in table I, the enterococci (1) hydrolyze bile esculin, (2) are CAMP negative, and (3) grow well in 6.5% NaCl broth.

Differentiation of the five species within this group involves nine or ten physiological tests.

Group D Streptococci (Nonenterococci)

The only medically significant nonenterococcal species of group D is *S. bovis.* This organism is found in the intestinal tract of humans as well as in cows, sheep, and other ruminants. It can cause meningitis, subacute endocarditis, and urinary tract infections. On blood agar, the organism is usually alpha hemolytic; occasionally, it is nonhemolytic. The best way to differentiate it from the group D enterococci is to test its tolerance to 6.5% NaCl. Note in table I that *S. bovis* will not grow in this medium, but all enterococci will.

F

As stated in Exercise 51, *Identibacter interactus* is a computer program designed to assist students in identifying unknown bacterial cultures. This CD-ROM program, which is distributed by WCB/McGraw-Hill Co. in Dubuque, is a powerful program that includes more than 50 tests to run on assigned bacterial unknowns. The organism data base includes about sixty species of chemoheterotrophic bacteria.

To run this program, you will select each test from pull-down menus. A color image of each test result will be displayed on the computer screen, and you must be able to correctly interpret the test result that is shown. Once you have tabulated sufficient information, you can identify your unknown by typing in the name of the organism. An audit trail of your choices can be saved to disk which can be evaluated by your instructor.

Before you attempt to use this program, read over the following pages of this Appendix. These twelve pages are the first portion of a fifty-nine page instructional manual that can be accessed from the CD-ROM. This information will explain more in detail how the program functions. A full copy of the manual should be available to you in the laboratory.

TABLE OF CONTENTS

PREFACE

Many microbiology laboratory courses use the identification of an unknown culture as an exercise to teach students about the types of characteristics that are used to distinguish bacterial species, and how classification schemes for bacteria are organized. This knowledge is of more than academic interest, because these identification strategies are also used in medical diagnostic laboratories.

A problem in teaching these concepts is that students generally identify only one unknown culture, because of time and/or money constraints. This is insufficient for them to learn the deductive reasoning skills involved in bacterial identification, or to understand how phenotypic information is organized in a reference such as *Bergey's Manual of Determinative Bacteriology*.

With this computer simulation, students can select among more than 50 tests (microscopic stains, growth on specific media, biochemical reactions) to run on their assigned unknown. Students select the test from a pull-down menu. A color image of the test result is displayed on the screen. The student must interpret this result. A print copy of a reference source (such as *Bergey's Manual of Determinative Bacteriology*) or the on-line help is then consulted to determine what additional test(s) are necessary to identify the unknown. When sufficient information has been collected, the student can choose to identify the unknown strain. An audit trail of the student's choices can be saved to disk and evaluated by the instructor.

First-time users can identify an unknown in 15 – 30 minutes, and experienced users can solve unknowns more quickly. Thus, students can repeat the exercise a number of times, and become familiar with the organization of bacterial identification schemes and the utility of specific tests for distinguishing species.

The color images of test results were made from assays described in the American Society for Microbiology's reference work *Methods for General and Molecular Bacteriology*. The organism database contains about sixty species of chemoheteroptrophic bacteria, including some human pathogens that would be difficult to use in an introductory microbiology laboratory. Neither photosynthetic or chemoautotrophic bacteria are included. In addition, virtually all of the included species will grow under aerobic conditions on typical enriched laboratory media (for example, nutrient agar). Thus, obligate anaerobes are generally not represented in the database.

We have also avoided the inclusion of strains which would require the use of tests such as serological reactions, bacterial virus sensitivity, or susceptibility to antimicrobial drugs in order to determine the identity of an unknown. The challenge to distinguish between phenotypically similar species is greatest within the 20 Enterobacteriaceae in the database. Within other groups, students are asked to distinguish among genera. The phenotypic characteristics for species are takenfrom data in *Bergey's Manual of Determinative Bacteriology*.

Dr. Allan Konopka
Department of Biological Sciences
Purdue University
West Lafayette, IN
June, 1996

450

SETUP INSTRUCTIONS

Installing *Identibacter.*

Identibacter can be run directly from the CD without installing anything from the CD, provided QuickTime is already installed.

Identibacter will run better if installed on the hard drive. To download ***Identibacter*** to the hard drive you will need 120 megabytes of free space. On the Macintosh drag (copy) the entire folder, "Identibacter interactus...", to the hard drive. In Windows 95 copy the folder, "Idbacter", to a harddrive. DO NOT drag or copy items from the folder one by one. These folders contain invisible files and folders that are needed by the application.

Installing QuickTime™ or QuickTime for Windows™.

You must have QuickTime™ or QuickTime for Windows™ installed. There are digitized video clips that are important. Installers for QuickTime 2.5 for Macintosh and QuickTime for Windows 2.1.1 are included on this CD.

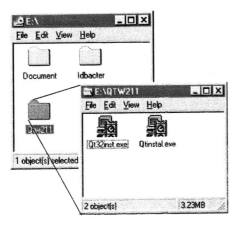

On the Macintosh open the folder, QuickTime 2.5, and double click on the installer to install QuickTime 2.5 in the current system.

In Windows 95 open the folder, "QTW211", and double click on the installer "Qt32inst.exe" to install QuickTime for Windows™.

Using Set Manager and Assigning Unknowns

Set Manager .PPC

Setman.exe

The 57 unknowns have been randomized into a hundred different sets.

To assign a particular unknown an instructor can use Set Manager to determine the ID number for the desired unknown in any set . Give the student the set number and the ID number for each unknown microorganism that you wish them to identify.

SetManager allows instructors to print a list of unknowns for any of the randomized sets.

Note: If you are running ***Identibacter*** from a harddrive, the SetManager can be removed to prevent students from accessing the information. Remove Setman.exe, or Set Manager.PPC and Set Manager.040 from the application folder.

Information in the manual

Appendix C of this manual contains species characteristics that are important for the identification of the microorganisms in *Identibacter*. You can access the information in Appendix C electronically from the Acrobat® pdf. The installer for Acrobat® Reader is on this CD (document folder). You may print all or any portion of this manual from Acrobat or Adobe Pagemaker 6.0 for any class that is using *Identibacter*. You can also find the information by consulting Bergey's Manual.

Installing Acrobat® Reader

On the Macintosh open the "Document" folder, and the "Acrobat Reader " folder. Double click on the "Acrobat Reader Installer" install the Acrobat Reader 2.1.

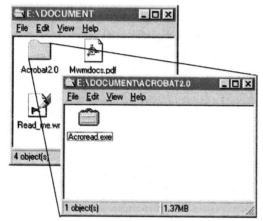

In the Windows environment open the "Document" folder, and then "Acrobat2.0" folder.

Use the installer, "Acroread.exe" to install the Acrobat® Reader 2.0.

Macintosh

- 68040 Macintosh or any PowerMac.
- Mac OS 7.1 , 7.5.5 preferred.
- 20 megs of RAM.[1]
- 14" color display capable of a 16 bit colordepth (thousands). A larger mutisync display or multiple monitors is preferrable for the multi-windowing environment.
- QuickTime 2.1™, 2.5 preferred.
- CD ROM player.[2]

Windows 95

- 486/66 or faster; prefer Pentium®.
- Windows 95™.
- 16 megs RAM.
- 13" color display capable of displaying at least 800 X 600 resolution with a depth of the monitor set at "High color (16-bit)."
- QuickTime for Windows 2.1.1™.
- CD ROM player.[2]

---

[1] On the Macintosh *Identibacter* requires a memory partition of 12MB RAM for powerMacs, (8MB for non-powerMacs). An additional 512K free RAM (1 Mb preferred) is required for the decompression of QuickTime files during operation.

[2] If you plan to play *Identibacter* from the CD, a 2X or better CD ROM player is recommended. If you are going to run identibacter from a harddrive, a CD ROM player is needed to to install identibacter on the harddrive.

Identibacter interactus can run directly from the CD or can be downloaded from the CD and run from the hard drive. The CD can be used on computers running either MacOS or Windows 95™.

Before you begin...

> Does your computer meet the minimium system requirements?   See page 10.
> Is QuickTime™ or QuickTime for Windows™ installed?   See page 9.
> Your instructor or computer administrator may have already taken care of this.

> Do you have a printed copy of this manual (Appendix C) or *Bergey's Manual,*
> or can you access Appendix C electronically?   See page 10.
> You will need some species information for the identification process.

> Do you have the numbers for each unknown microorganism that you are to identify?
> The first number is the set ID number, the second number is the unknown ID number
> for a particular species of microorganism in the given set.
> If you are practicing by yourself, you may randomly select a set and unknown ID at log-in time.

Insert the CD into the CD-ROM drive of your computer.

On the Macintosh...

identibacter

an icon "identibacter" for the CD will appear on the desktop.

Double click on this icon to open the CD.

This CD contains the following folders...

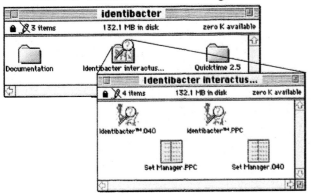

Open the folder, "Identibacter interactus..." by double clicking on the folder.

In Windows 95™ ...

double click on the icon "My Computer" to access the CD "Idbacter [E:]".

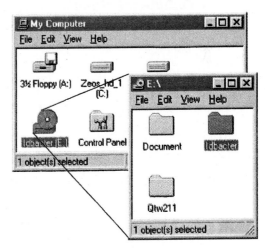

Open "Idbacter [E:]". You will find three folders in this window. The folder, "Idbacter", contains the application, "Idbacter.exe".

To start *Identibacter* double click on the icon, "Identibacter™.PPC" for power Macintoshes or "Identibacter™.040" for non-power Macintoshes, or "Idbacter.exe" for Windows 95.

QuickStart for Students

To identify an unknown microorganism...

double click on the icon,   tart *Identibacter*.

Identibacter™.PPC   Idbacter.exe

Log in
Type your name and ID number ➡

Select set number ➡
Select Unknown

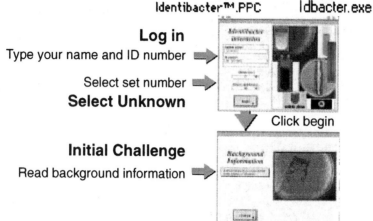

Click begin

Initial Challenge
Read background information ➡

Click continue

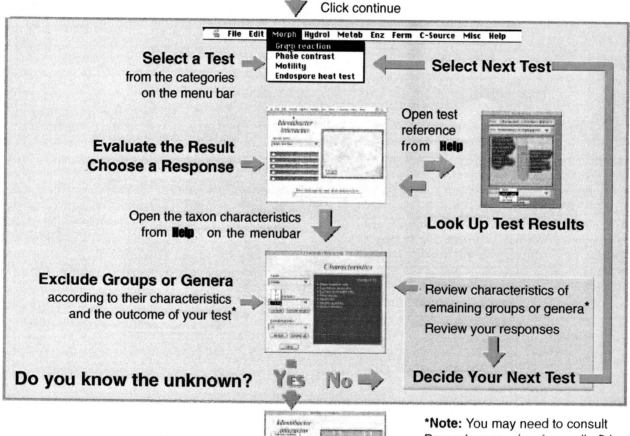

Select a Test
from the categories
on the menu bar

File Edit **Morph** Hydrol Metab Enz Ferm C-Source Misc Help

Gram reaction
Phase contrast
Motility
Endospore heat test

Select Next Test

Evaluate the Result
Choose a Response ➡

Open test
reference
from **Help**

Look Up Test Results

Open the taxon characteristics
from **Help** on the menubar

Exclude Groups or Genera
according to their characteristics
and the outcome of your test* ➡

Review characteristics of
remaining groups or genera*
Review your responses

Do you know the unknown? Yes No ➡ **Decide Your Next Test**

*Note: You may need to consult
Bergey's manual or Appendix C in
this manual

Identify the Unknown
Type its name in the response box ➡

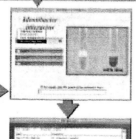

Print Report

QuickStart for Students

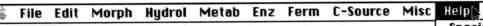

☀	File	Edit	Morph	Hydrol	Metab	Enz	Ferm	C-Source	Misc	**Help**

Help
- Species list
- Species by group

- Test reference
- Taxon characteristics

- Review responses

Access information from **Help** on the menubar.

Species List

Use the species list or species by group window to see the possible unknowns and check for proper spelling.

Test Reference

The test reference window illustrates possible results, background and procedural information about each test.

⬅ **Select the Desired Test**

⬅ **Select a Test Result**

Taxonomic Characteristics

The taxonomic characteristics window provides information on taxonomic groups.

Select group or genus ➡

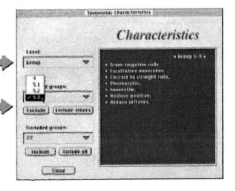

Exclude groups and genera ➡

Students can develop a strategy for identifying unknowns by reviewing taxonomic characteristics and eliminating groups and genera.

Reviewing Responses

Review responses in the main window. ➡

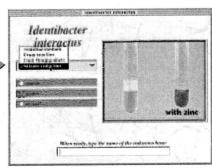

Review responses provides a history of the tests you chose and the responses you made.

CHAPTER I

THE IDENTIFICATION OF BACTERIAL ISOLATES

WHAT IS MEANT BY IDENTIFICATION?

The taxonomy of bacteria (or of any other group of organisms) consists of three interrelated areas. The task of arranging organisms into related groups is called **classification**. **Nomenclature** refers to the assignment of names to these groups, guided by a set of rules. **Identification** is the process of determining to which established taxon a new isolate or unknown strain belongs. This last task is the area of taxonomy which the computer program, *Identibacter interactus*, simulates.

THE RELATIONSHIP OF IDENTIFICATION TO BACTERIAL CLASSIFICATION AND NOMENCLATURE.

In order to identify an unknown bacterial isolate, the characteristics of the isolate must be compared to known taxa. In microbiology, the basic taxonomic unit is the **species**, and groups of related species are placed in the same **genus**. However, the term "species" does not have quite the same meaning in bacteriology as it does in the classification of plants and animals. In the latter cases, species are rigorously defined on the basis of individuals' capacity for interbreeding and geographical isolation. No such absolute tests are possible with asexual, globally-distributed bacterial strains.

Thus bacterial species are defined operationally — they are collections of similar strains. Classification schemes contain the criteria whereby the characteristics used to distinguish one species from other related ones are presented There is no "official" classification of bacteria, but reference sources such as *Bergey's Manual of Determinative Bacteriology* and *Bergey's Manual of Systematic Bacteriology* are the most commonly used resources in microbiology.

The identification of unknown cultures is a <u>practical application</u> of a classification scheme, so that a new isolate can be recognized as a member of an existing species.

WHY IS THE IDENTIFICATION OF BACTERIAL UNKNOWNS IMPORTANT?

Microbiologists must identify bacterial isolates for several practical reasons:

- Medical diagnostics — identifying a pathogen isolated from a patient.
- Food industry — identifying a microbial contaminant responsible for food spoilage.
- Research setting — identifying a new isolate which carries out an impor tant process.

A weakness of the classical classification scheme embodied in *Bergey's Manual* is that it is arbitrary; that is, it is weighted towards those characteristics that the community of microbiologists feel are most useful in distinguishing species. In addition, this classification scheme provides no insight into the evolutionary relationships among organisms, in the way that the Linnaean classification of plants and animals does. During the last decade, alternative means of characterizing bacteria have been developed based upon the chemical composition of the cells. Two of the methods are: (a) analysis of the pattern of fatty acids found in bacterial cell membranes, and (b) comparison of nucleic acid sequences. The analysis of 16s ribosomal RNA sequences has been especially useful in providing information about the evolutionary relationships among bacteria. See the bibliographic references for more information.

WHAT BACTERIA ARE REPRESENTED AS UNKNOWNS IN THE SIMULATION?

Bergey's Manual of Determinative Bacteriology contains information on approximately 4000 bacterial species. (Note that microbiologists believe the species known from cultures may represent only a few percent of all the bacterial species present on earth). The computer simulation contains a database of about 60 bacterial species. It is not representative of the breadth of species described in *Bergey's Manual*, but rather represents a set of species commonly used in undergraduate microbiology laboratories, and in addition includes some bacterial pathogens not usually assigned in introductory courses.

All of the species in the simulation are true Bacteria that make a cell wall. Therefore, neither Archaea nor mycoplasmas are represented. All of the species are chemoheterotrophic. Neither photosynthetic or chemoautotrophic bacteria are included. In addition, virtually all of the included species will grow under aerobic conditions on typical enriched laboratory media (for example, nutrient agar). Thus, obligate anaerobes are generally not represented in the database.

A list of the species used in *Identibacter interactus* is given in Appendix A.

WHAT TESTS DO I HAVE AVAILABLE TO ME?

A set of 52 tests can be accessed in the computer simulation. An overview of these tests is given in Appendix B. For a mor complete description of the tests, consult a general microbiology text or lab manual, or some of the references cited in the bibliography .

WHAT INFORMATION ON THE CHARACTERISTICS OF THE ORGANISMS IS AVAILABLE?

The computer simulation contains help on group and genus characteristics and information on a few organisms at the species level. Access these items from the **Help** menu. Information on the distinguishing characteristics of species in some groups is presented in Appendix C. You may also consult a more extensive reference source, such as *Bergey's Manual of Determinative Bacteriology* to obtain information.

THE STRATEGY OF IDENTIFICATION

Adapted from Cowan and Liston, "The Mechanism of Identification", in the 8th edition of *Bergey's Manual of Determinative Bacteriology*.

HOW DO I IDENTIFY AN UNKNOWN BACTERIUM?

Bacterial taxa are organized in groups. Each group is a cluster of genera. Your first task is to determine to which of these groups your unknown belongs. The titles of the groups provide hints as to what characteristics are most useful in making this determination (*e.g.*, "Facultatively anaerobic Gram-negative rods"). Thus, the true Bacteria are classified primarily on the basis of their morphology and physiology.

To distinguish broad categories, the oxygen requirement for growth is of special importance. Another important physiological characteristic to distinguish broad categories is whether the organism metabolizes carbohydrates (*e.g.*, glucose) by fermentation, respiration, or both modes.

Biochemical and physiological tests can then be used to distinguish among genera within a group, and between species once you have identified the correct genus. The specific tests that are most useful will differ depending upon the group you are investigating, but tests which are frequently useful are the catalase, oxidase, urease, and deaminase/decarboxylase reactions, as well as tests for fermentation products and nitrate reduction.

RULES OF THE GAME

1. Use all of the information available to you
2. Work from broad categories down to smaller, specific categories.
3. Apply "common sense" at each step.
4. Use the minimum number of tests to make the identification.

A goal of this simulation is to provide the opportunity to solve a number of unknowns. This will give you the experience to develop "common sense" about the organization and logic of bacterial classification.

PRACTICAL STEPS TO FOLLOW

1. Start with a pure culture.

   In the computer simulation this is a given, but in the laboratory one must be sure of this, because the reactions of mixed cultures are of no value.

2. Examine a Gram stain of the cells in the light microscope. Examination by phase contrast microscopy may also indicate a unique morphological property (for example, endospores).

   If unique morphological characteristics are present, confine your identification to groups having these characteristics.

3. Examine gross growth appearance on agar medium for pigments or other unique characteristics.

4. Test the oxygen requirements for growth.

5. Test the mode of carbohydrate metabolism — oxidative or fermentative.

6. Scan the characteristics of the genera to which your unknown may belong (based upon the data available from steps 2 – 5) to find specific tests that can distinguish between these genera.

WHAT IF I CANNOT IDENTIFY MY ORGANISM

1. In the real world, you would first check to see if you have a pure culture.

2. Review your interpretation of your test results.

3. Have you correctly used the diagnostic tables online or in *Bergey's Manual*?

4. Are there other tests that can distinguish between the remaining possible genera or species?

Reading References

General Information

Actor, Paul, et al. *Antibiotic Inhibition of Bacterial Cell Surface Assembly and Function.* Herndon, VA: ASM Press, 1988.

Alcamo, I. Edward. *Fundamentals of Microbiology,* 5th ed. Reading, Mass.: Addison-Wesley, 1997.

Atlas, R. M. and Bartha, R. *Microbial Ecology: Fundamentals and Applications,* 3rd ed. Menlo Park, Calif.: Benjamin/Cummings Publishing, 1993.

Baron, Samuel, ed. *Medical Microbiology.* 4th ed. Reading, Mass.: Addison-Wesley Publishing, 1996.

Freeman, Bob A. *Burrow's Textbook of Microbiology,* 22nd ed. Philadelphia: W. B. Saunders, 1985.

Gerhardt, Philipp, et al. *Methods for General and Molecular Bacteriology.* Herndon, VA: ASM Press, 1997.

Jakoby, W. B. *Methods in Enzymology.* New York: Academic Press, 1987.

Karam, Jim D., et al. *Molecular Biology of Bacteriophage T-4.* Herndon, VA: ASM Press, 1994.

Ketchum, Paul A. *Microbiology: Concepts and Applications.* New York: John Wiley and Sons, 1988.

Lederberg, Joshua, et al. *Encyclopedia of Microbiology.* New York: Academic Press, 1992.

Madigan, Michael T., John M. Martinko, and Jack Parker. *Brock Biology of Microorganisms,* 8th ed. Englewood Cliffs, NJ: Prentice-Hall 1997.

Madigan Michael T., and Barry L. Marrs. *Extremophiles.* New York: Scientific American Vol. 276, Number 4: pp 82–87, 1997

Myrvik, Quentin N., and Weiser, Russell S. *Fundamentals of Medical Mycology,* 2nd ed. Philadelphia: Lea and Febiger, 1988.

Norton, Cynthia F. *Microbiology.* 2nd ed. Reading, Mass.: Addison-Wesley Publishing, 1986.

Pelczar, M. J. and Chan, E. C. *Microbiology.* 5th ed. New York: McGraw-Hill, 1993.

Rippon, J. W. *Medical Mycology.* 3rd ed. Philadelphia: W. B. Saunders, 1988.

Ross, Frederick C. *Introductory Microbiology.* 2nd ed. Glenview, Ill.: Scott, Foresman, & Co., 1986.

Salyers, Abigail A. and Whitt, D. D. *Bacterial Pathogenesis.* Herndon, VA: ASM Press, 1994.

Talaro, K. and Talaro, A. *Foundations in Microbiology.* 2nd ed. Dubuque, IA: Wm. C. Brown Publishers, 1995.

Tortora, Gerard J.; Funke, B. R.; and Case, C. L. *Microbiology: An Introduction.* 4th ed. Menlo Park, Calif.: Benjamin/Cummings Publishing, 1992.

Volk, W. A., and Wheeler, M. F. *Basic Microbiology.* 6th ed. Reading, Mass.: Addison-Wesley Publishing, 1996.

Walker, Graham C. and Kaiser, Dale. *Frontiers in Microbiology: A Collection of Minireviews from the Journal of Bacteriology.* Herndon, VA: ASM Press, 1993.

Laboratory Procedures

American Type Culture Collections. *Catalog of Cultures,* 8th ed. Rockville, Md. n.d.

Atlas, R. M. and Snyder, J. W. *Handbook of Media for Clinical Microbiology.* Boca Raton, FL. CRC Press, 1996.

Chart, Henrik. *Methods in Practical Laboratory Bacteriology.* Boca Raton, FL. CRC Press, 1994.

Flemming, D. O., Richardson, J. H., Tulis, J. J., and Vesley, D. *Laboratory Safety: Principles and Practices.* 2nd ed. Herndon, VA.: ASM Press, 1995.

Garcia, Lynne S. and Brukner, David A. *Diagnostic Medical Parasitology.* 3rd ed. Herndon, VA: ASM Press, 1996.

Isenberg, Henry D., et al. *Clinical Microbiology Procedures Handbook,* Vols 1 and 2. Herndon, VA: ASM Press, 1992.

McGinnis, M. R. *Current Topics in Medical Mycology.* New York: Springer-Verlag, 1987.

Murray, Patrick R., et al. *Manual of Clinical Microbiology.* 6th ed. Herndon, VA: ASM Press, 1995.

Norris, John R., and Ribbons, D. W. *Methods in Microbiology.* Vols. 23 and 24. New York: Academic Press, 1991.

Shapton, D. A., and Shapton, N. F. *Principles and Practices of Safe Processing of Food.* New York: Academic Press, 1994.

Smith, Robert F. *Microscopy and Photomicrography.* Boca Raton, FL. CRC Press, 1994.

Murray, Patrick R., et al. *Manual of Clinical Microbiology.* 6th ed. Herndon, VA: ASM Press, 1997.

Bullock, G. R., and Petrusz, P. *Techniques in Immunocytochemistry,* vol. 3. New York: Academic Press, 1985.

Difco Laboratory Staff. *Difco Manual of Dehydrated Culture Media and Reagents,* 10th ed. Detroit, Mich.: Difco Laboratories, 1984.

Identification of Microorganisms

Anderson, O. Roger. *Comparative Protozoology: Ecology, Physiology, Life History,* Secausus, NJ: Springer-Verlag New York, Inc. 1987.

Balows, Albert et al. *Manual of Clinical Microbiology.* 5th ed. Bethesda: American Society for Microbiology, 1991.

Chandler, Francis W. and Watts, John C. *Pathologic Diagnosis of Fungal Infections.* Chicago, Ill.: American Society of Clinical Pathologists, 1987.

Goodfellow, M., and O'Donnell, A. G. *Handbook of New Bacteria and Systematics.* New York: Academic Press, 1993.

Holt, John G., Kreig, N. R., et al. *Bergey's Manual of Systematic Bacteriology,* vol. 1. 4th ed. Baltimore, Md.: Williams & Wilkins, 1984.

Jahn, Theodore L., et al. *Protozoa.* 2nd ed. Dubuque, Iowa: Dubuque, IA: WCB/McGraw-Hill, 1978.

Lapage, S. P., et al. *International Code of Nomenclature of Bacteria.* Herndon, VA: ASM Press, 1992.

Larone, Davise. *Medically Important Fungi: A Guide to Identification.* Herndon, VA. ASM Press, 1995.

Piggot, Patrick J., et al. *Regulation of Bacterial Differentiation.* Herndon, VA: ASM Press, 1993.

Sneath, Peter H. A., et al. *Bergey's Manual of Systematic Bacteriology,* vol. 2. Baltimore, Md.: Williams & Wilkins, 1986.

Staley, James T., et al. *Bergey's Manual of Systematic Bacteriology,* vol. 3. Baltimore, Md.: Williams & Wilkins, 1989.

Sanitary and Medical Microbiology

Balows, Albert et al. *Manual of Clinical Microbiology.* 5th ed. Herndon, VA: ASM Press, 1991.

Flemming, D. O., Richardson, J. H., Tulis, J. J., and Vesley, D. *Laboratory Safety: Principles and Practices.* 2nd ed. Herndon, VA. ASM Press, 1995.

Greenberg, Arnold E., et al. *Standard Methods for the Examination of Water and Wastewater,* 19th ed. Washington, D.C.: American Public Health Association, 1995.

Jay, James M. *Modern Food Microbiology.* 5th ed. New York: Chapman-Hall, 1996.

Kneip, Thedodore, and Crable, John V. *Methods for Biological Monitoring: A Manual for Assessing Human Exposre to Hazardous Substances.* Washington, D.C.: American Public Health Association, 1988.

Marshall, Robert T. *Standard Methods for the Examination of Dairy Products.* 16th ed. Washington D.C.: American Public Health Association, 1992.

Ray, Bibek. *Fundamental Food Microbiology.* Boca Raton, FL. CRC Press, 1996.

Vanderzant, Carl, and Splittstoesser, Don. *Compendium of Methods for the Microbiological Examination of Foods,* 3rd ed. Washington, D.C.: American Public Health Association, 1992.

Index